Advances in Intelligent Systems and Computing

Volume 1236

The series "Advances in Intelligent Systems and Computing" contains publications on theory, applications, and design methods of Intelligent Systems and Intelligent Computing. Virtually all disciplines such as engineering, natural sciences, computer and information science, ICT, economics, business, e-commerce, environment, healthcare, life science are covered. The list of topics spans all the areas of modern intelligent systems and computing such as: computational intelligence, soft computing including neural networks, fuzzy systems, evolutionary computing and the fusion of these paradigms, social intelligence, ambient intelligence, computational neuroscience, artificial life, virtual worlds and society, cognitive science and systems, Perception and Vision, DNA and immune based systems, self-organizing and adaptive systems, e-Learning and teaching, human-centered and human-centric computing, recommender systems, intelligent control, robotics and mechatronics including human-machine teaming, knowledge-based paradigms, learning paradigms, machine ethics, intelligent data analysis, knowledge management, intelligent agents, intelligent decision making and support, intelligent network security, trust management, interactive entertainment, Web intelligence and multimedia.

The publications within "Advances in Intelligent Systems and Computing" are primarily proceedings of important conferences, symposia and congresses. They cover significant recent developments in the field, both of a foundational and applicable character. An important characteristic feature of the series is the short publication time and world-wide distribution. This permits a rapid and broad dissemination of research results.

**** Indexing: The books of this series are submitted to ISI Proceedings, EI-Compendex, DBLP, SCOPUS, Google Scholar and Springerlink ****

More information about this series at http://www.springer.com/series/11156

Zuzana Kubincová · Loreto Lancia ·
Elvira Popescu · Minoru Nakayama ·
Vittorio Scarano · Ana B. Gil
Editors

Methodologies and Intelligent Systems for Technology Enhanced Learning, 10th International Conference. Workshops

Volume 2

Springer

Editors
Zuzana Kubincová
Department of Informatics Education
Faculty of Mathematics, Physics,
and Informatics
Comenius University
Bratislava, Slovakia

Elvira Popescu
Department of Computers
and Information Technology
University of Craiova
Craiova, Romania

Vittorio Scarano
Dipartimento di Informatica
Università di Salerno
Fisciano, Italy

Loreto Lancia
Department of Life, Health
and Environmental Sciences
University of L'Aquila
Coppito, Italy

Minoru Nakayama
Tokyo Institute of Technology,
School of Engineering
Tokyo, Japan

Ana B. Gil
BISITE Digital Innovation Hub
University of Salamanca
Salamanca, Spain

ISSN 2194-5357 ISSN 2194-5365 (electronic)
Advances in Intelligent Systems and Computing
ISBN 978-3-030-52286-5 ISBN 978-3-030-52287-2 (eBook)
https://doi.org/10.1007/978-3-030-52287-2

This Springer imprint is published by the registered company Springer Nature Switzerland AG
The registered company address is: Gewerbestrasse 11, 6330 Cham, Switzerland

Preface

Education, science, research and technology deployment in all areas of life are considered to be the basic pillars of the knowledge society. The International Conference in Methodologies and Intelligent Systems for Technology Enhanced Learning (Mis4TEL) serves as a forum for experts from all these fields, including not only education, information technology or computer science, but also such disciplines as psychology, medicine, social sciences, etc. It encourages multidisciplinary research and discussion on technology-enhanced learning promoting new intelligent and creative solutions for formal as well as informal learning and all types of learners. In addition to technological solutions, the technology-enhanced learning approach can be fostered by novel methods coming from different fields of research, and from diverse communities also including "fragile users," like children, elderly people or people with special needs.

The 10th edition of the conference expands the topics of the previous editions, highlighting the role of the most recent methods and technological opportunities (ranging from artificial intelligence and agent-based systems to robotics, virtual reality, Internet of Things and wearable solutions, among others) aiming to boost the discussion on how they can be employed to create novel approaches to TEL, innovative TEL solutions and valuable TEL experiences.

The conference program features also four selected workshops, which aim to provide participants with the opportunity to present and discuss novel research ideas on emerging topics complementing the main conference. In particular, the workshops focus on multidisciplinary and transversal aspects like TEL in nursing education programs, social and personal computing for Web-supported learning communities, interactive environments and emerging technologies for eLearning and TEL for future citizens. A total of 29 quality papers, with authors coming from various countries from Europe, Asia and America, have been selected for the workshops and included in the present volume.

This volume also includes three additional papers of the Student Competition which is an opportunity to showcase the student's creativity to leaders in the field, turn their ideas into reality and win fabulous prizes that will foster the development of their scientific interests and will help them with networking in the research community.

We thank the sponsor, the Armundia Group (https://www.armundia.com/), the support from national associations (AEPIA, APPIA, CINI and EurAI), the members of the Local Organization Team and the Program Committee members for their hard work, which was essential for the success of MIS4TEL'20.

Zuzana Kubincová
Loreto Lancia
Elvira Popescu
Minoru Nakayama
Vittorio Scarano
Ana Belén Gil

Organisation of MIS4TEL 2020

http://www.mis4tel-conference.net/

General Chair

Pierpaolo Vittorini	University of L'aquila, Italy
Tania Di Mascio	University of L'aquila, Italy

Technical Program Chair

Laura Tarantino	University of L'aquila, Italy
Marco Temperini	Sapienza University, Rome, Italy

Paper Co-chair

Rosella Gennari	Free University of Bozen-Bolzano, Italy
Elvira Popescu	University of Craiova, România
Ricardo Silveira	Universidade Federal de Santa Catarina, Brazil

Proceedings Chair

Fernando De la Prieta	University of Salamanca, Spain
Ana Belén Gil	University of Salamanca, Spain

Publicity Chair

Alessandra Melonio	Free University of Bozen-Bolzano, Italy

Demetrio Arturo Ovalle Carranza	National University of Colombia, Colombia
Nestor Dario Duque Mendes	National University of Colombia, Colombia

Workshop Chair

Zuzana Kubincová	Comenius University of Bratislava, Slovakia

Local Organizing Committee

Pierpaolo Vittorini	University of L'Aquila, L'Aquila, Italy
Tania Di Mascio	University of L'Aquila, L'Aquila, Italy
Giovanni De Gasperis	University of L'Aquila, L'Aquila, Italy
Federica Caruso	University of L'Aquila, L'Aquila, Italy
Alessandra Galassi	University of L'Aquila, L'Aquila, Italy

Organizing Committee

Juan M. Corchado Rodríguez	University of Salamanca, Spain, and AIR Institute, Spain
Fernando De la Prieta	University of Salamanca, Spain
Sara Rodríguez González	University of Salamanca, Spain
Javier Prieto Tejedor	University of Salamanca, Spain, and AIR Institute, Spain
Pablo Chamoso Santos	University of Salamanca, Spain
Belén Pérez Lancho	University of Salamanca, Spain
Ana Belén Gil González	University of Salamanca, Spain
Ana De Luis Reboredo	University of Salamanca, Spain
Angélica González Arrieta	University of Salamanca, Spain
Emilio S. Corchado Rodríguez	University of Salamanca, Spain
Angel Luis Sánchez Lázaro	University of Salamanca, Spain
Alfonso González Briones	University Complutense of Madrid, Spain
Yeray Mezquita Martín	University of Salamanca, Spain
Enrique Goyenechea	University of Salamanca, Spain, and AIR Institute, Spain
Javier J. Martín Limorti	University of Salamanca, Spain
Alberto Rivas Camacho	University of Salamanca, Spain
Ines Sitton Candanedo	University of Salamanca, Spain
Elena Hernández Nieves	University of Salamanca, Spain
Beatriz Bellido	University of Salamanca, Spain
María Alonso	University of Salamanca, Spain
Diego Valdeolmillos	AIR Institute, Spain
Roberto Casado Vara	University of Salamanca, Spain

Sergio Marquez	University of Salamanca, Spain
Jorge Herrera	University of Salamanca, Spain
Marta Plaza Hernández	University of Salamanca, Spain
Guillermo Hernández González	AIR Institute, Spain
Luis Carlos Martínez de Iturrate	University of Salamanca, Spain, and AIR Institute, Spain
Ricardo S. Alonso Rincón	University of Salamanca, Spain
Javier Parra	University of Salamanca, Spain
Niloufar Shoeibi	University of Salamanca, Spain
Zakieh Alizadeh-Sani	University of Salamanca, Spain

Contents

Workshop on TEL in Nursing Education Programs (NURSING)

Workshop on Interactive Environments and Emerging Technologies for eLearning (IEETeL)

Workshop on Interactive Environments and Emerging Technologies for eLearning, IEETeL

eLearning is a dynamic research area reflecting on the current requirements of all participants in the educational process for innovative teaching and meaningful learning combining existing knowledge with future perspectives.

The aim of IEETeL is to connect researchers, educators and technology experts giving them an opportunity to share and discuss new solutions, trends and realizations of eLearning environments and the adoption of emerging technologies in educational settings. These will draw the challenging problems in information gathering, processing and usage; development of web-based and mobile services; building intelligent and social-oriented applications in support of flexible, personalized, adaptable, in-demand learning.

The issue of this year, features 8 interesting accepted papers, contributing on the fundamentals of the future work in the scope of IEETeL.

Major topics of the accepted papers in IEETel workshop are Expert System, Machine Learning, Learning Performance, Security, Peer Assessment, Teamwork Evaluation, Information Retrieval, Knowledge Management, Gamification, Creativity and Blended learning. The common aspects of the various topics are about research on innovative approaches to Learning and Teaching using Technology.

Organization

Organizing Committee

Malinka Ivanova	Technical University of Sofia
Minoru Nakayama	Tokyo Institute of Technology
MarcoTemperini	Sapienza University of Rome

Program Committee

Fernando Albuquerque Costa	University of Lisboa, Portugal
Rachid Anane	Coventry University, UK
Katya Asparuhova	Technical University of Sofia, Bulgaria
Kay Berkling	Duale Hochschule Baden-Württemberg Karlsruhe, Germany
Maiga Chang	Athabasca University, Canada

Maria De Marsico	Sapienza University of Rome, Italy
Maya Dimitrova	Bulgarian Academy of Science, Bulgaria
Roumiana Ilieva	Technical University of Sofia, Bulgaria
Zuzana Kubincová	Comenius University in Bratislava, Slovakia
Luigi Laura	International Telematic University UNINETTUNO, Rome, Italy
Carla Limongelli	Roma Tre University, Italy
Matteo Lombardi	Griffith University, Australia
Victoria Marin	University of Oldenburg, Germany
Alexander Mikroyannidis	The Open University, UK
Daniela Minkovska	Technical University of Sofia, Bulgaria
Laurent Moccozet	University of Geneva, Switzerland
Elvira Popescu	University of Craiova, Romania
Ricardo Queiros	Politechnic of Porto, Portugal
Anna Rozeva	Technical University of Sofia, Bulgaria
Filippo Sciarrone	Roma Tre University, Italy
Andrea Sterbini	Sapienza University of Rome, Italy
Gemma Tur Ferrer	Universitat de les Illes Balears, Spain

Construction of Fuzzy-Classification Expert System in Cerebral Palsy for Learning Performance Facilitation

Malinka Ivanova[1](✉), Roumiana Ilieva[2], and Zhenli Lu[3]

[1] College of Energy and Electronics, Technical University of Sofia, 8 "Kliment Ohridski" boul., Sofia, Bulgaria
m_ivanova@tu-sofia.bg

[2] Faculty of Management, Department of Management and Business Information Systems, Technical University of Sofia, 8 "Kliment Ohridski" boul., Sofia, Bulgaria
rilieva@tu-sofia.bg

[3] School of Electrical Engineering and Automation, Changshu Institute of Technology, Chanshu 215500, People's Republic of China
zhenlilu@cslg.edu.cn

Abstract. The paper presents a novel method and conceptual architecture for implementation of fuzzy-classification expert system in the domain of rehabilitation methods for cerebral palsy. The expert system includes two blocks: Fuzzy block utilizing fuzzy algorithms for multi-criteria decision making and Machine learning block based on algorithms for tree classification and KMeans clustering. The proposed solution is designed for facilitation the learning performance of university students as well as for professionals who have to make decisions in the area of cerebral palsy and corresponding rehabilitation methods.

Keywords: Expert system · Learning performance · Cerebral palsy · Rehabilitation methods · Fuzzy theory · Machine learning

1 Introduction

The term "cerebral palsy" relates to permanent developmental disorders that occurred early in human biological development or in early childhood. These developmental disorders primarily refer to conditions of abnormal gross and fine motor functioning. It also may happen that children do not have proper spatial relationships and notion of themselves – they are not able to show where is his head, what is his left or right arm or leg, etc. Therapy of motor skill disorders in children is long-termed and should start as early as possible. Since cerebral palsy patients significantly differ among themselves, therapy should be adapted to a particular child. However, therapy may also be tiresome, uncomfortable or even painful for the child. Therefore, it is essential that the child is motivated to undergo therapy, which is a challenging therapeutic task.

Z. Kubincová et al. (Eds.): MIS4TEL 2020, AISC 1236, pp. 5–14, 2021.
https://doi.org/10.1007/978-3-030-52287-2_1

Recent research in the field of robot-assisted therapy for children with cerebral palsy suggests that a robotic system has a primary role of a facilitator – it may increase motivation and trigger social interactions between the children and the therapist.

It is a hot topic where researchers are developing a prototypical robotic system intended to be used as an assistive tool in therapy for children with cerebral palsy. To be used in such application robot must be able to fulfill some basic requirements: it has to be able to demonstrate requested exercises, its appearance should be attractive to children and it should be able to communicate (verbal and non-verbal) with patients. However, in contrast to other approaches, Prof. Branislav Borovac introduced a conceptual novelty: the robot's capacity to engage in a natural language dialogue may be of significant clinical benefit [1]. The anticipated benefit of the conversational robot is to support the therapist to conduct specific therapeutic exercises and to contribute to establishing affective attachment of the child to the robot. Dr. & A/Prof. Zhenli Lu et al. propose a training system for cerebral palsy rehabilitation that is developed on the human-computer interaction principles (Fig. 1) [2].

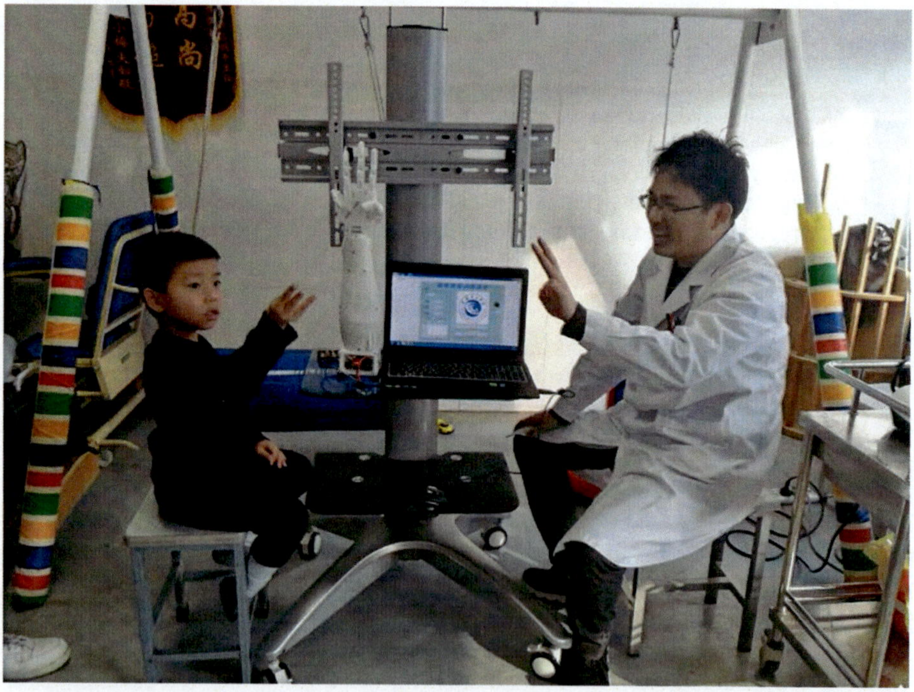

Fig. 1. Technical architecture of speech recognition based expert system in robotic assistant rehabilitation of cerebral palsy

So, the aim of this paper is to present a novel approach for expert system construction based on fuzzy theory and machine learning algorithms for educational purposes. It includes knowledge about rehabilitation methods suitable for applying at different types of cerebral palsy and it is a preliminary step for development of a more complex expert system for usage by therapists and patients.

2 The Expert Systems: State-of-the-Art

The term expert system is related to development of interactive software that collects expert and users knowledge in a given domain and emulates the human ability of decision making and problem solving. It could be driven by a pool of cases or based on a set of rules that together with the knowledge expert domain leads to the appropriate inference. Different techniques and machine learning algorithms are utilized for simulation the individual thinking, group brainstorming, reasoning and concluding.

In this section the expert systems are examined according to three criteria: (1) the domain area that is medicine and a very concrete topic on rehabilitation methods for cerebral palsy; (2) the used artificial intelligence methods for expert systems realization; (3) the utilization purpose that is focused on usage in educational settings.

2.1 Expert Systems in Medicine and Cerebral Palsy

An expert system for medical diagnostics of cerebral palsy is proposed by Borgohain et al. [3]. It works with determined rules and inferences are patients' diagnose according to the input symptoms and also classification of cerebral palsy in three groups: mild, moderate and severe. Its development is based on JESS (Java Expert System Shell). Another expert system for evaluating the patients with cerebral palsy and giving an advice about the suitable wheel chairs and devices for body support is created by Ni et al. [4]. It consists of knowledge base regarding the evaluation of seating and positioning, a set of rules, inference engine with guidelines pointing out the assistive technologies and specifications with seating/positioning devices, module with the authors' research in this topic and friendly and interactive user interface. A beginning step for implementation of intelligent expert system is done by Zammouri et al. who assess different reeducative therapies that could be applied to the children with cognitive difficulties [5]. Their work is focused on brain-computer interface for evaluation the brain workload during performance of a cognitive activity. A rule-based expert system for diagnosing patients with heart diseases and prescription of suitable treatments is proposed by Soltan et al. [6]. This medical expert system is developed through Visual Prolog.

2.2 Artificial Intelligence Methods in Expert Systems

A survey regarding expert systems in medicine that utilize artificial intelligence methods is performed by Singla et al. [7]. This work shows that medical expert systems are developed with purposes of patients' diagnostics, for giving advices concerning treatment and therapy, for producing guidelines for patients and doctors, and for educational training. The main technologies behind their implementations are pointed out as: fuzzy logic, Artificial Neural Networks and neuro-fuzzy approach. Another review done by Sheikhtaheri et al. reports that expert systems for clinical purposes are focused on: quality improvement of first aid, disease prediction, disease identification, diagnosing, therapy suggestion, giving advice, disease classification and their realization is driven by fuzzy theory, Artificial Neural Networks, support vector machine, Naïve Bayes classifier, Wavelet neural network [8]. An implemented variant of expert system for recognition of cerebral palsy at babies and small children and prediction for their future suffering is presented by Ojo et al. [9]. The input variables are related to the motor skills and output is possibility for cerebral palsy development. It is realized through applying Fuzzy theory and Fuzzy logic in MATHLAB. Another medical expert system based on Fuzzy logic, data mining techniques and machine learning algorithms is created in support of decision making and information delivery regarding diagnosis of vertebral diseases and appropriate treatment [10]. Its software platform is developed with functionality for knowledge sharing among doctors and patients through chat, forum and video and also its knowledge base could be extended with additional modules gathering knowledge for a wide variety of hospital services. Advantages of creation hybrid expert systems based on Fuzzy logic and Artificial Neural Networks in context of speech therapy are revealed in [11]. Such expert systems possess typical characteristics like pointing out the personalized disease treatment and giving expert opinion, but also possibility for self-learning.

2.3 Expert Systems for Educational Purposes

A review regarding the usage of expert systems in educational settings is done by Supriyanto et al. who outline the main purposes that we classify in the following groups: (1) for improvement teaching and learning – students characteristics analysis, student performance analysis and prediction, evaluation of student competency, realization of personalized learning, evaluation of teaching efficiency; (2) for improvement the educational process at all - eLearning evaluation, evaluation of requirements for technical education, evaluation of education, improvement the quality of learning lesson plans, giving an academic advice, academic programs evaluation, evaluation of master level criteria; (3) in support of librarians [12]. The main didactical concepts behind development of an expert system for purpose of mathematics teaching are presented by Salekhova et al. [13]. A new module of expert system is introduced Concept-Effect-Relationship that contains the relationship among subjects and it is used for providing individual educational strategy to every learner. The main aim of this expert system is to increase the effectiveness of the teachers' activities. The enhancement of students' learning is achieved through usage of an expert system in the database course that contains hard concepts for understanding [14]. The students are supported through immediate feedback and corrections when they have to decide a problem. Another expert system is developed to predict students' performance during the computer science course [15]. The input variables are related to information for students – their attitudes, study habits and ways for preparation and the output is the predicted outcome.

3 Expert System Construction

The expert system includes knowledge about the cerebral palsy (CP) types: Spastic, Athetoid, Ataxic and Mixed with their typical symptoms and rehabilitation methods

(RMs) for cerebral palsy. Six experts from China and Bulgaria (three academic professionals, one expert is professor and neurologist and two experts are doctors-neurologists) were asked to vote the importance of 13 rehabilitation methods that are classified in two groups: approaches without using any equipment (8 RMs) and approaches with using equipment for each type cerebral palsy (5 RMs). The goal is to select suitable RMs for a given CP type. The conceptual architecture of the constructed expert system is presented on Fig. 2. Experts and scientific papers are reliable information sources for knowledge base building. The collected knowledge in the domain of rehabilitation methods for cerebral palsy is used for decision making or problems solving by reasoning through two types of inference engines: fuzzy inference block and machine learning block. Fuzzy inference block is constructed on FuzzyTOPSISLinear algorithm for group decision making proposed by Chen [16] and FuzzyWASPAS method for alternatives ranking presented by Turskis et al. [17]. The ground of the both methods is creation of aggregated fuzzy decision matrix $\tilde{D}$ with m alternatives and n attributes:

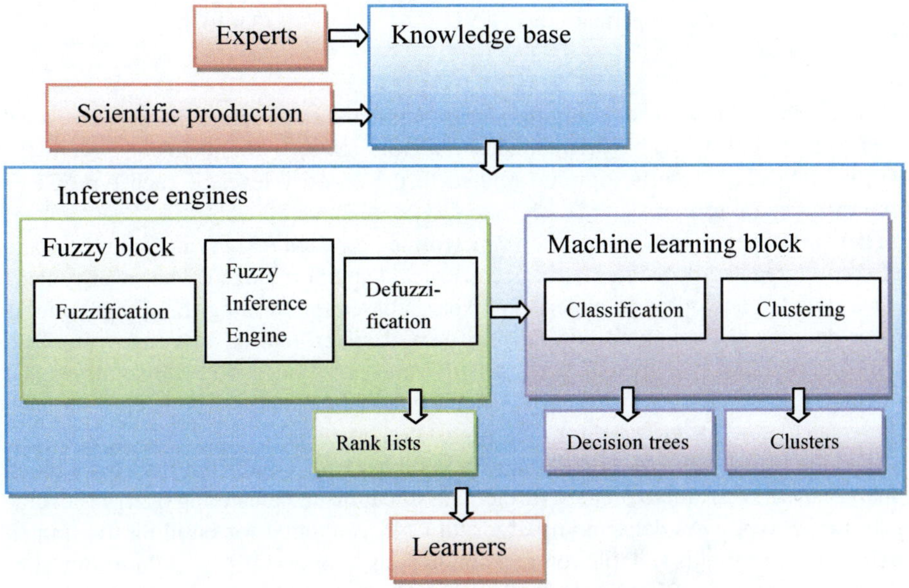

Fig. 2. Conceptual architecture of constructed expert system

$$\tilde{D} = \begin{vmatrix} \tilde{d}_{11} \ldots \tilde{d}_{1j} \ldots \tilde{d}_{1n} \\ \tilde{d}_{21} \ldots \tilde{d}_{2j} \ldots \tilde{d}_{2n} \\ \ldots\ldots\ \ldots\ldots\ldots \\ \tilde{d}_{m1} \ldots \tilde{d}_{mj} \ldots \tilde{d}_{mn} \end{vmatrix} \tag{1}$$

The fuzzy decision matrix contains the aggregated experts' vote that is in a 5-Likert scale from **1**-this RM is **not important** for this type of CP to this RM is **very important** for a given CP type, expressed in linguistic variables and fuzzy triangular numbers. Table 1 includes input data in linguistic variables before the experts 'vote aggregation. Table 2 summarizes the experts' opinion concerning the importance of every rehabilitation method for each cerebral palsy type.

Table 1. Rating scale and the corresponding triangular fuzzy numbers

Likert scale range	Meaning	Linguistic variable	Triangular fuzzy numbers
1	Not important	NI	(0,1,3)
2	Slightly important	SI	(1,3,5)
3	Moderately important	MI	(3,5,7)
4	Important	I	(5,7,9)
5	Very important	VI	(7,9,10)

Linguistic variables and triangular fuzzy numbers concerning the severity level are as follows: mild (M) – (0.1,0.3,0.5), moderate (MO) – (0.3,0.5,0.7), severe (S) – (0.5,0.7,0.9). The experts' opinion related to the severity level of each type CP is summarized in Table 3.

Then, the experts' vote aggregation is performed for each RM in fuzzy numbers and the fuzzy aggregated decision matrix is constructed and normalized. The severity level of each type CP according to experts is taken as bases for forming the group weights and for building the weighted normalized fuzzy decision matrix. The fuzzy numbers (1,1,1) and (0,0,0) are defined as fuzzy positive and fuzzy negative ideal solutions and the distance from each RM to them is calculated. The closeness coefficients for each RM are calculated that are used for ranking list creation.

Machine learning block uses all data from Tables 1, 2, 3 and from the Fuzzy block to analyze them with aim to prepare further classification and clustering of RMs according to the CP types. As data mining algorithm J48 is applied for building the pruned classification tree that is suitable for decision making support [18]. The algorithm SimpleKMeans is used for RMs clusters calculation, because of its proven effectiveness at solving clustering problems.

4 The Expert System Verification

The working capacity of the Fuzzy block from the proposed expert system is verified through usage of R software environment and running FuzzyMCDM package [19] with FuzzyTOPSISLinear and FuzzyMMOORA functions for realization of Multi-Criteria Decision Making algorithms as it is shown on Fig. 3.

The working capacity of the machine learning block is verified through usage of Weka software [20] and some of integrated machine learning algorithms for data classification and clustering (Fig. 4).

Table 2. Experts rating vote regarding the importance of every rehabilitation method for each cerebral palsy type

CP type	RM	Expert1	Expert2	Expert3	Expert4	Expert5	Expert6
Spastic	RM1	VI	I	I	VI	VI	VI
	RM2	I	I	I	VI	VI	VI
	RM3	VI	I	VI	VI	VI	VI
	RM4	MI	I	MI	I	I	VI
	RM5	I	I	I	VI	I	SI
	RM6	I	I	I	VI	I	VI
	RM7	MI	MI	MI	VI	I	VI
	RM8	MI	MI	MI	VI	I	SI
Athetoid	RM1	I	I	I	VI	VI	SI
	RM2	I	I	I	I	VI	VI
	RM3	I	I	I	VI	I	I
	RM4	I	I	I	VI	VI	MI
	RM5	I	I	I	VI	I	MI
	RM6	MI	I	I	VI	I	VI
	RM7	I	I	I	VI	I	VI
	RM8	MI	MI	MI	VI	I	SI
Ataxic	RM1	I	I	VI	VI	VI	VI
	RM2	I	I	I	I	VI	VI
	RM3	I	I	I	VI	I	VI
	RM4	MI	MI	MI	VI	MI	I
	RM5	I	I	I	VI	MI	SI
	RM6	MI	MI	MI	VI	I	SI
	RM7	I	I	I	VI	I	SI
	RM8	MI	I	MI	VI	I	SI
Mixed	RM1	VI	VI	I	VI	VI	VI
	RM2	VI	VI	VI	VI	VI	VI
	RM3	VI	VI	VI	VI	VI	VI
	RM4	I	I	I	VI	VI	VI
	RM5	MI	MI	MI	VI	I	VI
	RM6	I	I	I	VI	I	VI
	RM7	I	I	VI	VI	I	VI
	RM8	I	I	I	VI	I	SI

Table 3. The severity level of different CP types

	Expert1	Expert2	Expert3	Expert4	Expert5	Expert6
Spastic	M	MO	M	M	MO	M
Athetoid	MO	S	MO	MO	MO	MO
Ataxic	S	S	S	MO	S	S
Mixed	MO	MO	S	MO	S	MO

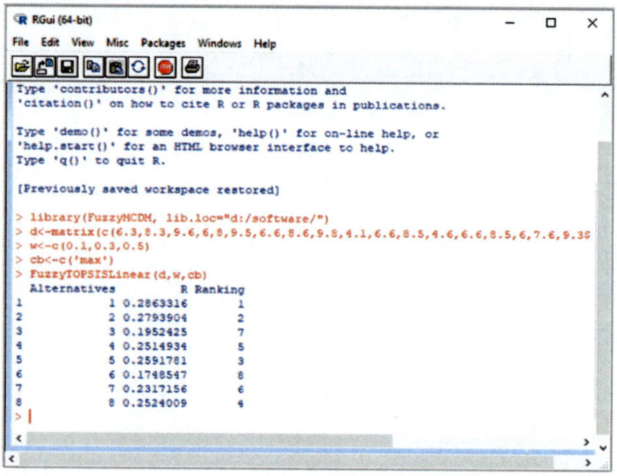

Fig. 3. The working capacity verification of the Fuzzy block

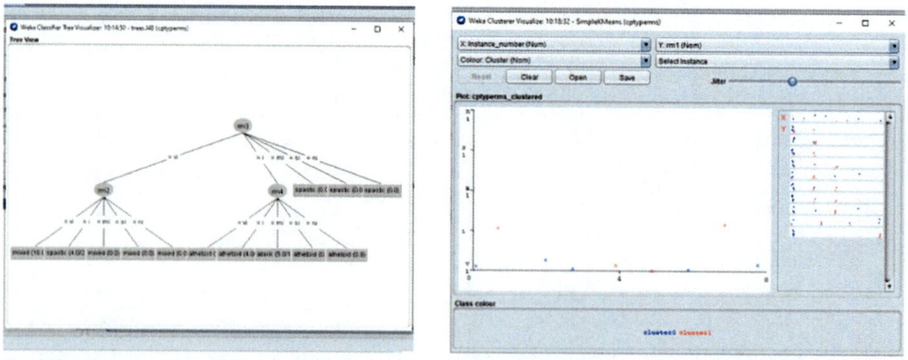

a). J48 algoritm for pruned decision tree construction

b). SimpleKMeans algorithm for clustering

Fig. 4. The working capacity verification of the machine learning block

5 Conclusions

IR (Intelligent Robotics) and HRI (Human Robot Interaction) related technologies are essential for expert system development, especially in the application of cerebral palsy rehabilitation. The novel method and conceptual architecture for implementation of fuzzy-classification expert system in the domain of rehabilitation methods for cerebral palsy are presented. Fuzzy block utilizing fuzzy algorithms for multi-criteria decision making and Machine learning block based on algorithms for tree classification and SimpleKMeans clustering are adopted to develop the expert system. Using the data by the surveying process and examined scientific production, the proposed solution designed for facilitation the learning performance of students, provided the construction of current expert system. The results of this work can give much help for the design of intelligent expert system for robotic assistant system in cerebral palsy rehabilitation.

Acknowledgments. The authors would like to thank the Research and Development Sector at the Technical University of Sofia for the financial support.

The previous work is funded by project of the 3rd Regular Session of the China-Serbia Inter-Governmental Scientific and Technical Cooperation Committee, project title "Study of human-robot interaction - Use of robot as assistive technology for cerebral palsy rehabilitation"; and the 12th Regular Session of the China-Slovenia Inter-Governmental Scientific and Technical Cooperation Committee, project title "Study on Key Intelligent Control Technology and Method of Robot Assistant System for Cerebral Palsy Rehabilitation".

The authors express their deep gratitude to all participants in the surveying process for the priceless engagement and the quick response. Special thanks to Prof. Dr. L. Haralanov, Head of the NCH Nervous Diseases Clinic, Dr V. Damyanov, Vice Head of 8 DCC, Senior Assist. Prof. O. Boyanova, Medical University-Sofia, Prof. Branislav Borovac, Department of Industrial Engineering and Management, Faculty of Technical Sciences, University of Novi Sad, Serbia; Prof. Marjan Mernik, Faculty of Electrical Engineering and Computer Science, University of Maribor, Slovenia; Prof. Yan Liu, Yu Zhao, Zhipeng Ma, Hong Wang and Changkao Shan with Changshu Institute of Technology, Xuanlin Shen, Head of Department of Rehabilitation Medicine, Changshu No. 2 People Hospital, P.R. China; and Dr. & A/Prof. Jun Liu with Faculty of Biomedical Engineering & Instrument Science, Zhejiang University, P.R. China for the enthusiastic support in the surveying process.

References

1. Mikov, A., et al.: Robot-assisted exercises in children with cerebral palsy-a case study. In: 9th International Conference on Children's Bone Health, 22–25 June, Salzburg, Austria (2019). https://doi.org/10.1530/boneabs.7.p206
2. Lu, Z., et al.: Face-expression and speech recognition based rehabilitation training system. Gao Ji Shu TongXun/Chin. High Technol. Lett. **29**(03), 287–294 (2019)
3. Borgohain, R., Sanyal, S.: Rule Based Expert System for Cerebral Palsy Diagnosis. ArXiv abs/1207.0117 (2012)
4. Ni, B.-N., et al.: An expert system in specialized seating/positioning for severe cerebral palsy. Chin. J. Biomed. Eng. **18**(3), 113–122 (1999)
5. Zammouri, A., Moussa, A., Mebrouk, Y.: Brain-computer interface for workload estimation: Assessment of mental efforts in learning processes. Exp. Syst. Appl. **112**, 138–147 (2018)

6. Soltan, R.A., Rashad, M.Z., El-Desouky, B.: Diagnosis of some diseases in medicine via computerized experts system. Int. J. Comput. Sci. Inf. Technol. (IJCSIT) **5**(5), 79 (2013)
7. Singla, J., Grover, D., Bhandari, A.: Medical expert systems for diagnosis of various diseases. Int. J. Comput. Appl. **93**(7), 36–43 (2014)
8. Sheikhtaheri, A., et al.: Developing and using expert systems and neural networks in medicine: a review on benefits and challenges. J. Med. Syst. **38**(9), 110 (2014)
9. Ojo, A.H., et al.: Fuzzy expert system for the intelligent recognition of cerebral palsy. J. Comput. Sci. Appl. **21**(1), 59–72 (2014)
10. Keles, A.: Expert doctor verdis: integrated medical expert system. Turk. J. Elect. Eng. Comput. Sci. **22**(4), 1032–1043 (2014)
11. Schipor, O., et al.: From fuzzy expert system to artificial neural network: application to assisted speech therapy. Artif. Neural Netw. Models Appl. (2016). https://doi.org/10.5772/63332. Joao Luis G. Rosa, IntechOpen
12. Supriyanto, G., et al.: Application of expert system for education. In: 3rd Annual Applied Science and Engineering Conference, AASEC 2018. IOP Publishing (2018). https://doi.org/10.1088/1757-899x/434/1/012304
13. Salekhova, L., et al.: The principles of designing an expert system in teaching mathematics. Univers. J. Educ. Res. **1**(2), 42–47 (2013). https://doi.org/10.13189/ujer.2013.010202
14. Gerald, V., et al.: An expert system helps students learn database design. J. Innovative Educ. **3**(2), 273–293 (2005). https://doi.org/10.1111/j.1540-4609.2005.00070.x
15. Kuehn, M., et al.: An expert system for the prediction of student performance in an initial computer science course. In: 2017 IEEE International Conference on Electro Information Technology (EIT), 4–17 May, Lincoln, NE, USA (2017)
16. Chen, C.T.: Extensions of the TOPSIS for group decision-making under fuzzy environment. FuzzySets Syst. **114**, 1–9 (2000)
17. Turskis, Z., et al.: A hybrid model based on fuzzy AHP and fuzzy WASPAS for construction site selection. Int. J. Comput. Commun. Control **10**(6), 873–888 (2015)
18. Kapoor, P., Rani, R.: Efficient decision tree algorithm using J48 and reduced error pruning. Int. J. Eng. Res. Gen. Sci. **3**(3), 1613–1621 (2015)
19. Package 'FuzzyMCDM' (2016). https://ftp.uni-sofia.bg/CRAN/. Accessed 10 Feb 2020
20. Frank, E., et al.: The WEKA Workbench Online. Appendix for "Data Mining: Practical Machine Learning Tools and Techniques, 4th edn. Morgan Kaufmann, Burlington (2016)

Security in Multimedia Information Systems: Analysis and Prediction

Daniela Minkovska[1] and Malinka Ivanova[2](✉)

[1] Faculty of Computer Systems and Technologies, Technical University of Sofia, 8 "Kl. Ohridski" boul., Sofia, Bulgaria
daniela@tu-sofia.bg

[2] College of Energy and Electronics, Technical University of Sofia, 8 "Kl. Ohridski" boul., Sofia, Bulgaria
m_ivanova@tu-sofia.bg

Abstract. Nowadays, a big part of eLearning applications is web-based, designed with integrated interactive multimedia objects and extensive information exchange among clients and servers. This requires appropriate security measures to be taken to protect users, infrastructure and databases from a variety of threats and attacks. The paper presents the developed multimedia information system for eLearning purposes to train students from the Informatics course. It is realized through object-oriented language C# and Microsoft Visual Studio, ASP.Net, HTML5, CSS3, SQL Server Manager Studio and ToolBook and it is constructed taking into account the importance of security issues for protection the participants in educational process and university resources. The literature overview based on bibliometrics approach is performed to outline the current state in the field of security of multimedia information systems. The findings point out that the security topic is still not in the center of research in the explored in the paper context that exposes the educational assets at risks of attackers' malicious activities. Also, a predictive model, based on machine learning algorithms is created to present trending issues.

Keywords: Security · Multimedia information system · eLearning · Machine learning · Predictive model

1 Introduction

Multimedia information systems adopt the typical characteristics from information systems and multimedia applications. Boell and Cecez-Kecmanovicis duscuss that the definition of the term *information systems* could be very complex when it is presented from different aspects: technical, social, socio-technical, operable [1], but it can be said that these systems should possess some main technical functions related to information storage, retrieval, transmission, management and presentation. Integration of suitable media objects like: text, graphics, pictures, images, audio and video files, animations that allow programmed interactions to be performed is an important issue for achieving

Z. Kubincová et al. (Eds.): MIS4TEL 2020, AISC 1236, pp. 15–24, 2021.
https://doi.org/10.1007/978-3-030-52287-2_2

reach user experience in a multimedia application. *Multimedia* is characterized with several advantages: complex information could be presented in simple way, the information could be delivered in the form of dynamic animations or interactive simulations, adding audio or video objects could contribute to better idea understanding, different levels of interactivity could be planned according to the users' preferences, several hierarchical models of course content could be designed [2].

Nowadays, a big part of information systems assisted through multimedia is web-based, distributing information and services with support of web and Internet technologies and using the principles of hypermedia. Their applications are in continuous progress, including more and more different areas like: tourism – for explaining routes, destinations, events, for marketing purposes, for travelers' management [3], cultural heritage – for presenting, studying, exploring a wide variety of cultural artifacts [4], business – for planning, controlling, forecasting, decision making, products/services demos. Also, multimedia information systems are well accepted for educational purposes as eLearning tool in order to facilitate students' learning and to support educators' activities. For example, Saadé and Galloway present a web-based multimedia information system for material learning by students involved in the course of management information systems [5].

It can be seen that multimedia information systems propose information and services to many users, transmitting a big data array through Internet and using database storage on servers. Then, the system security appears to be a very important feature and must be taken into account at its development and utilization. This exploration tray to understand whether the educational systems are secure, what kind of techniques are applied for data securing and what are the criteria that evaluate a system for secure.

So, the aim of the paper is to present the developed multimedia information system for students training in the Informatics course and the utilized security measures as well as to summarize and analyze the contemporary issues regarding security in multimedia information systems. A predictive model is created to outline the prognosis about the published scientific papers that treat the security problems in multimedia information systems.

2 The Method

The used method for outlining the relevant issues to security in multimedia information systems is based on analysis of bibliographic data, corresponding to scientific papers indexed in the database SCOPUS. The submitted queries on 13 February 2020 include keywords: *multimedia*, *information systems*, *security* and the search is limited to the last ten years: from 2010 to 2019. The returned result includes 155 documents that additionally are sorted on relevance. The extracted bibliographic dataset in.bib format is analyzed through application Biblioshiny that is part of the Bibliometrix package and is started in the R Studio platform (Fig. 1) [6]. Such approach allows a preliminary exploration about given topic to be performed as well as the "big picture" to be described. Then, the detailed examination of selected literature sources with highest impact is performed for better understanding the security issues in multimedia information systems.

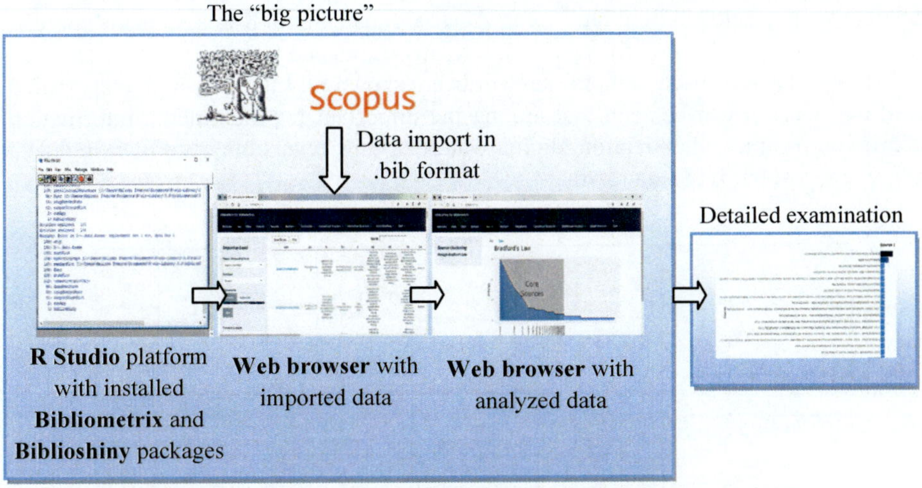

Fig. 1. The used method for scientific literature analysis

3 The-State-of-the-Art

The current state of the scientific research in the area of *security of multimedia information systems* is outlined through the method of bibliometric analysis as it is above described. The interest to the examined topic from the researchers' side is summarized through published by them scientific production. The exploration shows that the number of the published papers from 2010 to 2018 year obtains minimum value in 2018 year. Then, the line increases from 2018 to 2019 year, but not so noticeable. It means that the researchers' interest to the topic is returned slowly again, placing in the focus the importance of security issues for multimedia information systems. The published papers by countries are depicted on Fig. 2.

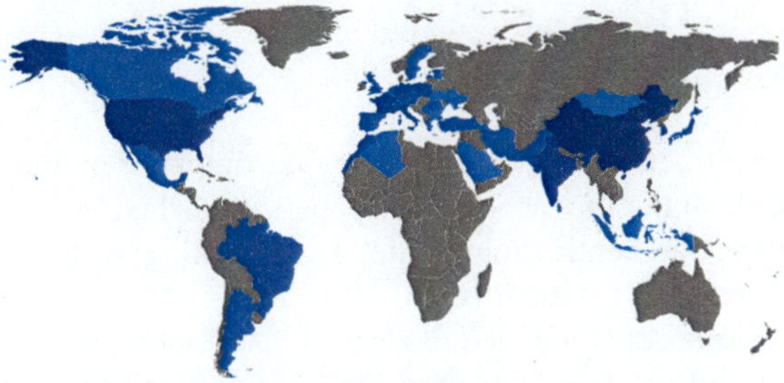

Fig. 2. Country scientific production

In dark blue color are countries with the biggest volume of scientific production and they are: China, USA, and India, then are countries marked in blue color: France, Germany, Italy, Taiwan, Brazil, Poland and the rest countries with smaller contribution to the topic are presented in light blue color: Austria, Japan, South Korea, Czech Republic, Greece, Morocco, Pakistan, Switzerland, UK, Argentina and Canada.

The twenty most used authors' keywords are shown on Fig. 3 through constructed word tree map. It can be seen that among the important for multimedia information security terms are: authentication, digital watermarking, reversible watermarking, data hiding, encryption, ETSI standards.

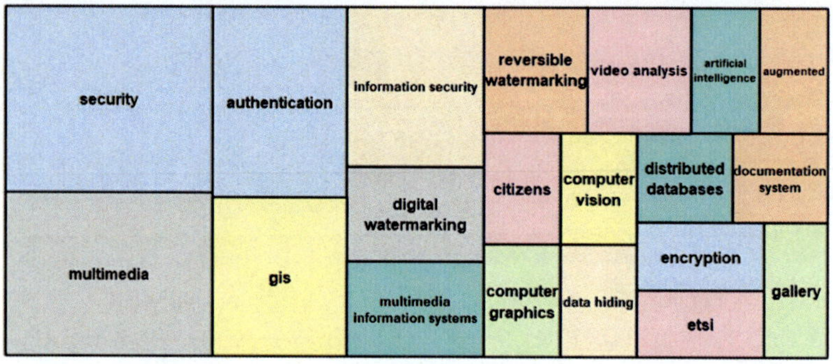

Fig. 3. The word tree map

The sources of published scientific production according to the value of h-index we divided into three groups: (1) journals: IEEE Multimedia, Multimedia Tools and Applications, (2) book series: Advanced Materials Research, Advances in Intelligent Systems and Computing, Lecture Notes in Electrical Engineering, and (3) conference proceedings: Communications in Computer and Information Science, International Conference on Intelligent Information Hiding and Multimedia Signal Processing, Military Communications Conference, Conferences of international society for optics and photonics, Conference on network and information systems security, European conference on Information Systems, others. The detailed information of selected literature sources with highest impact is presented in Table 1.

3.1 Determined Security Methods

In this section the identified security methods in the context of multimedia information systems are discussed taking into account the papers published in the literature sources with the highest impact: authentication, digital watermarking, reversible watermarking, data hiding, encryption, ETSI standards.

Authentication is a process for recognizing the user' identity and typically for this purpose a username and password are used [7]. They are compared with the database information and if they match the user get access to the multimedia information system. Other methods are: multifactor authentication through smart card usage and PIN code and through biometrical method that uses human physical or behavioral characteristics.

Digital watermarking is a process of embedding digital sign in a digital media that is used to copyright it and to show the owner [8]. Another application is related to data authentication that takes into account the sent watermarked digital media and received

Table 1. Literature sources with the highest impact and number of scientific papers

	Security	Authentication	Watermarking	Data hiding	Encryption	ETSI standards
Multimedia Tools and Applications	3455	1057	1081	1605	998	21
Advanced Materials Research	24637	2033	371	30	1894	41
International Conference on Intelligent Information Hiding and Multimedia Signal Processing	61	14	11	58	14	0
Military Communications Conference	661	62	2	14	65	3
European Conference on Information Systems	1094	147	1	54	91	3

one – they must be identically. Such techniques are cyclic redundancy check and parity check.

Reversible watermarking of digital media is a technique for full extraction of the watermark that gives possibility for complete restoration of this media [9]. In digital watermarking sometimes the watermark in its extraction could damage the digital media that then the media cannot be recovered. Several types of watermarking exist: robust – the digital media processing will not break the watermark, fragile – the digital media modification will break the watermark and this technique is suitable for media authentication, semi-fragile – only minor modifications on digital media will not break the watermark.

Data hiding is a technique that allows identification, copyright and annotation of digital media to be done through embedding data [10]. It is applicable for copyright protection and as proof for content integrity. The secret picture is integrated in the main one and could be read after its extraction. The Steganography is the science that examines methods and techniques for digital media protection through hidden data embedding.

Encryption is a process for conversion the plaintext (multimedia data) into cipher code through usage of appropriate symmetric or asymmetric cryptography algorithms

[11]. It protects digital media from unauthorized disclosure; it is used for keeping multimedia data integrity and for copyrighting the owner, when it is transmitted via Internet and when they are stored in databases.

ETSI is a European Standards Organization that is responsible for standardization of electronic communications and services, including in the field of cyber security and digital signatures [12]. Standards are used to describe when systems are secure and for what kind of requirements they have to respond.

4 Development of Secure Multimedia Information System

In this section the development of multimedia information system with several measures for data security is presented. The aim of the system is to deliver knowledge to students enrolled in the Informatics course that includes eleven topics. The system is realized in the form of web-based application through utilization of object-oriented language C# and Microsoft Visual Studio, ASP.Net, HTML5, CSS3, SQL Server Manager Studio and ToolBook authoring environment. The system recognizes three types of users: students, educators and administrators. The main page allows a registration process by unregistered students to be performed and then login into the system. Registered students have access to the course material, to additional information with contacts and to the Facebook profile of all course participants. ToolBook is used as an authoring tool for creation the course multimedia content. The educators have possibility to develop/edit course content, to create course learning strategies and to see statistical data regarding the content usage. Administrators have access to the system through Admin panel with functionality to manage registration queries (to accept or reject a given registration), to manage database and course data, to control the learning analytics data, to manage all registered participants and to change the users' access level. The function architecture of

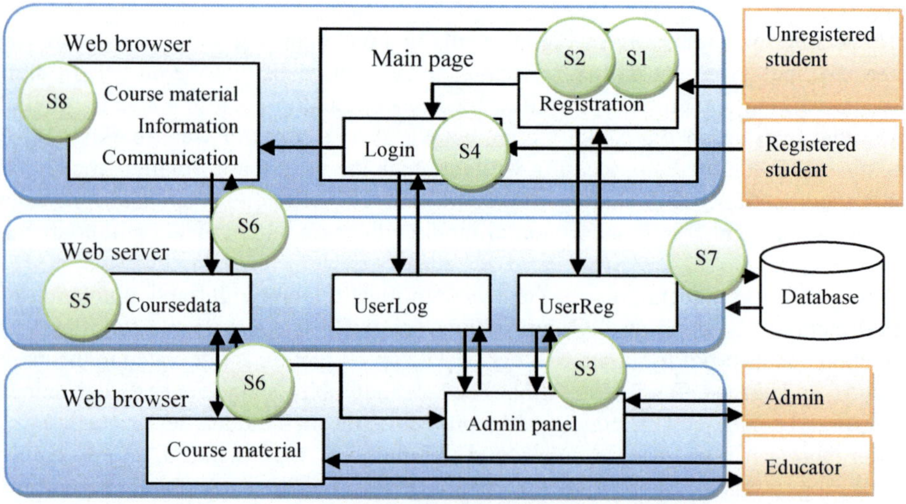

Fig. 4. Functional architecture of the multimedia information system with security points

the multimedia information system with the main security points is presented on Fig. 4. The security points are marked through symbols from S1 to S8 and are below explained.

***S1* – Registration security** – Unregistered users (students/educators) must be registered through password with high strength and unique username. It will hamper the *brute force attack, dictionary attack, password spraying attack, keyword logger attack.* Another possible way for registration is unregistered user to send a request for access to the administrator, filling up several fields with basic information. Then, the administrator has to decide whether to allow this registration query or not. A limited number of registration queries from one IP are allowed to protect the system form different kinds of *DNS DoS attacks* that are reasons for unavailability of a given service.

***S2* – Two-factor authentication** – The access to the system is possible after students' registration on the registration page. This security measure guarantees the system to be used only by authorized students and protects the course content from the *copyright aspect.* Additional measure is realized through the element *I'm not a robot* captcha clicker for protection from *abusive traffic.* Also, the administrator is facilitated, because will not administrate multiple false queries. In the same way, the database is protected from overload.

***S3* – Privileged access management** – After user's registration through password and username, the system automatically does not give the access to the system resources. The administrator must prove the query realizing privilege access strategy, giving full or limited access to the course. It will decrease the risk from credentials stealing and from access obtaining to important resources.

***S4* – Pattern matching** – The login is performed after two-factor authentication – username and password and captcha clicker. Regular strings are used for checking the allowed symbols for the fields from the registration form with aim the database to be protected from harmful commands. The used regular strings guarantee that SQL injection attack could not occur and records in the database will be not changed, stolen or damaged.

***S5* – Secure programming code** – ToolBook environment supports developers giving them possibilities to create secure programming code that will be very hard for changing from the side of unauthorized users who could provoke HTML injection or Cross-site scripting attack. The course lessons are prepared in different ToolBook files that also contribute to the increased security, because every file must be damaged one by one.

***S6* – Encrypted communication** – ToolBook created files are converted in HTML/CSS pages when the application is published on the web server. Then, the transmitted data among users and web server uses secure HTTPS protocol. HTTPS is responsible for secure authentication and for protection the integrity of transmitted data, hampering the man-in-the-middle attack. The encrypted communication between user and server functions against eavesdropping and tampering attacks.

***S7* – Database security** – Development of suitable database structure and organization as well as usage of admin username and strong password will lead to higher database security. For example, the security is improved, because the table with requested queries for registration is different from the table with the registered users. If an attacker sends harmful registration query this will not damage the data about registered users and the system will continue to perform its normal functions.

***S8* – GDPR (General Data Protection Regulation) security** – The requirements and recommendations of GDPR are followed with aim the users' private data to be protected. The students at registration process must give only basic information for him/herself that will confirm his/her status in the university. The redundant data is not required that will protect students if there is stealing, leakage or changing of information. Through the option *Settings*, the students can manage his/her profile that give possibility for observation or change the content or visibility of some types shared data.

The prototype of the developed multimedia information system from students' and administrators' point of view is presented on Fig. 5.

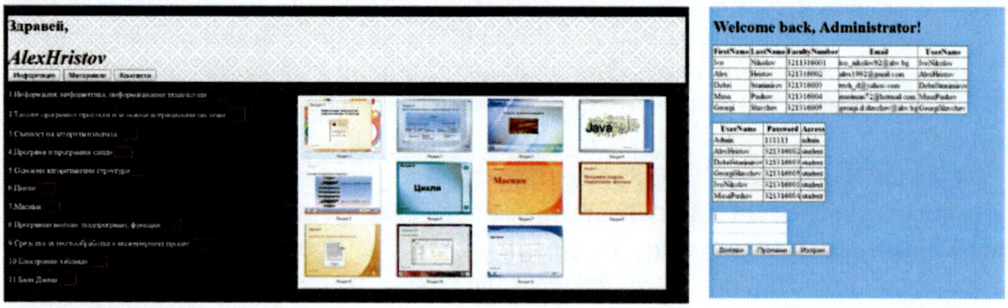

a). Students' view b). Administrator panel

Fig. 5. Prototype of the multimedia information system

5 Predictive Model

The predictive model based on the gained experience at development of secure multimedia information system and literature review conductance regarding the importance of security measures is created with aim to depict the future trends in the discussed topic. The predictive model is constructed on the data, taken from SCOPUS database for 10 year period (from 2010 to 2019 year) regarding the number of scientific papers that in their title, abstract and keyword include the terms: security, authentication, watermarking, encryption, https, multimedia, information system and eLearning. To select the best algorithm for predicting the future number of scientific papers that present security issues in multimedia information systems for eLearning purpose, three algorithms for prediction are evaluated: linear regression, additive regression and regression by discretization accorting to their accuracy, as it shown in Table 2. The data is trained for each algorithm in Weka software and the chosen test option is cross-validation of 10 folds. Linear regression algorithm is selected for predictive model construction, because of its better accuracy and the model in shown on Fig. 6.

Table 2. Comparison the algorithms according to their accuracy

	Linear regression	Additive regression	Regression by discretization
Correlation coefficient	0.9708	0.595	0.6431
Relative absolute error	23.8226	99.9551	66.8307
Root relative squared error	22.7038	92.028	77.0739

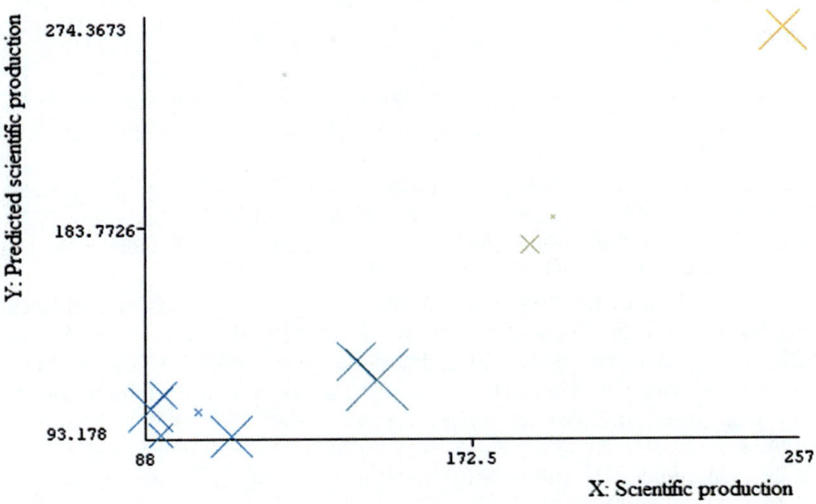

Fig. 6. The created predictive model

6 Conclusion

The exploration of the scientific literature according to defined above research method shows that the security topic grasps again the scientist interests, because of the increased multimedia production, emerging new technologies and extremely intensive information transfer via Internet. And it is applicable for educational sector too; where multimedia gives possibilities for achieving high interactivity, multi-format presentation and complex hierarchical content structures. Also, the exploration shows that among the most used author keywords in the scientific papers are: authentication techniques, digital watermarking, reverse watermarking and data hiding methods, encryption algorithms and application of ETSI standards for cyber security. And it means that more attention requires these techniques and methods at systems development and utilization. One part of them is taken into consideration at prototype development of multimedia information system for eLearning purpose in the Informatics course like authentication and encryption methods. Other part of secure measures such as pattern matching techniques, privilege access management strategy, secure programming code, and GDPR policy are considered as very important for the system realization in educational context and this is the reason for their implementation.

Also, the security issues in multimedia information systems are taken into account at creation the predictive model through usage of machine learning algorithms. The model predicts the number of scientific papers indexed in SCOPUS that discuss the security problems in multimedia information systems for eLearning purposes.

Acknowledgements. The authors would like to thank the Research and Development Sector at the Technical University of Sofia for the financial support.

References

1. Sebastian Boell, K., Cecez-Kecmanovic, D.: What is an information system? In: Proceedings of 48th Hawaii International Conference on System Sciences, pp. 4959–4968 (2015). https://doi.org/10.1109/hicss.2015.587
2. Introduction to Multimedia and Hypermedia, Chap. 1. https://ftms.edu.my/v2/wp-content/uploads/2019/02/MMGD0101-chapter-1_052015.pdf. Accessed 18 Feb 2020
3. Kanellopoulos, D.N.: Current and future directions of multimedia technology in tourism. Int. J. Virtual Technol. Multimedia **1**(2), 187–206 (2010)
4. Cucchiara, R., Grana, C., Borghesani, D., Agosti, M., Bagdanov, A.D.: Multimedia for cultural heritage: key issues. In: Grana, C., Cucchiara, R. (eds.) MM4CH 2011. CCIS, vol. 247, pp. 206–216. Springer, Heidelberg (2012). https://doi.org/10.1007/978-3-642-27978-2_18
5. Saadé, R.G., Galloway, I.: Understanding intention to use multimedia information systems for learning. Issues Informing Sci. Inform. Technol. **2**, 287–296 (2005). https://doi.org/10.28945/828
6. Aria, M., Cuccurullo, C.: Bibliometrix: an R-tool for comprehensive science mapping analysis. J. Informetr. **11**(4), 959–975 (2017)
7. Georgescu, A.: Vulnerabilities in a two-factor user authentication in multi-server networks protocol. In: de la Puerta, J.G., et al. (eds.) International Joint Conference SOCO'14-CISIS'14-ICEUTE'14. AISC, vol. 299, pp. 495–504. Springer, Cham (2014). https://doi.org/10.1007/978-3-319-07995-0_49
8. Wang, Y., Yang, C., Zhu, C.: A multiple watermarking algorithm for vector geographic data based on coordinate mapping and domain subdivision. Multimedia Tools Appl. **77**(15), 19261–19279 (2017). https://doi.org/10.1007/s11042-017-5358-6
9. Kim, C., Yang, C.-N.: Watermark with DSA signature using predictive coding. Multimedia Tools Appl. **74**(14), 5189–5203 (2013). https://doi.org/10.1007/s11042-013-1667-6
10. Ri, Y., Dong, J., Wang, W., Tan, T.: Adaptive histogram shifting based reversible data hiding. In: Pan, J.-S., Tsai, P.-W., Watada, J., Jain, L.C. (eds.) IIH-MSP 2017. SIST, vol. 82, pp. 42–50. Springer, Cham (2018). https://doi.org/10.1007/978-3-319-63859-1_6
11. Mao, X., Liu, Y., Lu, S., Li, Y.: A new secure communications solution for network application. Adv. Eng. Forum **6–7**, 932–936 (2012). https://doi.org/10.4028/www.scientific.net/AEF.6-7.932
12. ETSI European Organization for Standardization. https://www.etsi.org/. Accessed 18 Feb 2020

Estimating Student's Performance Based on Item Response Theory in a MOOC Environment with Peer Assessment

Minoru Nakayama[1(✉)], Filippo Sciarrone[2], Masaki Uto[3], and Marco Temperini[4]

[1] Tokyo Institute of Technology, Meguro, Tokyo 152-8552, Japan
nakayama@ict.e.titech.ac.jp

[2] Rome Tre University, Rome, Italy
sciarro@ing.uniroma3.it

[3] The University of Electro-Communications, Chofu, Tokyo, Japan
uto@ai.is.uec.ac.jp

[4] Sapienza University, Rome, Italy
marte@diag.uniroma1.it

Abstract. Peer Assessment is a powerful strategy to support educational activities and the consequent learners' success. Learning performance of participating is often estimated in a peer assessment setting using Item Response Theory. In this paper, a feasibility of estimating individual performance is examined for a simulated data set representing a MOOC environment, where one thousand students are supposed to perform a Peer Assessment session, where each peer assesses three other peers' work. For each student the modeling traits "ability", "consistency", and "strictness" are evaluated using Generalized Partial Credit Model, and the validity of such calculation is confirmed. While taking into consideration the limits of the synthetic sample production, this experiment provides an evidence of the possibility to predict learning performance in the large scale learning conditions of a MOOC.

Keywords: Peer Assessment · MOOC · Item Response Theory

1 Introduction

Massive Open Online Courses (MOOCs) have been developing for several years, with the aim of delivering learning contents to a numerous and worldwide audience [6]. They have well known and long studied problems with the dropout rates, which are in part due to the nature itself of these courses (still quite usually easy to enrol in and inexpensive), and in part for issues related to maintaining motivation and continuing engagement.

Peer Assessment (PA) is a powerful strategy to support educational activities and the consequent learners' success. With PA the learners can be exposed to

Z. Kubincová et al. (Eds.): MIS4TEL 2020, AISC 1236, pp. 25–35, 2021.
https://doi.org/10.1007/978-3-030-52287-2_3

different cognitive experiences: on the one hand they answer to a request (such as a question, or a task to be performed); on the other hand they are requested to assess other learners' work, being then involved in cognitive activities of higher level [3].

To encourage participant's learning activity, PA has been sometimes introduced in a MOOC [1,17], although the reliability of the assessments, and in general the applicability of the strategy are still under verification.

PA has been proven reliable in applications where the teacher mediation and participation in the grading process are supported. In this approach the teacher grades a fraction of the learners' answers, and all the remaining answers have automated grading, based on the learners' models calibrated by the PA activity and by the available teacher's grades.

In particular, the learners modeling, deemed to represent the ability of the learner to answer a question, and the different ability to assess others' answers, has been studied in several research work. In [4] a Bayesian Network based modeling is defined, for the two learner's traits mentioned above, and a teacher mediated approach to PA is described. For this approach, involving a certain amount of teacher's grading (although just a fraction of the whole class of learners), an alternative has been proposed in [15], based on a modified version of K-NN algorithm [12], that is expected to be lighter from the computational point of view. This kind of teacher activity, though, is difficult to be granted if the number of students goes into the thousands, as is the case of MOOCs.

In [19] Item Response Theory (IRT) has been studied for similar purposes, conducting to a modeling of the learner's ability to assess. This method allows for a lighter computational cost of the process of modeling and automated assessment, but has the limit of having been experimented in too limited contexts (as far as the number of students is concerned).

The problem of having these approaches experimented on limited numbers of students is general, and in this paper we try and show a way of overcoming it, by the production of a "simulated MOOC" (i.e. a set of, say, 1000 simulated students, representing learners respecting a statistical distribution of their models' features).

This paper confirms the feasibility of determining the individual ability by using the IRT modelling technique applied to a huge PA data-set, based on the above simulation.

The remainder of the paper is structured as follows. In Sect. 2 the process of validation is shown, while in Sect. 3 the PA data-set building process is depicted. IRT is showed in Sect. 4. Our experimental results are in Sect. 5 while in Sect. 6 conclusions and future works are illustrated.

2 Process of Validation for the IRT Approach to PA

Basically, in the application of PA we are dealing with in this paper, we imagine that the PA system (PAS) will manage the data coming from a session of PA,

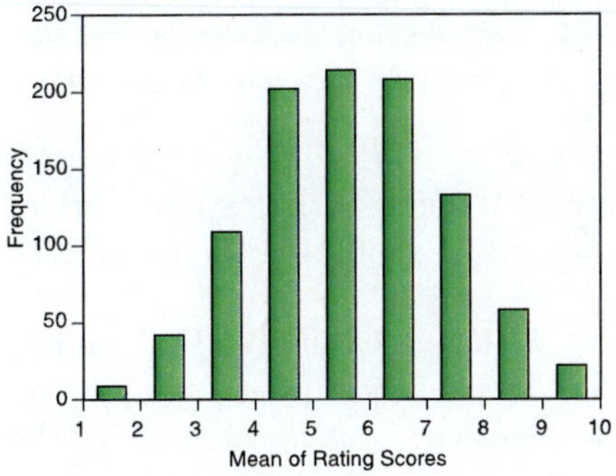

Fig. 1. Histogram of mean rating scores.

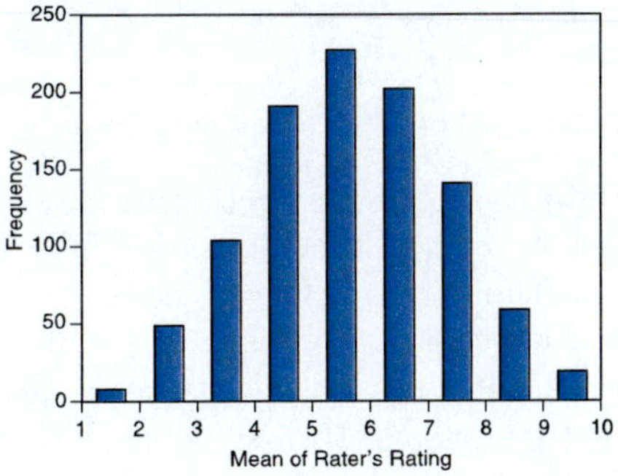

Fig. 2. Histogram of mean peer's ratings.

and the grades provided by the teacher, to produce automated grading of the non-teacher-graded answers.

In a PAS there are three main data-sets of interest, described in the following paragraphs: the result of an experiment is in evaluating how well they are correlated, that is how well the PAS is able to compute valid learner's models and correct grading.

First there is the set of all the learners' models, describing the actual skills of the MOOC's members. We call this set Real (learners') Models (*RMs*). These data are generated according to a statistical distribution underlining them, and will be used to estimate the good behavior of the PAS.

Then there is the set of data coming from the PA activity (PA data-set). From the point of view of the data originated by a PA session, we can imagine that mainly two data-set are produced by Gaussian simulation, as follows: 1) the grades given by the learners to their peers (e.g. 3 grades for each peer, given and received); 2) the teacher's grade for each student's work. Notice that in real applications the data-set in 2) would contain only a fraction of the whole set of grades, whereas in our (simulated) experiments it covers the whole class (this is obtained by simulating a Gaussian distribution of the teacher's grades, so not involving real work from a real teacher).

Based on the PA data-Set, the PAS produces a new set of learners' grades. (Computed Models *CMs*). If the PAS is doing a good job, the new grades in *CMs* will be close to the ones in *RMs*. A more formal definition of the experiment we are going to present is as follows.

Definition 1. *An experiment of PAS consists of the following steps:*

1. *The RM data-set creation, based on a given statistical distribution;*
2. *The creation of the PA data-set, based on a given statistical distribution;*
3. *The application of the PAS algorithms by managing the creation and update of the grades in CM;*
4. *The computation of the grades provided by the PAS on those answers for which the teacher's grade in the PA data-set was not used;*

5. *Evaluation of the results, whereas the results can be the analysis of the correlation between the real and computed grades, that is between the teacher's real grades Vs. those computed by the PAS.*

In the following Section we discuss the creation steps of the experiment. In Sect. 4 we show the application of IRT with the aim of validating the use of IRT to a (simulated) MOOC context.

In Sect. 4 we will use a modeling consistent with the features of IRT, and in particular with the features of a *Generalized Partial Credit Model* with Rater Parameters. By this modeling technique, the learner j is represented by the following features:

- θ_j, latent ability of j
- α_j, consistency of rater j
- β_{jk}, strictness of the rater j for rating category k

3 The Peer Assessment Data-Set

In this section we briefly report on the process of creation of the *RM* and *PA data-set*, provided by our simulation system [16].

3.1 The Data Generation Procedure

First data to be generated is the PA data-set, based on a specific graphical interface used by the experimenter (typically a teacher). The interface allows to generate the grades given by each peer to her/his n peers (in our use of the simulation system $n = 3$).

These grades are are not actually given directly by the teacher, rather they are generated randomly, according to the classic bell-shaped Gauss distribution, which has proven to be very reliable in representing grades distribution in a learning context. For such distribution, the parameters μ (the value of the mean), and σ^2 (variance) are set by the experimenter. In our case $\mu = 5.5$ and $\sigma^2 = 1.63$

The student models are then implicitly defined based on the PA data-set, as we suppose that each student in the experiment is behaving accordingly to her/his model.

3.2 The Features of the Generated Data

The simulated data-set consists of 1000 participants, where three of them assess another participant's performance using a 10 points scale, i.e. a participant's performance i, r_i, is computed by the grades given by other three participants such as $i + 1, i + 2, i + 3$, here $i = 1, \ldots, 1000$.

Simple statistics of generated data are regulated by the algorithm. Both means of rating scores and peer's ratings are summarized in Figs. 1 and 2. Also, the relationship between two means is illustrated in Fig. 3.

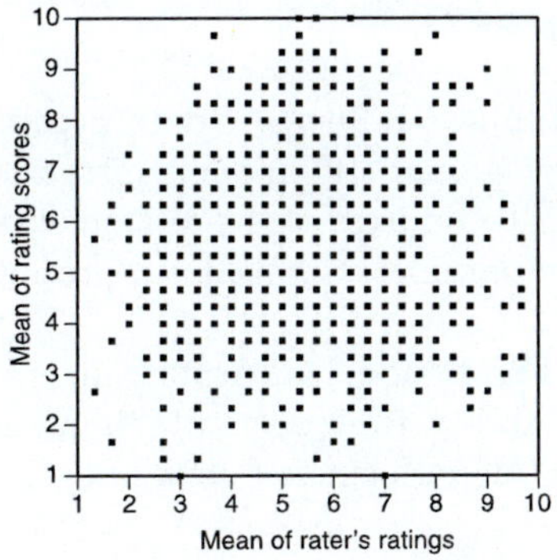

Fig. 3. Scatter gram between peer's ratings and rating scores for each participant

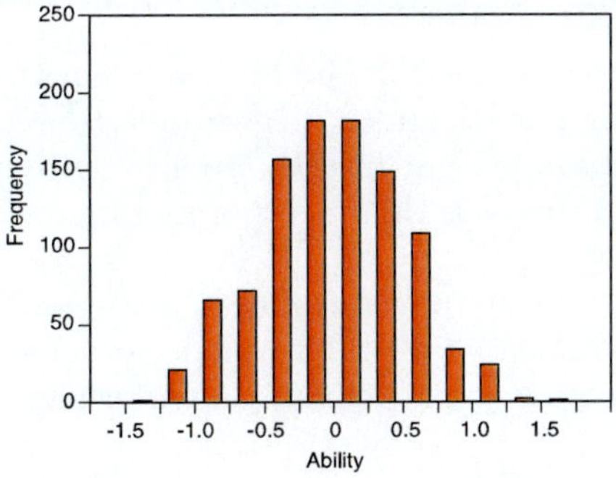

Fig. 4. Histogram of calculated ability of participants.

4 The Item Response Theory

This study attempt to apply IRT [10], a test theory based on mathematical models, to the PA data. IRT models provide an item response function that specifies the probability of a response to a given test item as a function of latent participant's ability and the item's characteristics such as difficulty and discrimination. IRT enables to estimate participant ability while considering item characteristics.

IRT also has the following advantages: i) participant responses to different test items can be assessed on the same scale; ii) missing data can be easily estimated.

IRT has traditionally been applied to test items for which responses can be scored as correct or incorrect. In recent years, however, there have been attempts to apply polytomous IRT models to performance assessments, including PA.

4.1 The Generalized Partial Credit Model

A representative polytomous IRT model is the generalized Partial Credit Model (GPCM) [13]. The GPCM gives the probability that participant j receives score k for test item i as

$$P_{ijk} = \frac{\exp \sum_{m=1}^{k} [\alpha_i(\theta_j - \beta_{im})]}{\sum_{l=1}^{K} \exp \sum_{m=1}^{l} [\alpha_i(\theta_j - \beta_{im})]}, \tag{1}$$

where:
α_i is a discrimination parameter for item i, β_{ik} is a step difficulty parameter denoting difficulty of transition between scores $k-1$ and k in the item, and θ_j is the latent ability of participant j. Here, $\beta_{i1} = 0$ for each i is given for model identification.

4.2 The GPCM with Rater Parameters

As described in the Introduction, this study applies IRT to PA data comprising participants × peer-raters. However, the traditional IRT models introduced above are not directly applicable to such data. To address this problem, many IRT models that incorporate rater characteristic parameters have been proposed [20].

This study introduces a state-of-the-art GPCM model incorporating rater parameters [21]. This model provides the probability that peer-rater r assigns score k to participant j's performance for item (or performance task) i as

$$P_{ijrk} = \frac{\exp \sum_{m=1}^{k} [\alpha_r \alpha_i(\theta_j - \beta_i - \beta_r - \beta_{rk})]}{\sum_{l=1}^{K} \exp \sum_{m=1}^{l} [\alpha_r \alpha_i(\theta_j - \beta_i - \beta_r - \beta_{rk})]}, \quad (2)$$

where, α_r reflects the consistency of rater r, β_{rk} is a step difficulty parameter denoting difficulty of transition between scores $k-1$ and k in the rater r. β_r is represents the strictness of rater r, and β_i is the difficulty of item i. Here, $\prod_i \alpha_i = 1$, $\sum_i \beta_i = 0$, $d_{r1} = 0$, and $\sum_{k=2}^{K} d_{rk} = 0$ are given for model identification.

This study applies this model to the PA data described in Sect. 3. It is noteworthy that, in the data, the number of performance tasks is fixed to one. In this case, α_i and β_i are ignorable because the model identification constraints restricts the value of $\alpha_{i=1} = 1$ and $\beta_{i=1} = 0$.

4.3 The Parameter Estimation

To estimate IRT model parameters, marginal maximum likelihood estimation using an EM algorithm has been commonly used [2]. However, for complex models like that used in this study, EAP estimation, a form of Bayesian estimation, is known to provide more robust estimations [5]. EAP estimates are calculated as the expected value of the marginal posterior distribution of each parameter. For the EAP estimation, Markov Chain Monte Carlo (MCMC), a random sampling–based estimation method, is generally used [5,18].

The Metropolis-Hastings-within-Gibbs sampling algorithm has been used as a MCMC algorithm for IRT models [14]. The algorithm is simple and easy to implement, but it requires long times to converge to the target distribution [8]. The Hamiltonian Monte Carlo (HMC) is an alternative MCMC algorithm with high efficiency. In recent years, the No-U-Turn (NUT) sampler [8], an extension of HMC that eliminates hand-tuned parameters, has been proposed. Because the "Stan" software package makes implementation of a NUT-based HMC easy, this algorithm has recently been used for parameter estimations in various statistical models, including IRT models [9,11].

For the aforesaid reasons, we use a NUT-based MCMC algorithm for parameter estimations. We calculate EAP estimates as the mean of parameter samples obtained from 500 to 1,000 periods of three independent MCMC chains. Furthermore, the standard normal distribution is used as the prior distributions.

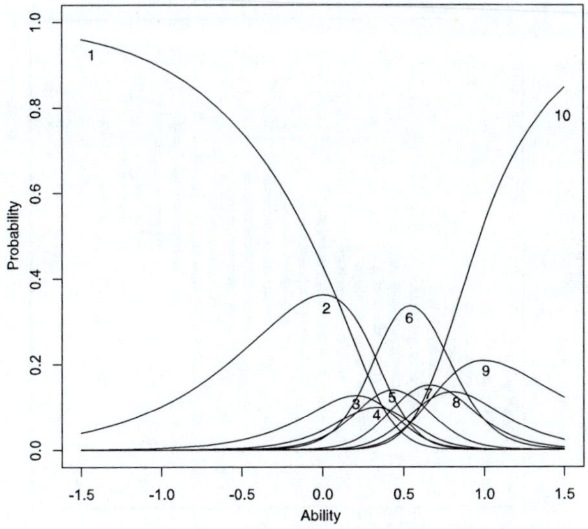

Fig. 5. Curves of item category response function (ICRF) for the 4th rater.

5 Results

We estimated the IRT parameters from the PA data described in Sect. 3.

Table 1. Simple statistics of ratings and estimated parameters.

Variable	N	Mean	STD
Rating scores	1000	5.50	1.63
Peer's rating	1000	5.50	1.66
Ability	1000	0.00	0.51
Consistency	1000	0.90	0.46
Strictness	1000	0.00	0.48

Though each rater gave scores to three participants only, such as rates "1", "4" and "6", the MCMC algorithm estimated the IRT parameters. The MCMC was run using 2.5 GHz 14 core Intel Xeon W processor. The calculation time was 1628.57 s. We confirmed the Gelman–Rubin statistic $\hat{R}$ [7], which is generally used as a convergence diagnostic. Values for these statistics were less than 1.1 for all parameters, indicating that the MCMC runs converged.

Table 1 shows the overall simple statistics for learner's model features ability (Fig. 4), rater consistency and strictness. Furthermore, as an example, Fig. 5 depicts the curves of item category response function (IRCs) for the 4-th rater. In the Figure, the horizontal axis represents the latent ability θ, while the vertical

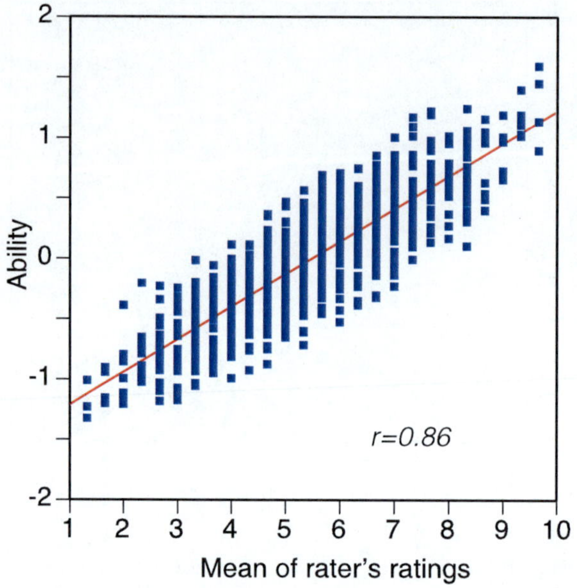

Fig. 6. Relationship between mean peer's ratings and the estimated ability.

axis shows the response probability for each rating category. The figure indicates a characteristic of overall rating behaviors.

To illustrate a validation of estimated ability using our model, the relationship between the ability and means of rater's ratings is summarized in Fig. 6. The Figure shows a strong correlation, with $r = 0.86$. In the Figure, a regression line is overlapped to the scatter-plot. Since rating behaviour affects to estimate the ability, some deviations for the ability are observed.

In order to examine the rating behaviours, the relationships between rating score and consistency, and rating score and strictness are summarized in Fig. 7 and 8 respectively. The consistency is again deviating over the rating scores and the behaviour may be controlled by the data generation procedure. The strictness decreases with rating score, the correlation coefficient is $r = -0.89$. The strictness indicates severeness of rating so that there is negative relationship between them. In addition to the relationship, the data generation may influence some deviation of the strictness which are observed in the Figure.

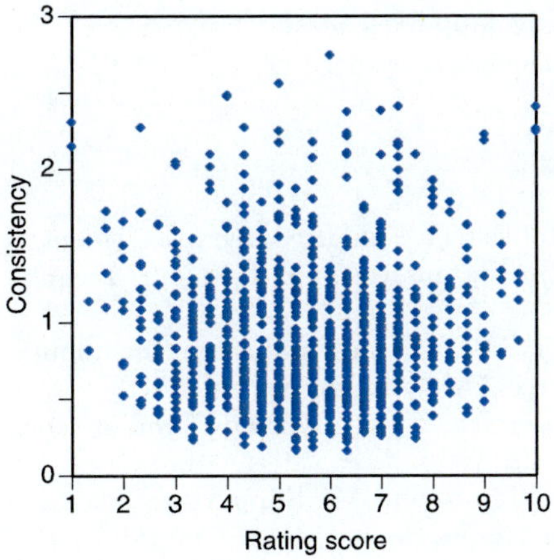

Fig. 7. Relationship between mean rating scores and the estimated consistency.

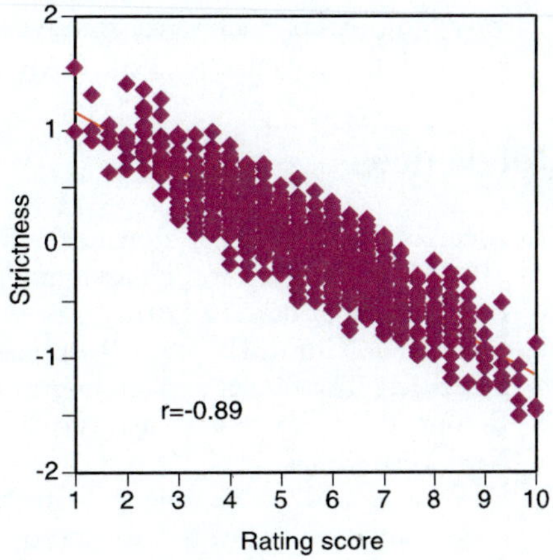

Fig. 8. Relationship between mean rating scores and the estimated strictness.

6 Conclusions and Future Work

This paper examined a possibility of estimating student's performance which was based on a simulated data set using the IRT. The simulated data-set is generated with 1000 students and with PA based on the requirement of having 3-peer's grading per each student, where a 1–10 point scale is used for the grades. The performance was estimated by using the GPCM and some detailed parameters were calculated with EAP estimation and MCMC techniques.

As a result, we had the opportunity to estimate values of ability, consistency and strictness for each participant, in a learning context characterized by one thousand students.

We modeled the students by means of the above mentioned parameters, and the feasibility of the estimation for a large data set was examined, confirming the validity of the simulated approach for the trial.

In particular, the limited experimentation we presented allowed us to plan a more extensive use of simulation of PAS in a MOOC context, and the enrichment of the PAS capabilities by means of IRT. Such future work will have the intent of checking on PAS aspects, that can have influence on the PAS performance, and see how their improvements could make the PAS better. By PAS aspects above, we mean, for instance, the number of peers in the class, the number of peer assessments required by each peer, the applied rating scale, the tuning of teacher's grading, by recommendation of what peer's work would be more beneficial to grade by the teacher, in order for the PAS to produce a better automated grading, and in general the algorithms to actually computing the automated grading in a PA session.

Acknowledgement. This research was partially supported by the Japan Society for the Promotion of Science (JSPS), KAKEN (17H00825).

References

1. Alcarria, R., Bordel, B., deAndrés, D.M., Robles, T.: Enhanced peer assessment in MOOC evaluation through assignment and review analysis. Int. J. Emerg. Technol. Learn. **13**(1), 206–219 (2018)
2. Baker, F., Kim, S.H.: Item Response Theory: Parameter Estimation Techniques. Statistics, Textbooks and Monographs. Marcel Dekker, New York (2004)
3. Bloom, B.S.: Taxonomy of Educational Objectives. David McKay Company Inc., New York (1964)
4. De Marsico, M., Sciarrone, F., Sterbini, A., Temperini, M.: Supporting mediated peer-evaluation to grade answers to open-ended questions. EURASIA J. Math. Sci. Technol. Educ. **13**(4), 1085–1106 (2017)
5. Fox, J.P.: Bayesian Item Response Modeling: Theory and Applications. Springer, Heidelberg (2010)
6. de Freitas, S.I., Morgan, J., Gibson, D.: Will MOOCs transform learning and teaching in higher education? engagement and course retention in online learning provision. Brit. J. Educ. Technol. **46**(3), 455–471 (2015)
7. Gelman, A., Rubin, D.B.: Inference from iterative simulation using multiple sequences. Stat. Sci. **7**(4), 457–472 (1992)
8. Hoffman, M.D., Gelman, A.: The No-U-Turn sampler: adaptively setting path lengths in Hamiltonian Monte Carlo. J. Mach. Learn. Res. **15**, 1593–1623 (2014)
9. Jiang, Z., Carter, R.: Using Hamiltonian Monte Carlo to estimate the log-linear cognitive diagnosis model via Stan. Behav. Res. Methods **51**(2), 651–662 (2019)
10. Lord, F.: Applications of Item Response Theory to Practical Testing Problems. Erlbaum Associates, Mahwah (1980)
11. Luo, Y., Jiao, H.: Using the stan program for bayesian item response theory. Educ. Psychol. Meas. **78**(3), 384–408 (2018)
12. Mitchell, T.M.: Machine Learning, 1st edn. David McKay, New York (1997)
13. Muraki, E.: A generalized partial credit model. In: van der Linden, W.J., Hambleton, R.K. (eds.) Handbook of Modern Item Response Theory, pp. 153–164. Springer, Heidelberg (1997)
14. Patz, R.J., Junker, B.: Applications and extensions of MCMC in IRT: multiple item types, missing data, and rated responses. J. Educ. Behav. Stat. **24**, 342–366 (1999)
15. Sciarrone, F., Temperini, M.: K-openanswer: a simulation environment to analyze the dynamics of massive open online courses in smart cities. Soft Computing (2020). In Press
16. Sciarrone, F., Temperini, M.: Simulating massive open on-line courses dynamics. In: Proceedings of iTHET 2019, Magdeburg, Germany, pp. 1–9 (2019)
17. Sun, D.L., Harris, N., Walther, G., Baiocchi, M.: Peer assessment enhances student learning: the results of a matched randmized crossover experiment in a college statistics class. PLoS ONE **10**(12), 1–7 (2015)
18. Uto, M.: Rater-effect IRT model integrating supervised LDA for accurate measurement of essay writing ability. In: Proceedings of the International Conference on Artificial Intelligence in Education, pp. 494–506 (2019)
19. Uto, M., Ueno, M.: Item response theory for peer assessment. IEEE Trans. Learn. Technol. **9**(2), 157–170 (2016)

20. Uto, M., Ueno, M.: Empirical comparison of item response theory models with rater's parameters. Heliyon **4**, 1–32 (2018)
21. Uto, M., Ueno, M.: Item response theory without restriction of equal interval scale for rater's score. In: Proceedings of the International Conference on Artificial Intelligence in Education, pp. 363–368 (2018)

Retrieving Relevant Knowledge from Forums

Guy Betouene(✉) and Laurent Moccozet

University of Geneva, CUI, 7 route de Drize, Battelle, 1227 Carouge, Switzerland
{Guy.Betouene,Laurent.Moccozet}@unige.ch

Abstract. There are many instances of online content management systems where users collectively produce knowledge. Some of them are used for educational purposes such as forums in Massive Open Online Courses MOOCs. However, the lack of structuring of the useful content reduces the possibility of further reusing them. We believe that reorganizing the knowledge available in a forum can make it more easily reusable as a pedagogical resource. In this article, we present the first part of our analysis, namely developing a method for sorting the information and thus for assessing the quality of the information. We examine the effectiveness of our method on the Stackoverflow forum thread examples. The results are relevant and show that it is possible to assign weights to the information contained in the discussion forums in order to sort them out.

Keywords: Forum summary · Reuse discussion forums · Information retrieval

1 Introduction

Discussion forums are widely used to exchange knowledge between learners and teachers, between novices and experts or between peers. Their omnipresence in MOOCs in particular has generated renewed interest and led to numerous works to analyze their content with various objectives.

1.1 Context

Many online services and resources available on the Web are widely used to find answers to specific questions. These questions can be answered using tools such as search engines, which analyze published content to build indexes. With the emergence of the Semantic Web, we also see that there is a need to restructure the data published on the Web, with for example the Google's knowledge graph[1] together with the Schema microdata format: http://schema.org that improve the response capabilities of search engines.

Part of the content can be published on the Web by third party, by using tools that facilitate the publication of content, such as wikis or discussion forums. Some

[1] Knowledge Graph - Wikipedia: https://en.wikipedia.org/wiki/Knowledge_Graph. Knowledge Graph is a knowledge base used by Google and its services to improve the results of its search engine with information from various sources.

Z. Kubincová et al. (Eds.): MIS4TEL 2020, AISC 1236, pp. 36–46, 2021.
https://doi.org/10.1007/978-3-030-52287-2_4

systems such as Stackoverflow: stackoverflow.com/are very popular for getting accurate answers on technical topics. However, it is difficult to reuse the knowledge stored in these systems once the discussion is over. One of the reasons for this is related to the structuring of the data. They are organized according to the mechanism of user interaction: questions/answers sorted chronologically and according readers' votes.

We aim at reorganizing the available knowledge in a forum and transform it into learning resources so that it can be easily reused. A first step to achieve is to identify and sort the relevant knowledge from the forum threads.

1.2 Related Work

Forums are very old tools for exchange between users and they have aroused a lot of interest in analyzing their content. For example: selection of answers relevant to the topics discussed, summary of discussions, extraction of data.

Messages in the discussion forum are acts of dialogue, tagging messages to identify their roles in the discussion helps determine which messages are relevant [1, 2]. These tags can be used to summarize the discussion thread [3, 4]. The summary can be done by extracting relevant sentences or relevant concepts [5, 6].

A thread is stored as an HTML web page whose structure varies depending on the forum website. The extraction of the data allows to obtain various information such as the user name, the date of the message, the text of the message [7–9].

1.3 Contribution

Our objective is to discover relevant information and to sort them according to their weight of relevance. Message labelling identifies the type of message in the discussion so gives an idea of the content. However, this method does not determine the relevance of the messages. In this case, other approaches must be used. To do this, we draw on the method of syntactic emotion analysis [10] to identify the type and relevance of the content and also the method of the document summary to establish a relationship between the information and the topic of the discussion. We propose an approach to measure the level of relevance of an answer.

Next, we present the first part of the analysis, namely the development of a method for sorting information and thus for assessing the quality of the information. This document is organized as follows. Section 2 presents the strategy used to sort the information. Section 3 discusses the results of the analysis and Sect. 4 the conclusion and future work.

2 Sorting Information from a Discussion Thread

To sort the information, one must first assess its quality according to criteria, then weight and combine the criteria to obtain an overall indicator of quality.

2.1 Assess the Quality of Information

In order to assess the quality of the information contained in a forum answer, it is first of all necessary to identify the criteria that can be extracted from it and to define how they should be interpreted. Then assign a weight to these criteria and find a way to combine them in order to assign a value to the quality of the answer.

Criteria for Assessing the Quality of Information
Following an analysis of forum threads, we have identified the following criteria: the repetitions, the users vote and the feedbacks. We assume that if a solution comes up several times, it is relevant; a message with a high score is probably relevant; and the feedback validates the answer. The following section presents the strategy used to assign weights to these criteria.
Weight the Criteria of Evaluation

Repetition
The weight of answer according to the repetition is the number of occurrences of this answer in relation to the total number of answers. Thus, the weight of answer according the repetition is:

$$r_{a_i} = \#a_j/A$$

where a_j is a distinct answer; $\#a_j$ is the number of occurrences of a_j and A is the total number of answers

Vote
The weight of answer according to the vote is the vote score divided by the total vote:

$$v_{a_i} = \frac{v_i}{\sum v_i}$$

where v_i is the vote of the answer i.

Feedback
Sentiment analysis [11–14] and syntactic emotions analysis [10] make it possible to determine whether the feedback validates the response (positive polarity) or not (negative polarity). The weight of the polarities oscillates between -1 and 1.

Polarity is determined from expressions e.g., "this solution works". However, a feedback can contain several validation expressions and can be both negative and positive. In this case, the emerging polarity is the sum of the polarities encountered, e.g. "Well, **this solution works. Your answers are good; it's a good thing to do; only the last line is incorrect**". An answer can have several feedbacks, to find the emerging polarity; the average polarity of all the feedbacks is calculated. The weight of answer according all feedbacks:

$$f_{a_i} = \left\{ \frac{\sum_1^n pol_{a_{ij}}}{n} \right\}$$

where $pol_{a_{ij}}$ is the polarity of answer according to one feedback.

Combination of Criteria

The most important criterion is the explicit evaluation of the response (a priori tested by the responder) in the readers' feedbacks. It is sufficient to determine the quality of the response. The problem is that not every response receives systematic feedback and the feedback does not always contain an explicit evaluation of the response. In these extreme situations, it is necessary to refer to the other criteria. However, it is important to ensure that the explicit evaluation of the response, whether positive or negative, in a feedback has a greater impact than the other criteria. In a discussion thread, several responses may provide different solutions to the same question. If a solution appears more than once, this reinforces its relevance. The second time a solution occurs, its weight will become so significant that it will impose itself over the other criteria. The answer with the most votes is the one displayed at the top of the answers. It is therefore the most consulted and appears the most relevant. However, the distribution of votes does not systematically reflect the quality of the answers. The votes therefore have less weight in the final decision. An answer is relevant if the sum of the weights of the criteria is greater than the sum of the thresholds of those criteria.

$$w_{a_i} = f_{a_i} + v_{a_i} + r_{a_i}$$

where w_{a_i} is the sum of weights of criteria; f_{a_i} is the weight of answer according feedbacks; v_{a_i} is the weight of answer according to the vote; r_{a_i} is the weight of answer according to the repetition.

2.2 Example 1

To illustrate the effect of our method, we have selected a thread based on the following question: "***What is going wrong with web deployment from Visual Studio and App service?***" Table 1 shows the different messages in the discussion thread. The answers containing expression "*delete settings WEBSITE_RUN_FROM_ZIP*" are repetitive.

Table 1. Answers

Type	Post
Question	Suddenly Web Deployment started failing. `Could not find file 'D:\home\site\wwwroot\App_Offline.htm'`
Answer1	All of a sudden VSTS default deployment mode became Run-From-Zip. The solution is setting *Select deployment method* checkbox in VSTS deploy and be sure *Web Deploy* is selected. To "unlock" the service you need to delete settings *WEBSITE_RUN_FROM_ZIP* from *Application settings* page Note: The new name of this settings is WEBSITE_RUN_FROM_PACKAGE
Answer2	I hade the same problem today with net core 2.2 and the solution was to remove setting "WEBSITE_RUN_FROM_PACKAGE" in azure appsettings
Answer3	We got hit by this same issue - the file system becomes readonly when WEBSITE_RUN_FROM_PACKAGE = 1. **Azure App Service** seems to be adding this app setting automatically during recent platform upgrades. I do suggest using Run-From-Package over web deploy - but you can easily revert their forced updates by setting WEBSITE_RUN_FROM_PACKAGE = 0. If you are on Azure DevOps - the latest version of App Service Deploy v4 supports Run-From-Package
Answer4	Delete the appp settings WEBSITE_RUN_FROM_ZIP then retry deploy from VS. This worked for me
Answer5	If you choose the deployment method by yourself in the release step this should not appear:
Answer6	My deployment was failing with the same message and stack trace, but was caused by a policy assignment that disallowed the location for my resource. (The resource had been created in US West, and then a policy was applied at the tenant root level that only allowed US West 2.). After the policy was updated, the deployment succeeded

Table 2 shows the feedback and voting on the answers. The first answer has the highest vote and the last one has zero votes. Only answers 1, 2 and 5 received feedback. Answer 1 received five feedbacks from five different users; the username is to the right of the answer.

Table 2. Feedback

Answer	Feedback	Vote
1	This has been killing me, **thank you so much** for posting the answer!! **It does not work for me**. I have not WEBSITE_RUN_FROM_ZIP in **It's not in** appsettings, it's in Azure App settings. – Rambalac WEBSITE_RUN_FROM_PACKAGE **is the new name** – devlord **Lifesaver**, I have no idea how you figured this out… – The Senator	**38**
2	**Did the job for me**. Thanks – VRPF	**21**
3		10
4		2
5	**This solution works for me**. – AsValeO	1
6		0

Table 3 presents the results of the feedback analysis. For each answer, we have the number of the feedback, the expression that indicates the polarity and the average weight. The negative feedback of answer 1 reduces its weight.

Table 3. Feedback analysis

Answer	Feedback			Weight
	No	Sentiments' expressions	Weight	
1	1	Thank you so much	1	(4 – 1)/5 = **0.6**
	2	It does not work for	−1	
	3	It's not in	1	
	4	Is the new name	1	
	5	Lifesaver	1	
2	1	Did the job for me	1	1/1 = **1**
5	1	This solution works for me	1	1/1 = **1**

Table 4 calculates the weighting of responses according to the criteria and the different thresholds. Responses 1, 2 and 4 can be considered repetitive because they share the same block of data. The voting threshold is equal to the harmonic mean 3.86, which is between the first three responses with values >= 10 and the last three responses with values >= 2. The answer that received no feedback has a weight of 0.5.

Table 4. Weight on criteria

Answers	Vote	Repetition	Feedback	Total
1	38/72 = 0.53	3/6 = 0.5	0.6	1.63
2	21/72 = 0.29	3/6 = 0.5	1	1.79
3	10/72 = 0.14	1/6 = 0.17	Empty - > 0.5	0.81
4	2/72 = 0.03	3/6 = 0.5	Empty - > 0.5	1.03
5	1/72 = 0.01	1/6 = 0.17	1	1.18
6	0/72 = 0	1/6 = 0.17	Empty - > 0.5	0.67
Threshold	3.86/72 = 0.05	1/6 = 0.17	0.5	0.72

Table 5 gives the rank of the responses according to total weight. Although response 1 has the highest vote, response 2 has the highest total due to feedback. Answer 4 has a good repetition score but has no feedback, so answer 5 is significantly better than answer 4. In this example, only answer 6 is not relevant, it has no vote, no repetition and no feedback, the answer is probably not interesting.

Table 5. Sort according the total

Answers	Total	Vote	Repetition	Feedback
2	1.79	0.29	0.5	1
1	1.63	0.53	0.5	0.6
5	1.18	0.01	0.17	1
4	1.03	0.03	0.5	0.5
3	0.81	0.14	0.17	0.5
6	0.67	0	0.17	0.5

2.3 Example 2

Another thread based on the following question: "***Is there a method to clone an array in jQuery?***" Table 6 shows the different messages in the discussion thread.

Table 6. Answers

Type	Post

question	This is my code:
	`var a = [1,2,3]`
	`b = $.clone(a)`
	`alert(b)`
	Doesn't jQuery have a 'clone' method? How can I clone an array using jQuery?

Answer1	`a = [1];` `b = a.slice();`
Answer2	`copy = $.merge([], a);`

Answer3	**b = $.clone(a)** to **b = $(this).clone(a)** but it some time dont work. A great alternative is
	`var b = jQuery.extend({}, a);`
	`//Deep copy`
	`var b = jQuery.extend(true, {}, a);`

Answer4	`var newArray = JSON.parse(JSON.stringify(orgArray));`
Answer5	`var a = [1,2,3]` `var b = [].concat(a);`

Answer6	`if (!Array.prototype.clone) {`
	`Array.prototype.clone = function () {arr1 = new Array(); for (var property in this) {arr1[property] = typeof (this[property]) == 'object' ? this[property].clone() : this[property]}`
	`return arr1; }`

Answer7	`var a = [1,2,3]` `b = JSON.parse(JSON.stringify(a));`
Answer8	`let arrayCopy = [...myArray];`

Table 7 shows the feedback and voting on the answers. The first answer has the highest vote and the last two have no vote, no feedback.

Table 7. Feedback

Answer	Feedback	Vote
1	This is synonymous with slice(0) ? – Peter Ajtai @Peter Ajtai - yeah. – meder omuraliev **Thanks meder.** .... **The only thing to watch out for**, is that Arrays are objects. **But this is pretty much** a corner case in the context of this quest **Still keeping reference. Doesn't work**. – neoswf **Just watch out**: **This does a shallow copy**. **Not a deep copy**. So any ob**jects in the array are not copied**, they are simply referenced. – Ariel	**151**
2	**This also does a shallow copy**, and therefore is equivalent to the accepted answer. – rych @rych **This is not equivalent, as it also works on** array-like objects that don't have a .slice method (e.g. NamedNodeMap). – janek37	10
3	Japan- Extend is meant for Objects. Ur first func was also meant for Object. **Extend works great for an array too, just change empty object** { } into an empty array []. – Chris Leonard	6
4	**This is obviously not the right option** for a big, complex structure, **but it's a pretty nice** one-liner, and actually the approach I ended up taking.	5
5	**That's doesn't do the trick** – neoswf **Basically the same problem** as using slice. – podperson	1
6	**This is just in case you've** attached other properties to the array in addition to the values in the array? (as opposed to just using slice(0))? – Peter Ajtai @Peter - **slice(0) is good**. **I'm just showing another way of solving it**. **Great man! Thanks**! – neoswf	1
7		0
8		0

Table 8 presents the results of the feedback analysis. For each answer, we have the number of the feedback, the expression that indicates the polarity and the average weight.

Table 8. Feedback analysis

Answer	Feedback			Weight
	No	Sentiments' expressions	Weight	
1	3	Thank meder, …	1	**(2 – 1)/4 = 0.25**
	4	Doesn't work	−1	
	5	Just watch out This does a …	1	
	6	Jects in the array are not …	0	
2	1	This also does a shallow copy	0.5	**(0.5 + 1)/2 = 0.75**
	2	This is not, as it also works on	1	
3	1	Extend works great for an …	1	**1**
4	1	This is obviously not … pretty nice	0.75	**0.75**
5	1	That's doesn't do the trick	−1	**(−1–1)/2 = −1**
	2	Basically the same problem	−1	
6	1	Great man! Thanks!	1	**1**

Table 9 calculates the weighting of responses according to the criteria and the different thresholds. The voting threshold is very low because the first value is much higher than the others. Answer 5 has a negative feedback, which influences its total score. Only answers 4 and 7 are repetitive, so despite the absence of a vote, answer 7 is relevant.

Table 9. Weight on criteria

Answers	Vote	Repetition	Feedback	Total
1	151/174 = 0.87	1/8 = 0.13	0.25	1.25
2	10/174 = 0.06	1/8 = 0.13	0.75	0.94
3	6/174 = 0.03	1/8 = 0.13	1	1.16
4	5/174 = 0.03	2/8 = 0.25	0.75	1.03
5	1/174 = 0.01	1/8 = 0.13	−1	−0.76
6	1/174 = 0.01	1/8 = 0.13	1	1.14
7	0/174 = 0	2/8 = 0.25	0.5	0.75
8	0/174 = 0	1/8 = 0.13	0.5	0.63
Threshold	3.6/1 74 = 0.02	1/8 = 0.13	0.5	0.65

3 Discussion

The examples of analysis on actual data shows how relevant the proposed criteria are. The choice of the harmonic mean as a voting threshold is an elegant solution. The harmonic mean defines a threshold that considers the differences between the values. It seems that the differences between the votes can bias the judgement. Votes attributed to lower-rated responses are more meaningful, because they require users to go beyond the first response in the forum thread. This observation plays in favor of a reorganization of the forum's knowledge, but also of its linear presentation sorted according to the votes. The "reorganization of the data in the form of a graph" from the discussion could be stated in the future work section. In contrast to the other criteria, the weighting of the answers in relation to the feedback is independent to the feedback of the other answers. Thus, the values can be close to 1 respectively −1. This favors the feedback because the feedback is the most relevant criterion.

The default value when a response has no feedback is 0.5, so that it does not influence the final decision.

4 Conclusion

The objective of this work is to propose a method to evaluate the quality of the answers in relation to the topic of the discussion. For this purpose, we studied criteria that can be used for assessment, namely: feedback, repetition and voting. We defined weighting mechanisms for these criteria and defined thresholds from which these criteria become relevant. The study of the examples illustrates the capabilities of the method. The method allows an assessment of the content considering the context, i.e. voting, repetition, feedback. It considers the intuitive behavior of users, the reasons why they read and vote the answers. It allows a good analysis of the feedback. It highlights the influence of the different validation criteria and the impact of each one in relation to the others. Most importantly, it gives a weight of relevance to each answer, which allows you to order or reorganize them. It has yet to be thoroughly tested and evaluated to formally validate it.

In order to reuse the contents of the forum, the selection of relevant answers allows the threads to be filtered to provide only the useful content or to proceed to a summary of the answers afterwards. This filtering operation is as interesting for a learner who will not need to review the entire thread as it is for a teacher who will be able to easily analyze the answers and reuse them in his teaching. This operation also brings a valorization of the correct answers which can play an additional incentive for the forum contributors. It is also possible to reorganize all the relevant answers to a question in a different form (e.g. a concept map) in which learners can explore the answers according to the concepts and knowledge used in the answers.

4.1 Future Work

We expect to test our method on a large scale and to set up other analysis mechanisms inspired by the analysis of syntactic emotions and the summary of documents. We are also interested in reorganizing the data in order to facilitate access to it and hence its reusability in learning or training contexts. The goal is to improve the presentation of the data as well as the interactions in order to explore them in an optimized way (for example, without having to browse through unnecessary data). Our next study will focus on approaches to data structuring.

References

1. Kim, S.N., Wang, L., Baldwin, T.: Tagging and linking web forum posts. In: Proceedings of the Fourteenth Conference on Computational Natural Language Learning, pp. 192–202. Association for Computational Linguistics, Uppsala (2010)
2. Ramesh, A., Goldwasser, D., Huang, B., Daumé III, H., Getoor, L.: Understanding MOOC discussion forums using seeded LDA. In: Proceedings of the Ninth Workshop on Innovative Use of NLP for Building Educational Applications, pp. 28–33 (2014)
3. Bhatia, S., Biyani, P., Mitra, P.: Summarizing online forum discussions - Can dialog acts of individual messages help? In: EMNLP 2014 - 2014 Conference on Empirical Methods in Natural Language Processing, Proceedings of the Conference, pp. 2127–2131. Association for Computational Linguistics (ACL) (2014)
4. Jeong, M., Lin, C.-Y., Lee, G.G.: Semi-supervised speech act recognition in emails and forums. In: Proceedings of the 2009 Conference on Empirical Methods in Natural Language Processing, vol. 3, pp. 1250–1259. Association for Computational Linguistics, Singapore (2009)
5. Tarnpradab, S., Liu, F., Hua, K.A.: Toward extractive summarization of online forum discussions via hierarchical attention networks. In: The Thirtieth International Flairs Conference (2017)

6. Carbonaro, A.: WordNet-based summarization to enhance learning interaction tutoring. J. e-Learn. Knowl. Soc. **6**, 67–74 (2010)
7. Pretzsch, S., Muthmann, K., Schill, A.: FODEX – towards generic data extraction from web forums. In: 2012 26th International Conference on Advanced Information Networking and Applications Workshops, pp. 821–826 (2012)
8. He, G., Zhang, Y., Wu, X.: Information extraction of forum based on regular expression. In: 2013 5th International Conference on Intelligent Human-Machine Systems and Cybernetics, pp. 118–122 (2013). https://doi.org/10.1109/IHMSC.2013.175
9. Sarencheh, S., Potdar, V., Yeganeh, E.A., Firoozeh, N.: Semi-automatic information extraction from discussion boards with applications for anti-spam technology. In: Taniar, D., Gervasi, O., Murgante, B., Pardede, E., Apduhan, B.O. (eds.) Computational Science and Its Applications – ICCSA 2010, pp. 370–382. Springer, Heidelberg (2010)
10. Ganascia, J.-G.: Extraction automatique de motifs syntaxiques. In: TALN 2001, Tours, France (2001)
11. Waltinger, U.: Sentiment analysis reloaded: a comparative study on sentiment polarity identification combining machine learning and subjectivity features. In: Proceedings of the 6th International Conference on Web Information Systems and Technologies (WEBIST 2010) (2010)
12. Wiebe, J., Riloff, E.: Creating subjective and objective sentence classifiers from unannotated texts. In: Gelbukh, A. (ed.) Computational Linguistics and Intelligent Text Processing, pp. 486–497. Springer, Heidelberg (2005)
13. Wilson, T., Wiebe, J., Hoffmann, P.: Recognizing contextual polarity in phrase-level sentiment analysis. In: Proceedings of Human Language Technology Conference and Conference on Empirical Methods in Natural Language Processing, pp. 347–354 (2005)
14. Socher, R., Perelygin, A., Wu, J., Chuang, J., Manning, C.D., Ng, A.Y., Potts, C.: Recursive deep models for semantic compositionality over a sentiment treebank. In: Proceedings of the 2013 Conference on Empirical Methods in Natural Language Processing, pp. 1631–1642 (2013)

Negative Badges in Teamwork Evaluation – Preliminary Results

Zuzana Kubincová(✉)

Comenius University in Bratislava, Bratislava, Slovakia
kubincova@fmph.uniba.sk

Abstract. Already several years ago, we integrated team-based projects into our university courses. Because we perceive grading all team members with the same mark or the same number of points as inappropriate, we tried to find another way to evaluate the teamwork. In an effort to fairly evaluate the contributions of particular team members to a joint project, we started to use a peer review within the team. Students evaluated the contribution of each team member in percentage and were also supposed to justify their rating with verbal comments. As they were not really willing to write comments or open answers to teachers' questions, we had to find another way to allow them to express their opinion of the teammate's work. Therefore, we involved gamification in the team assessment and started to use badges. In the last course run, we expanded our badge set to include negative badges. This paper presents the preliminary results of our research on how students awarded negative badges, whether they did not misuse them, whether giving negative badges was related to the team relationships, and whether the negative badges fit in team evaluation.

Keywords: Team project · Teamwork evaluation · Team review · Badges

1 Introduction

The ability to work in a team is one of the competences currently highly valued by employers. Therefore, it is important for students to have the opportunity to train the skills needed for teamwork as soon as possible, preferably during their studies.

Assessing the teamwork is a common problem faced by teachers when introducing team projects into the classroom. An egalitarian approach in which all team members receive the same rating for a project has been flagged as inappropriate in several publications as it does not take into account the actual contributions of individual team members to the project outcome and thus discourages students from active participation in teamwork and can even lead to freeloading [1–3].

Many educational professionals have addressed this topic in their publications and have proposed different solutions to this problem [4–6]. One of them is the involvement of the students themselves in the evaluation of teamwork. This process can be accomplished using various methods, e.g. evaluation of team colleagues by allocating points,

Z. Kubincová et al. (Eds.): MIS4TEL 2020, AISC 1236, pp. 47–55, 2021.
https://doi.org/10.1007/978-3-030-52287-2_5

distributing percentages, category-based approach, etc. [7–9]. Numerical evaluation of the teammate's contribution seems to be the easiest to translate into the overall evaluation of the team project by the teacher. However, it is not sufficient to give a rating of team colleagues only by points or percentages. It is very useful if the evaluation also contains a verbal commentary, which serves both to justify the numerical evaluation and also as a motivation for the recipient (colleague who is evaluated), who can, based on the commentary, find out what specifically needs to be improved.

However, in our experience, students are reluctant to write verbal reviews and comment on the work of team colleagues. They need to be properly motivated to be willing to do so.

As several studies [10–12] show, one way of increasing motivation and willingness to engage in certain activities may also be gamification, i.e. the use of gaming elements in non-game contexts. As examples of gamified systems can serve a framework for group work evaluation enhanced by gamification components – user points and scores [10] or a gamified peer assessment model to deal with the lack of student engagement and motivation, where several gamification elements, such as points, rankings, badges, medals, and missions, were used [13].

For several years, team projects and team reviews have been part of our university courses for students of Applied informatics. Trying to employ certain gamification elements in the coursework assessment we introduced badges as part of the team evaluation three years ago. In this paper we bring a short report on the use of negative badges in the last run of the course.

2 Badges in Team Review

The course in question is an optional web-design course which is embodied in the master study program in Applied informatics. Development of a web application carried out through team projects is an integral part of the course already for several years as well as mutual evaluation of the teammates' work, so called team review.

In the earlier course runs the team review was performed by every student by dividing 100% among all the other teammates and providing each teammate with comments on their contribution to the teamwork. The comments were led by three questions prepared by the teacher and focused on (a) evaluation of the teammate's contribution in the given phase of the project, (b) the most valuable thing the teammate has done for the team, and (c) what should the teammate improve to more effectively help the team.

However, although the students recognized this type of evaluation to be useful, they grudged writing comments on their colleagues. As we better preferred to motivate students than to force them to do something they were not willing to do, we started to search for other ways for students to express their opinion of their teammates' work. We resorted to gamification and decided to use badges as the representation of particular skills needed for the team work.

At the beginning, only badges representing skills useful in team work were used (positive badges) [14]. Nevertheless, some of the students expressed their interest in using also badges reflecting negative behavior of some their colleagues in the team work. Therefore, we integrated the badge 'lazy' in the badge set two years ago. However, as the 'lazy' label was perceived as too negative (also by teachers), in the last course run, the 'lazy' badge was replaced by four other negative badges: Freeloader, Follower, Last Minute and Solo Player.

3 Project Settings and Team Review Methodology

The project development constituted a crucial and mandatory component of the course. It was divided into three phases, each of them evaluated independently. Students worked on the project in 3–4-member teams. Since the course is optional, the number of the enrolled students is usually not very high in last years. There were 15 students enrolled in the last course run, who formed four teams. The teams were created by students based only on their preferences.

The whole procedure of the project development was as follows: After the students formed the teams their first task was to find a suitable topic for web application (both desktop and mobile) they were about to develop using the iterative web design process. Then, in the first project phase, they created the specification for the application, prepared a prototype for a use-case scenario with two personas, and tested it. Based on the testing results they finalized the improved prototype and submitted it for the teacher evaluation. In the second phase they developed the actual web application based on their prototype, tested it and after finalizing it they submitted it for the evaluation. In the third phase the students received additional requests on their application from the teachers. Their task was to use the iterative design process (trained in previous two phases) to produce version 2 of their application based on these requests and submit it (together with the required documentation) for the teacher evaluation.

Each project phase was followed by the team review. During this activity, all students evaluated the contribution of their teammates to the project. Besides dividing 100% among all the other members of the team (Fink method, see [8]), and optional writing a comment to the particular teammates they also could award them badges, expressing this way their opinion of the teammates' work and their skills. The same set of badges was used in all team review activities during all three project phases. The whole badge set is depicted in Fig. 1 and the meaning of particular badges is described in Table 1.

The main course activities such as the team creating, submitting the project in each phase (including the improved submission in the first phase) as well as team-reviewing were administered using the tools integrated with our own learning management system courses.matfyz.sk [15].

As it was mentioned before, each project phase was evaluated by the teachers independently based on the quality of the submitted work and also based on the team review. Every student could earn 60 evaluation points for the whole project – 20 points per project phase. The particular student's score in each phase could be increased or decreased depending on the percentage evaluation of the student's work by their teammates.

4 Results

The original purpose of digital badges was to serve as "a validated indicator of accomplishment, skill, quality or interest that can be earned in various learning environments"

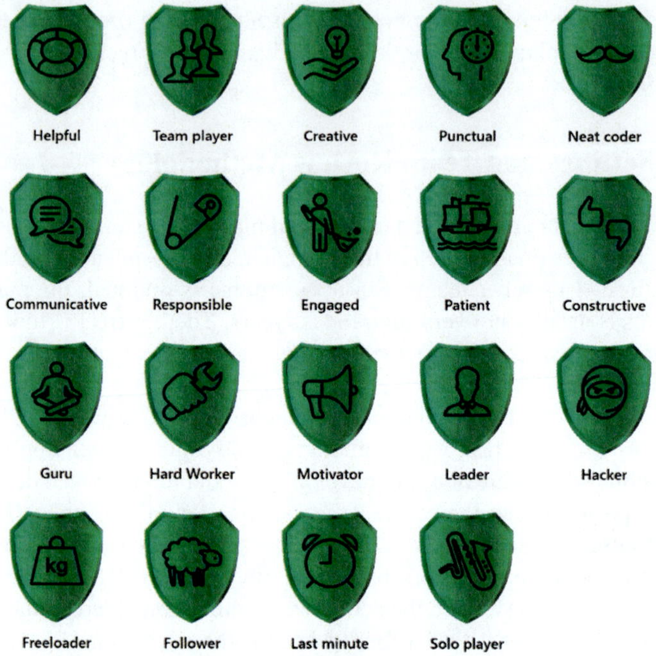

Fig. 1. Set of badges.

[16]. Thus badges were perceived as positive. In our courses, we also took them along the same line when we started to use them in team-reviewing. However, the situation changed in the recent course run, as also badges reflecting an undesirable behavior of teammates could have been allotted.

We wondered how students would use this type of badges, whether they would grant them appropriately, and whether there was a risk of their misuse.

4.1 Awarding Badges

Although the evaluation of the teammates by badges was an optional part of team review, the students used this evaluation option to a rather large extent. Altogether they awarded a badge 408 times, which creates an average of more than 27 badges allotted/earned by a student. The largest number of badges awarded by one student was 49. On the other side, there was one student, who awarded only two badges (one positive and one negative, to two teammates in the same phase of team review). However, not using badges was an exceptional behavior.

Negative badges were allotted in 7 out of 408 instances which makes the share only 1.72% of all badges the students gave to their colleagues.

The most frequently used negative badge was 'Follower' that was allotted 4 times, by four different reviewers to four different recipients. Each of the remaining three negative badges ('Freeloader', 'Last minute', 'Solo player') was used only once.

Table 1. Meaning of badges

Badge	Meaning
Communicative	Has strong communication skills
Constructive	Gives constructive feedback
Creative	Proposes creative solutions
Engaged	Always eager to put effort into the project
Guru	Understands the project thoroughly
Hacker	Solves any programming problem like a piece of cake
Hard worker	Puts a lot of hard work into the project
Helpful	Constantly helps colleagues
Leader	Shows excellent leadership skills
Motivator	Motivates others to deliver the best work
Neat coder	Writes clean code that is easy to read
Patient	Is patient with others, especially in stressful situations
Punctual	Does things on time
Responsible	Can be relied upon
Team player	Puts effort into working with colleagues
Freeloader	Just rides along, contributes little or nothing to the team
Follower	Blindly follows the lead of others, does not bring anything new to the team
Last minute	Irresponsible, disorganized, does everything at the last moment
Solo player	Does everything by herself/himself, does not really work with the team

Altogether, five students evaluated the teammate's work by negative badges while only one of them allotted more than one negative badge (in fact, he gave a negative badge three times out of which twice to the same student).

From the other side, there were five students who received a negative badge plus one student who received two negative badges. Two of these students did not pass the course.

Analyzing the data related to the negative badges we came to an interesting finding: apart from the above mentioned student who allotted only two badges during the whole semester and another student giving a negative badge to the teammate who dropped out the course already during the first project phase, all the remaining students who evaluated their teammates' work by negative badges, awarded the same teammates simultaneously also several positive badges. This fact brings us to the belief that students really used the negative badges with consideration and sensibly, just trying this way to give notice to the teammates what they needed to improve while giving them also praise for the matters they mastered well.

The negative badges were allotted evenly in all three rounds of team review.

4.2 Survey

After the course and exams were over, we conducted a survey among the students aiming to find out the students' opinion of the team review system in the course. The students were asked to fill in a questionnaire consisting of 15 questions mostly focused on evaluating by badges, some of them dealing with negative badges.

Since two of the students dropped out the course shortly after the semester started, the questionnaire was handed out to 13 students and collected from 12 of them. However, not all 12 students answered all of the questions.

In some questions, students selected one or more answers from the options offered or wrote down their own answers. In other questions they answered on a scale of 1–5, with 1 being the worst option and 5 the best one.

The most frequent answers to the question of what do the students think about our team-review system were: 'It helped us to split project points fairly' and 'It helped us to improve teamwork' which we were satisfied with as these were the main purposes of the team reviewing.

According to the students' answers, the badges they and their teammates received represented their work on the project as well as the work of their teammates quite well, with slightly more positive answers in the former case (see Fig. 2).

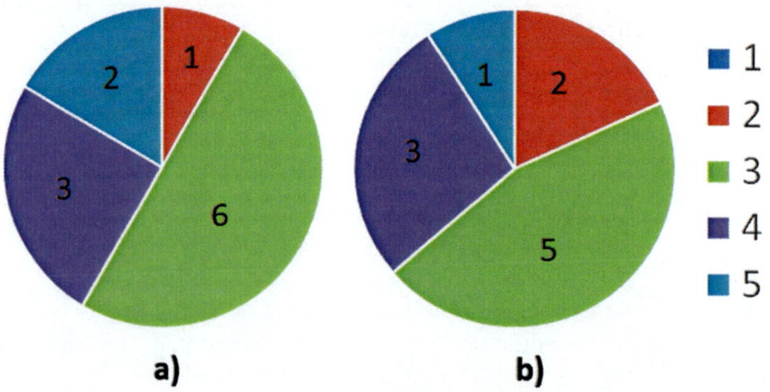

Fig. 2. a) Do you think that the badges you received during team-review represent your team-work? (Scale: 1 I strongly disagree – 5 I totally agree) b) Do you think that the badges your teammates received represent the work they did on the project? (Scale: 1 I strongly disagree – 5 I totally agree)

Although 5 out of 12 students stated that they tried to improve their team-work to earn badges in the next rounds (values 4 and 5), the average answer was 2.75 which is less than average value on the scale 1–5. Even a mite less positive were the beliefs of students in willingness of their teammates to improve their team-work after being evaluated by badges (average value 2.42 on the scale 1–5, with modus 1). Answers to both questions are depicted in Fig. 3.

According to the students' claims the badges helped them realize their strengths in team-work better than realize what they need to improve in team-work with average values 3.42 and 3.17 respectively on the scale 1–5 (Fig. 4).

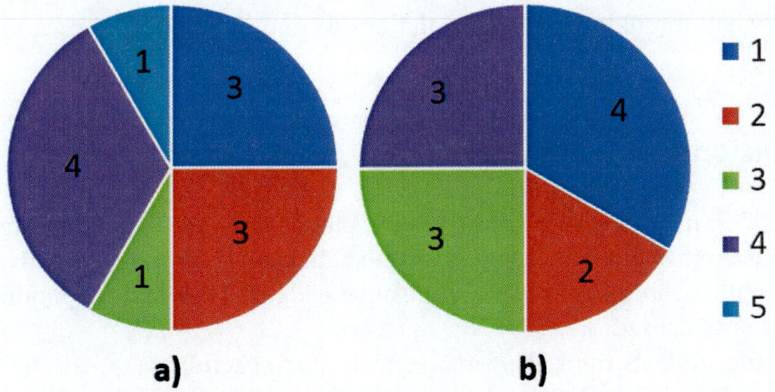

Fig. 3. a) Did you try to improve your team-work to earn badges in the next rounds? (Scale 1: I did not try at all – 5 I tried very hard) b) Do you think your colleagues improved their team-work after being evaluated by badges? (Scale: 1 I strongly disagree – 5 I totally agree)

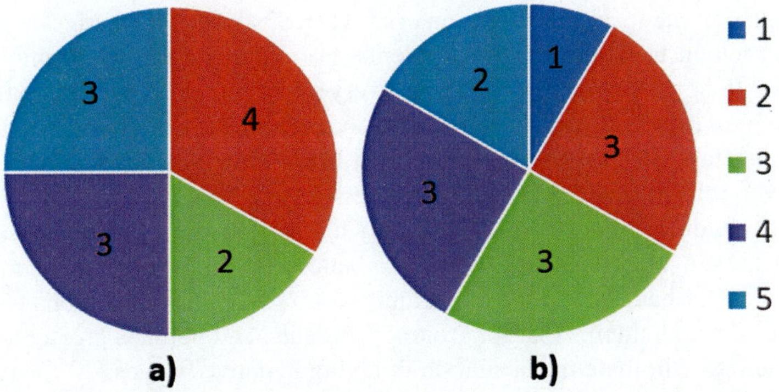

Fig. 4. a) Did the badges help you realize your strengths in team-work? (Scale: 1 I strongly disagree – 5 I totally agree). b) Did the badges help you realize what you need to improve in teamwork? (Scale: 1 I strongly disagree – 5 I totally agree).

Answering the question of what was the impulse related to badges that helped them realize what they need to improve the students most commonly commented that it was both positive badges which they did not receive and negative badges which they received.

We also tried to reveal the students' opinion about negative badges, such as Freeloader, Follower, Last-minute and Solo player in general. The number of students who declared they do not mind them, but they would not use them, was the same as the number of students who thought that negative badges are needed, and more types of them should be used. However, the students most frequently answered they do not insist on them, but they would use them in a justified case.

Students answered also the question 'If you did not award any negative badge, why?'. Besides claiming that no one in the team deserved it, the most common answer was that the student did not feel comfortable to award negative badges in general, even if someone would deserve it. Nevertheless, it seems that allotting negative badges is not related to bad relations in the team, because two-thirds of students rated the relations in their teams to be friendly and for all the others the relations were professional, correct, and not bad in any case.

5 Conclusions

The results of team review in all three project phases showed that the negative badges were not used extensively. The students tried to balance out allotting a negative badge to a teammate with awarding them several positive badges. Thus in our opinion, negative badges were not misused.

Most of the students think the negative badges are useful to express the discontent with the teammate's behavior or performance in the team-work. On the other hand, many students grudge using them even in the case a teammate deserves them. It seems that their hesitation is not connected with the relations in the team, plenty of them just do not feel comfortable to award negative badges in general. Furthermore, we also saw negative badges allotted in teams where the relations were friendly.

As certain comments in the questionnaire revealed, some of the badges we designed as negative seemed to be puzzling to students. For example, the 'Solo player' badge does not have to be perceived as negative in every situation, but in the context of a team project we regard it as such. One of the students commented on the 'Follower' badge that it could be also taken as positive, because "…it is not always bad if you have someone to do what you tell them and do not solve uselessness."

Since we conducted our research with only a small sample of students, we need to verify the results in a larger group. We plan to better explain to students the meaning of individual negative badges and the opportunity to express their objections to the work of teammates through them. A larger group of students, and perhaps greater anonymity, could also help to eliminate the hesitation in giving negative badges.

Acknowledgement. The author would like to thank to her colleagues Martin Homola and Ján Kľuka who served as the course lecturers. This work was supported by Slovak national project VEGA 1/0797/18.

References

1. LaBeouf, J.P., Grffith, J.C., Roberts, D.L.: Faculty and student issues with group work: what is problematic with college group assignments and why? J. Educ. Hum. Dev. **5**(1), 13–23 (2016)
2. Bacon, D.R., Stewart, K.A., Silver, W.S.: Lessons from the best and worst student team experiences: how a teacher can make the difference. J. Manage. Educ. **23**(5), 467–488 (1999)
3. George, J.M.: Extrinsic and intrinsic origins of perceived social loafing in organizations. Acad. Manage. J. **35**(1), 191–202 (1992)
4. Gibbs, G.: The assessment of group work: lessons from the literature. Assess. Stand. Knowl. Exch., 1–17 (2009)
5. Hindle, B.P.: The 'project': putting student-controlled, small-group work and transferable skills at the core of a geography course. J. Geogr. High. Educ. **17**(1), 11–20 (1993)

6. Lejk, M., Wyvill, M.: Peer assessment of contributions to a group project: a comparison of holistic and category-based approaches. Assess. Eval. High. Educ. **26**(1), 61–72 (2001)
7. Levine, R.E.: Peer evaluation in team-based learning. In: Team-Based Learning for Health Professions Education: A Guide to Using Small Groups to Improve Learning, pp. 103–116. Stylus (2008)
8. Michaelsen, L.K., Fink, L.D.: Calculating peer evaluation scores. In: Team-Based Learning: A Transformative Use of Small Groups in College Teaching, pp. 241–248. Stylus (2004)
9. Lejk, M., Wyvill, M.: Peer assessment of contributions to a group project: student attitudes to holistic and category-based approaches. Assess. Eval. High. Educ. **27**(6), 569–577 (2002)
10. Moccozet, L., Tardy, C., Opprecht, W., Léonard, M.: Gamification-based assessment of group work. In: 2013 International Conference on Interactive Collaborative Learning (ICL), pp. 171–179. IEEE (2013)
11. Šuníková, D., Kubincová, Z., Navrátil, M.: How to use open badges? Inf. Commun. Technol. Educ., 222–230 (2015)
12. Hamari, J.: Transforming homo economicus into homo ludens: a field experiment on gamification in a utilitarian peer-to-peer trading service. Electron. Commer. Res. Appl. **12**(4), 236–245 (2013)
13. Tenório, T., Bittencourt, I.I., Isotani, S., Pedro, A., Ospina, P.: A gamified peer assessment model for on-line learning environments in a competitive context. Comput. Hum. Behav. **64**, 247–263 (2016)
14. Šuníková, D., Kubincová, Z., Homola, M.: A badge for reducing open answers in peer assessment. In: Hancke, G., Spaniol, M., Osathanunkul, K., Unankard, S., Klamma, R. (eds.) ICWL 2018. LNCS, vol. 11007, pp. 14–24. Springer, Cham (2018). https://doi.org/10.1007/978-3-319-96565-9_2
15. Homola, M., Kubincová, Z., Čulík, J., Trungel, T.: Peer review support in a virtual learning environment. In: Li, Y., et al. (eds) State-of-the-Art and Future Directions of Smart Learning. LNET, pp. 351–355. Springer, Singapore (2016). https://doi.org/10.1007/978-981-287-868-7_43
16. Carey, K.: A future full of badges. Chronicle High. Educ. **58**(32), A60–A60 (2012)

What Do Higher Education Students Have to Say About Gamification?

Fernando Albuquerque Costa(✉), Joana Viana, and Mónica Raleiras

Instituto de Educação, Universidade de Lisboa, Lisbon, Portugal
{fc,jviana,mraleiras}@ie.ulisboa.pt

Abstract. Gamification has been a strategy widely used in recent years, particularly at higher education level. The idea is to improve the teaching and learning process through the use of different mechanics and other elements based on games, hoping to get more motivation and commitment from students. As a contribution to the discussion on the value that this approach can bring in terms of its effectiveness of students' achievement, it seems particularly relevant to hear what students have to say. The study here presented was developed in the context of a gamification experience in an engineering course. We will present the results of the content analysis of the focus-group interviews carried out before academic activity ended. Preliminary results show that although students resume a globally positive experience about gamification, it is important to ensure that the "rules of the game" are clear, that the articulation between the elements makes sense and that educational feedback is not neglected.

Keywords: Gamification · Students' perceptions · Students' engagement

1 Introduction

With the purpose of promoting the involvement of students in learning, gamification has been a strategy widely used recently, particularly at the level of higher education. In fact, the majority of studies reported in the literature relate invariably to the idea of improving the teaching and learning process through the use of different mechanics and other elements based on games [1–6], counting on a greater motivation of the student in learning [7–9] and, consequently, their greater involvement and participation in the different activities of a given course [10].

Although an important part of the studies carried out do not yet offer a clear balance in terms of students' academic performance that allows a categorical conclusion on the effectiveness of the gamification of learning, from the point of view of the curriculum and curriculum development it seems to be a very promising strategy, therefore deserving the attention of many researchers [11].

As Yildirim [12] points out, if from the theoretical point of view it is possible to defend the idea that gamification positively affects the pedagogical work in which it is inserted, it is necessary to consider the students' opinion, asking them if from their point of view

Z. Kubincová et al. (Eds.): MIS4TEL 2020, AISC 1236, pp. 56–65, 2021.
https://doi.org/10.1007/978-3-030-52287-2_6

this happens in practice. In a recent study, the author seeks to understand precisely how gamified processes are perceived by students, whether there is a convergence of opinions around a common basis regarding the concept of gamification and what elements of the gamification process they highlight.

In this line and as a contribution to the discussion on the value that this pedagogical organization strategy can bring in terms of its effectiveness on students' success, it seems particularly relevant to hear what these students have to say. These questions make particular sense in the study presented here, given its integration in a broader investigation with the purpose of conceiving and developing a gamification model that allows students to follow alternative paths, adapting automatically, for example, in function on their individual characteristics and preferences, or in function of their behavior throughout the process (an adaptive gamification model).

Before presenting results, we start with an overview of other studies aiming at understanding students' perceptions on this new way of organizing the teaching and learning process.

2 Related Work

As Rapp [13] points out, in the gamification of learning there are not many studies in which qualitative approaches predominate, although the quantitative metrics do not allow a complete representation of what he calls "subjective user experiences" and the effective quality of their experiences as a player (student-player).

As mentioned earlier, actually several authors refer the need for studies with a qualitative focus, of a longitudinal character and developed in natural environments [14], in order to be able to make a more complete and effective assessment of the gamified teaching and learning strategies, allowing us to conclude, for example, about which game elements are preferred by students, or what the use of different game strategies means for them, including narratives or scenarios [15].

If we add the fact that the gamification of learning is a very generic construct, allowing great variations, both in terms of the elements used, and of applications and configurations, it is easy to recognize the importance of studying the diversity of specific contexts and pedagogical designs [11] that only a more naturalistic approach can guarantee. In a recent study on the lessons learned from two years of gamification experiences, Kermek et al. [16] highlight precisely the extent and time span as important variables for assessing the effects of gamified learning experiences, since they end up providing elements that otherwise would not be possible to obtain.

In this line and as an example, the authors note the decrease in students' attention after a few weeks, observing a dilution of the enthusiasm shown initially. The disappearance of the novelty effect that students experience underlines the relevance of studying the implementation of gamification processes in the long term, since they will allow observing to what extent there is continuity of the effects attributed to the game elements in terms of motivation and consequent involvement by students.

According to Domínguez et al. [3] the analysis of the gamification experiments suggests that this type of pedagogical approach can have an emotional and social impact on students, namely through the use of reward systems and social mechanisms that stimulate competition. The desire for sociability is also presented by students as an important driver for their participation in the gamified learning processes [17], highlighting the way how elements of the game are organized influences their actions.

Assuming an innovative, fun and encouraging way to represent the progress in carrying out activities and their results, reward systems, such as leaderboards, are referred to by many students as a source of motivation, not just because they can see their work instantly published and recognized, but also because they can compare their progress with other classmates [11, 17].

In the study done by Aldemir et al. [11] the majority of participants state that the leaderboard promote a competitive environment, considering this positive as a stimulus in itself, although they mainly highlight the issue of reputation which, from a social point of view, represents to integrate a ranking with the best.

According to Ejsing-Duun and Karoff [17] although the playful dimension of competitiveness is appreciated by most students, encouraging greater activity, for some students this approach is referred to as a source of stress and retraction and consequent discomfort and loss of motivation [3], ending up to decrease the pleasure inherent in the playful factor inherent to the gamification of learning proposals. For many, this fact can also occur when, as Robertson [18] refers, the gamification becomes a mere "pontification". Some more critical students even claim to prefer traditional classes because they feel that leaderboard is not a good way to represent and distinguish who gets the best results in terms of learning.

Although in most studies of gamification, the challenges constitutes the incentive and the basis of all the activity that students are expected to do - some students confess that without the challenges they would not read the proposed content – [11], the possibility of choice is an aspect highlighted by some students, which may be an element to consider in the design of new learning gamified experiences. Because they are differentiated and adapted to each one, they are perceived as more significant and more motivating. This is what students point out in the study of Domínguez et al. [3] when they emphasize the importance students attach, for example, to the fact that they can have the content presented in different formats.

The possibility for students to be challenged in different ways, either by using quizzes, enhancing self-assessment and self-regulation, or by being involved in tasks that appeal to creativity, or others in which students may have an effective role in terms of individual control and decision, contributes to the sustainability of the learning path an increases the levels of commitment and interest in learning [19].

On the other hand, the way challenges are perceived by students will depend above all on their nature, that is, on the type of task to be performed. For instance, the challenges that required students to write critical reflections were not the most appreciated by students, even being referred to as less engaging [11]. In fact, for the students surveyed, being mandatory, requiring reflective thinking, written reflections ended up causing stress and demotivation.

In the study by Gomes et al. [19], the vast majority of students surveyed considered that gamified design was in general a motivating factor, due to different factors. Among them, students denote great appreciation to the existence of a narrative that helps them to structure and understand the process, or the existence of rules of the game, that is, the mechanics of the gamified system and, or finally, the use of feedback, in particular that one provided by students themselves.

Moreover, authors make a set of suggestions that seems to be important to underline here. For the narrative, they suggest the care to be taken in creating something that is really stimulating, using for example the video as a way to contextualize and "translate visually" the story in which the students would be involved. Facilitating in this way the understanding of the tasks to be performed. On the other hand, in addition to defending the idea that feedback is crucial to create and maintain optimal levels of participation and commitment on the part of students, they suggest that its realization goes beyond the forms commonly used, presenting feedback among peers as a stimulating factor, namely in terms of interpersonal relationships, which will enhance the increase in personal satisfaction and intrinsic motivation in performing tasks. Finally, they suggest that particular attention should be paid to the mechanics of the gamified system, which should be clarified with the students, so that the rules underlying the design and functioning of the "game" are known and appropriate from the beginning.

3 Context of Study

The context of the study is the gamification implemented in a curricular unit on Multimedia Content Production (MCP), from the 4th year of a MS.c. course on Information Systems and Computer Engineering, at Instituto Superior Técnico, the School of Engineering of the University of Lisbon, Portugal.

Students started with 500 experience points (XP) and acquire additional XP by completing activities, which included activities carried out in laboratory classes (15%, 3000 XPs), a multimedia presentation (15%, 3000 XPs) to publicly present at the end of the semester, mini-tests (30%, 3000 XPs) done regularly throughout the semester after of some theoretical classes and a set of tasks and challenges that they can perform throughout the semester on the platform (continuous assessment, 40%). The percentage of forty includes, in terms of evaluation, the points obtained through the realization of the challenges proposed in the skill tree (25%, 5000 XPs) and the accumulation of points and achievements (badges) obtained through the tasks accomplished, corresponding to 15% (3000 XPs + 1000 extra).

The entry point of the gamified experience was the leaderboard, a public webpage linked from the forums, which allowed students to track both theirs and others' progress and where scores were sorted in a descending XP order. By clicking a row, a dedicated page for the respective player is displayed, which include a rich dashboard with several charts portraying student progress and a list of the completed achievements and badges earned so far. Extra experience points were allocated throughout some of the challenges, which would earn XP above the maximum grade, although the final grade would always be clamped to 100%. The main goal of this decision is to allow students to get the same amount of grade through different paths, thus enabling them to do more of what they like or to earn the XP lost on failed assignments, thus potentiating learning from trial and error.

4 Methodology

The study here presented is part of a broader investigation for which the main purpose is to design and develop an adaptive gamification model. Considering that in this development process it is particularly significant to hear the students, we opted for collecting and analyzing qualitative data [20] using focus-group interviews. The option for conducting focus-group interviews is in line with the need for conducting qualitative studies [11] in order to explore the perceptions of students about the gamified learning experience that, in a given context, is provided to them [12]. In particular, two focus-group were carried out with a total of 15 students, in a universe of about 110, among those who responded favorably to the challenge launched by the interviewers (a convenience group [20]). The interviews took place even before the teaching activities ended, thus ensuring that the interviewees were very aware about the lived experience.

For the interviews, carried out by two researchers, a structured script was built based on the research question - *What are the students' perceptions about the configuration and implementation of the gamification experience of MCP*? -, and on the following research goals: (i) to characterize their motivation to perform the proposed tasks and challenges and of the gamified learning experience in general; (ii) to characterize their opinion about the different gamification elements; (iii) to identify factors that influence the gamified learning experience (positive highlights, obstacles, etc.); (iv) to collect suggestions and proposals for reformulating the pedagogical design of the course, both from the point of view of content and gamification strategies. The interview topics were prepared based on the literature on gamification and development of this type of frameworks, having been previously examined by two specialist members of the project team.

After the audio recording of the interviews with the permission of the participating students, and their transcription, these data was then analyzed based on a thematic categorical analysis [21], following closely the data reduction procedures suggested by Miles, Huberman & Saldaña [20]. The transcripts were read several times to get a general sense of the participants' thoughts, then proceeding with the use of *Nvivo12* to the identification of units of meaning and respective coding based on a system of categories and subcategories created from the objectives previously stated. A calculation of the reliability of coding was carried out using the coding performed by two researchers ($r = 85$).

5 Results

Table 1 shows the distribution of number of units of meaning for the 4 dimensions arising from the research goals ($N = 597$) and from which a presentation will be made.

5.1 Perceptions of the Gamified Learning Experience

The preliminary results show that students have a positive perception of their experience in the gamified discipline. The experience of "playing the game" is described as easy to onboard, with positive effects on their motivation and learning.

Table 1. Content analysis results: dimensions and categories

Dimensions	Categories		N	%
G.1. Perceptions of the gamified learning experience	1.1. Effects on motivation		21	38.18
	1.2. Effects on learning		18	32.73
	1.3. Player experience		16	29.09
		Subtotal	***55***	***9.21***
G. 2. Appreciation of the gamification design	2.1. Game elements		136	38.86
	2.2. Assessment		39	11.14
	2.3. Progression		75	21.43
	2.4. Collaboration		21	6.00
	2.5. Pedagogical feedback		36	10.29
	2.6. Rules of the Game		43	12.29
		Subtotal	***350***	***58.63***
G. 3. Conditioning factors	3.1. Learner characteristics		33	48.53
	3.2. Training context		22	32.35
	3.3. Interaction		13	19.12
		Subtotal	***68***	***11.39***
G.4. Suggestions and proposals for reformulating the pedagogical design of the course	4.1. On the gamification strategy		48	38.71
	4.2. On the assessment methods		13	10.48
	4.3. On the organizaton of the individual learning process		9	7.26
	4.4. On the Rules of the Game		39	31.45
	4.5. On the platform		15	12.10
		Subtotal	***124***	***20.77***
		Total	***597***	***100.00***

They characterize this discipline as being more motivating, interesting and dynamic than the non-gamified ones. Motivation can decrease during the semester, as stated by some students, and it's not clear for them if the effects are about achieving the best position in the final grade raking in competition with colleagues (extrinsic motivation) or about the achievement of personal learning goals (intrinsic motivation).

The advantages pointed out for gamification in the learning process are about the organization of students' work and attention. Students refer the constancy of the challenges, gradual progression, flexibility of paths and the record of their activities as positive aspects that help them better organize and regulate their own learning process, and manage their time and deadlines.

5.2 Appreciation of the Gamification Design

Game Elements. The skill-tree is very appreciated by students for its flexibility and the possibility to resubmit the work after receiving feedback. The different ways to get experience points and badges also allow a diversity of opportunities to focus on the elements of the game and learning challenges that attract each student the most. But in some students' opinion, experience points or achievements badges can be obtained with behaviors they consider irrelevant to the learning process. If not carefully designed, badges can promote a certain "cheating" in the game. The leaderboard divides students' opinions. For some, seeing their name on the ranking has a positive effect and leads them to engage in the discipline constantly throughout the semester. For others it is a meaningless element as they have no interest in comparing their score and not being well positioned in the ranking may have a demobilizing effect. However, they value the possibility of checking what each colleague did to achieve that position in the ranking and what they can do if they wish to increase their score. The references to the discussion forums where students ask for feedback and explain their doubts on assignments are mostly about their configuration and the lack of tailored notifications. The feedback given by their peers to the assignments receives a positive appreciation as being useful, specific and helping them to understand where they could do better. Quizzes are perceived as the element of gamification over which students have little control and that can introduce some arbitrariness into the game. The references are mostly about the questions on factual knowledges, which students consider as not being relevant for their learning process, and the coverage of details given during the classes.

Assessment. The existence of a greater number and variety of assessment elements makes them feel more in control of their final grade, since this will reflect the work they developed during the course. On the other hand, students frequently mention the lack of explicit criteria that could help them improve as one of the major difficulties.

Progression. The criticisms pointed out by both groups are mostly about the integration of the gamified component within the rest of the curriculum design. Students refer that learnings from classes, labs and skill-tree are not integrated and run in parallel. They mention the need of different mindsets for each element of the same discipline and a certain discrepancy between some of the challenges rewards and the difficulty and time they demand.

Collaboration. One element that contributes to the general positive feeling about their experience is to the interaction with a greater number of colleagues, making connections beyond their small group of close friends, and having to resort to them to ask for tips on solving challenges. They believe these interactions contribute to increase the sense of belonging to the class group.

Feedback. Students consider that face-to-face feedback, carried out in the labs, is more useful since they feel it is easier to expose doubts and don't have to wait for answers in the forums. Some mention avoiding to ask complex questions in the forums due to their belief that written communication introduces lack of clarity and ambiguity.

Rules of the Game. Despite mentioning that they had no difficulty entering the game mechanism, students are unanimous in reporting difficulties regarding what we may call the "rules of the game", i.e., the descriptions of the challenges and instructions for their completion, and the evaluation criteria used to score the schoolwork. They mention a certain frustration and discomfort when facing a challenge (from skill tree or lab activity) and not knowing exactly what is being asked or where to start. They consider that the lack of precision in some challenges' instructions and guidelines can lead to unfair penalties if they don't deliver what the teacher expected. When referring to situations in which, given the lack of more specific guidelines and instructions for carrying out the challenges, they did autonomous research and teach themselves how to solve the proposed challenge, they consider having learned more, in a deeper and more lasting way and gained more skills. These students state their preferences for video instructions and guidelines with examples.

5.3 Conditioning Factors

The factors pointed out by the students as being able to condition the gamification experiences are, essentially, of three dimensions. First, the characteristics of the students involved regarding their receptivity to a pedagogical strategy that promotes competition and requires constant work throughout the semester. Students from both groups believe gamification may not be suitable for less competitive students. In fact, in both groups some students are more cautious about their enthusiasm. They justify this feeling with them not enjoying the competitive behaviors promoted by the leaderboard and the quest for achievements badges. They also believe that this way of organizing students' work can harm those who, despite their commitment and skills, prefer sparse high concentration periods of assessments for which they prepare intensely but only in the previous days. Second, the training context on which gamification is applied. Students frequently refer the time demanding characteristic of the course as a limit to apply gamification to other disciplines. Finally, students refer the quality of the platform used in terms of performance and possibilities of interaction as a conditioning factor for a motivating learning experience.

5.4 Suggestions for Improvement

Despite the positive perception of the gamified discipline, students have considerations and recommendations for improving its dynamics. The students discuss among themselves suggestions for the discipline's gamification framework design and its articulation with other curricular elements. These suggestions refer to: i) the gamification strategy, adjusting the progression of challenges and integrating learning outcomes from classes, labs and skill tree (some students mention the desire for an introductory game level where one could learn "how to play the game": how to answer the challenges and what tools to use); ii) the assessment methods, with quizzes and badges that assess relevant knowledge and authentic learning products; iii) the learning paths available, with more flexibility so that gamification becomes adaptive to each student; iv) the rules of the game, improving the feedback and directing it towards the work processes, with an explanation of the evaluation criteria (within the game the rules of maximum score are very clear - behaviors and the frequency that correspond to each badge, except in the evaluation of the challenges and labs work); v) and, finally, greater interactivity with the platform creating reminders, relevant notifications and recommendations for learning paths and activities.

6 Final Synthesis

In general terms, we can conclude that students' perceptions seem to confirm the motivational effect identified in the literature on the use of gamification as a strategy for organizing and developing curriculum, even though this motivating effect may slow down over time [16]. In particularly, students highlighted the contribution to the self-regulation of the learning process, and the possibility they had of learning by trial and error. On the other hand, they confirm the idea that their participation in gamified learning processes is somehow driven by the desire for sociability [3]. The social component of the experience is, in fact, an element referred by the students as much positive.

Globally, students show preference in the gamification component specifically done through the platform, highlighting the possibility of creating their own learning paths and the possibility of differentiation of paths according to their personal learning objectives, giving several suggestions of interactive aspects that they would like to see implemented and that could assist them in managing their activity during the semester, such as reminders about the activities to be carried out, explicit recommendations on alternative paths and selective notifications about the activity in the forums.

With regard specifically to the aspects that may facilitate their effective participation, students highlight the need for articulation of the different components of the gamification, the importance of explicit and clear game rules, and the existence of pedagogical feedback spaces specially oriented to effectively support their work.

In short, the students' perspective makes clear that, regardless of the gamification and the characteristics of the strategy adopted, aspects related to the curricular dimension are determinant for the success of their pedagogical experience of gamification. The internal coherence and the articulation of the curricular elements, the quality of pedagogical feedback, among others, in their view are crucial aspects if the goal is the effective improvement of the teaching and learning process.

Although valuing the practical interest of the conclusions here presented for the improvement of the gamification design, we remark the need to carry out other similar studies, giving the voice to the students, the ultimate beneficiaries of gamified teaching and learning models.

Acknowledgement. This paper is supported by national funds through Fundação para a Ciência e a Tecnologia (FCT), Portugal, in the frame of the Project GameCourse – Improving College Learning with Gamification (PTDC/CCI-CIF/30754/2017).

References

1. Ar, N.A.: The Effects of Gamification on Academic Achievement and Learning Strategies Usage of Vocational High School Students. Sakarya University, Sakarya (2016)

2. Buckley, P., Doyle, E.: Gamification and student motivation. Interact. Learn. Environ. **24**(6), 1162–1175 (2014)
3. Domínguez, A., Saenz-de-Navarrete, J., de-Marcos, L., Fernández-Sanz, L., Pagés, C., Martínez-Herrálz, J.: Gamifying learning experiences: practical implications and outcomes. Comput. Educ. **63**, 380–392 (2013)
4. Faghihi, U., et al.: How gamification applies for educational purpose specially with college Algebra. BICA 2014, Procedia Comput. Sci. **41**, 182–187 (2014)
5. Rouse, K.E.: Gamification in Science Education: The Relationship of Educational Games to Motivation and Achievement. The University of Southern Mississippi, USA (2013)
6. Sanmugam, M., Abdullah, Z., Mohamed, H., Aris, B., Zaid, N.M., Suhadi, S.M.: The affiliation between student achievement and elements of gamification in learning science. In: 4th International Conference on Information and Communication Technology (ICoICT), pp. 1–4. IEEE (2016)
7. Bell, K.R.: Online 3.0-the rise of the gamer educator the potential role of gamification in online education. Retrieved from ProQuest Dissertations and Theses database (2014)
8. Measles, S., Abu-Dawood, S.: Gamification: game–based methods and strategies to increase engagement and motivation within an e-learning environment. In: Society for Information Technology & Teacher Education International Conference, pp. 8319–8324 (2015)
9. Wongso, O., Rosmansyah, Y., Bandung, Y.: Gamification framework model, based on social engagement in e-learning 2.0. In: 2nd International Conference on Technology, Informatics, Management, Engineering & Environment, Bandung, Indonesia, pp. 10–14 (2014)
10. Dicheva, D., Dichev, C., Agre, G., Angelova, G.: Gamification in education: a systematic mapping study. Educ. Technol. Soc. **18**(3), 75–88 (2015)
11. Aldemir, T., Celik, B., Kaplan, G.: A qualitative investigation of student perceptions of game elements in a gamified course. Comput. Hum. Behav. **78**, 235–254 (2018)
12. Yildirim, Y.: Students' perceptions about gamification of education: a Q-method analysis. Educ. Sci. **42**(191), 235–246 (2017)
13. Rapp, A.: A qualitative investigation of gamification: motivational factors in online gamified services and applications. IJTHI **11**(1), 67–82 (2015)
14. Nacke, L.E., Deterding, S.: The maturing of gamification research [Editorial]. Comput. Hum. Behav. **71**, 450–454 (2017)
15. Hew, K.F., Huang, B., Chu, K.W.S., Chiu, D.K.W.: Engaging Asian students through game mechanics: findings from two experiment studies. Comput. Educ. **92–93**, 221–236 (2016)
16. Kermek, D., Novak, M., Kaniski, M.: Two years of gamification of the course - lessons learned. In: Proceedings of the 2018 41st International Convention on Information and Communication Technology, Electronics and Microelectronics, MIPRO 2018, pp. 754–759 (2018)
17. Ejsing-Duun, S., Karoff, H.S.: Gamification of a higher education course: what's the fun in that? In: Proceedings of the European Conference on Games-Based Learning, pp. 92–98 (2014)
18. Robertson, M.: Can't Play, Won't Play. Hide & Seek: Inventing New Kinds of Play (2010)
19. Gomes, C., Pereira, A., & Nobre, A.: Desenho instrucional gamificado no ensino superior online: a perceção e experiência dos estudantes. RE@D – Rev. Educ. Distância Elearn. **2**(1), 97–119 (2019)
20. Miles, M.B., Huberman, A.M., Saldaña, J.: Qualitative Data Analysis. A Methods Sourcebook. Sage, Arizona (2014)
21. Bardin, L.: Análise de Conteúdo. Edições 70, Lisboa (1977)

Analysis of Relationship Between Students' Creative Skill and Learning Performance

Malinka Ivanova(✉) and Tsvetelina Petrova

College of Energy and Electronics, Technical University of Sofa, 8 Kl. Ohridski boul., Sofia, Bulgaria
{m_ivanova,tzvetelina.petrova}@tu-sofia.bg

Abstract. The aim of the paper is to explore the relationship between students' creative ability and learning performance through solving classification problems. The data is gathered via survey tool that is online delivered to students from College of Energy and Electronics at Technical University of Sofia. Dataset is used for decision-making models creation through utilization of three machine learning algorithms: J48, AdaBoost.M1 and RandomTree that are compared and evaluated concerning their performance. The models point out as the best predictors for evaluation the relationship between creativity and learning performance: creation of course work according to the first idea, working pace, the frequency for realization the new idea in practice, the ability for combination existing ideas to produce something new and factors that influence on the course work quality.

Keywords: Creativity · Learning performance · Machine learning · J48 · AdaBoost.M1 · RandomTree · Classification models

1 Introduction

Creativity is a cognitive ability related to the generation of original ideas or mixing several ideas to create new products [1]. Curiosity and creativity are connected and their relationship gives possibilities for making experiments based on emerged ideas [2]. The educators possess an important role because they could stimulate the students' curiosity and creativity through the organization of appropriate teaching/learning environments. It means that creativity is a process that includes idea coming, idea development and idea realization through utilization of explorative environment and results in artifact origination.

Creative skill is considered as an important part of the key competences for 21st century [3] and also important competence for lifelong learning [4]. Zhang et al. say that the lack of creative abilities of students could impact on their learning performance [5]. A strong relationship between creativity and learning styles is found by Eishani et al. [6], showing that students' individual differences reflect on the learning process and academic performance.

Learning performance is a term that explains how students organize their learning to reach the learning goals and what learning activities are taken for achieving

Z. Kubincová et al. (Eds.): MIS4TEL 2020, AISC 1236, pp. 66–75, 2021.
https://doi.org/10.1007/978-3-030-52287-2_7

successful learning outcomes. Nakayama et al. research the influence of students' characteristics and their learning behavior in the context of note-taking during an online course on learning performance [7]. Also, Nakayama et al. prove the possibility to predict learning performance taking into account the note-taking activities of students enrolled in a blended-learning course [8]. Shell et al. conclude that creative competency is connected to the students' knowledge retention and to the strategy for self-regulation [9]. Connection between creative competency and students' grades is not confirmed. Pretz and McColum discuss the meaning of self-perception creativity and global self-perception creativity and conclude that the first one is closer to the students' specific task conductance and the second one is related to the personality [10].

The prediction of learning performance attracts researchers' attention, because of its significant relationship with final students' outcomes as well as the possibility for pushing students into the direction of successful results through optimization of their learning. Predictive models regarding learning performance are developed considering individual learning characteristics of students, their learning activities and created artifacts by them. For example, Popescu and Leon prove that students' behaviour in a social learning environment is a good predictor for learning performance through using proposed by them Large-Margin Nearest Neighbor Regression algorithm [11]. Durga and Thangakumar compare several predictive models regarding the learning performance of engineering students according to standard metric factors [12]. The predictions are created through utilization of machine learning algorithms and optimization techniques.

The paper aims to present the findings related to the connection among personal characteristics, creative ability and learning performance of students during their course work preparation. The classification models are created in Weka environment that outline the main predictors for evaluation creativity and learning performance. Three machine learning algorithms are utilized and their accuracy is compared taking into account the small dataset, prepared after students' surveying.

2 Classification Machine Learning Algorithms

In this section three classification algorithms are explored which are used for construction the decision-making models in Weka environment (open-source software for machine learning) [13]. Also, the meaning of parameters for classifiers' evaluation is described.

AdaBoost.M1 is an algorithm for creation of boosting trees [14]. The algorithm forms a strong model through a combination of many "weak" classifiers. In the beginning, all instances in the training dataset have equal weights $\omega(i)$ and the first "weak" classifier is created. Then, at each iteration, the weights are updated as the weights of correctly classified instances receive lower value: $\omega(i) = e/(1-e)$, where e is the classifier error. The process is accomplished when the error e becomes zero or obtains pre-defined value. The result is a set of classifiers with given errors e, which values are utilized for identification of the predicted classes. The final predicted class is this one which receives the largest sum of classifiers' vote. The vote of each classifier has weight, calculated through: $-\log(\frac{e}{1-e})$.

C4.5 algorithm (in Weka it is called **J48** algorithm) is applied for construction pruned classification trees that consist of a root node (the starting point for tree building), end nodes-leaves (containing the decisions) and internal nodes (containing the questions) [15]. The training instances are used for tree building and for classifying the test data. At each node, the data is sorted with aim the best splitting attribute to be identified that is performed according to value of normalized information gain. Usually, the created trees are influenced by sample noise and the final result is over-fitted with data trees – the trees become larger and more complex. Complex trees include an increased number of splits and also new parameters are added into the model. To be reduced the trees' complexity they are pruned. In this case, the number of tree' nodes are reduced as the algorithm removes these with unsatisfied benefit for instances classification.

RandomTree algorithm uses training data to produce a set of trees as each tree is constructed through a random subset of variables [16]. It means that at each node, the best split is found after usage of the random subset of variables, instead of all variables utilization. The algorithm takes input variable subset and performs classification for every tree. The class with major vote is the classification result. Additional estimation procedures (cross-validation, bootstrap or test subset) are not applied to estimate the training error. The algorithm during the data training internally sets up the error. This algorithm produces a full free without applying any pruning techniques.

One approach for evaluating the classification algorithms' performance is to found their accuracy taking into account a set of parameters [17–19]:

- The **percentage of correctly classified instances** – this is the ratio of correctly predicted instances to the total number of made predictions.
- **Kappa statistic** – it is used to describe the accuracy of multi-class and imbalanced class problems and it is expressed through: (observed agreement-expected agreement)/(1-expected agreement).
- **True Positive (TP) rate** (Sensitivity) – this is the proportion of instances which are classified as class A and all instances which truly belong to class A i.e. TruePositiveRate = TruePositive/(TruePositive + FalseNegative). This parameter is equivalent to the parameter **Recall**.
- **False Positive (FP) rate** (Specificity) – this is the proportion of instances which are classified as class A, but actually, they belong to other class B i.e. FalsePositiveRate = FalsePositive/(FalsePositive + TrueNegative).
- **Precision** – this is the proportion of these instances which truly belong to the class A to all instances that are classified as class A i.e. Precision = TruePositive/(TruePositive + FalsePositive).
- **F-measure** – it calculates through: 2*Precision*Recall/(Precision_Recall) and shows how precise is a given classifier i.e. how many instances are correctly classified and how robust is this classifier i.e. it does not miss a larger number of instances.
- **MCC** (Matthews correlation coefficient) – it evaluates the quality of binary classifications and it is calculated through: MCC = (TruePositive*TrueNegative-FalsePositive*FalseNegative)/SQRT((TruePositive + FalsePositive)(TruePositive + FalseNegative)(TrueNegative + FalsePositive)(TrueNegative + FalseNegative)).
- **ROC Area** (Receiver Operating CharacteristicArea) – measures the entire area under the ROC curve. The ROC curve is constructed as x axis plots False Positive Rate and y axis plots True Positive Rate.

- **PRC Area** (Precision-Recall Curve Area) – measures the entire area under the PRC curve. The PRC curve is created as x-axis plots Recall and y-axis plots Precision.

3 Experiment and Models

Thirty two students from Technical University of Sofia, College of Energy and Electronics were asked to share their opinion in an online survey about their learning performance at course works preparation and to self-evaluate their creative abilities. The survey includes questions in the form of Likert scale, yes/no and free answers. The questions are classified in three groups: (1) related to personal data – gender and ages, (2) for better understanding whether they are creative persons, (3) regarding the connection between creativity at course work preparation and learning performance.

The respondents are divided to the following six groups according to their ages: (1) 21–25 ages – 9 students, (2) 26–30 ages – 6 students, (3) 31–35 ages – 3 students, (4) 36–40 ages – 9 students, (5) 41–45 ages – 4 students, (6) 46–50 ages – 1 student. 72% of the surveyed students are male and 28% of them are female.

The first model, constructed through **J48** algorithm, reveals whether the students possess creative skill (Fig. 1). It is created through the following attributes:

@attribute *frequencynewideas* with values {veryrarely1, rarely1, notsooften1, often1, veryoften1} – indicates how often the students come up with new ideas;

@attribute *realizationonpractice* with values {veryrarely2, rarely2, notsooften2, often2, veryoften2} – indicates how often these new ideas are realized in practice;

@attribute *ideascombination* with values {veryrarely3, rarely3, notsooften3, often3, veryoften3} – it indicates how often the students combine existing ideas to create something new;

@attribute *likeexperiments* with values {yes1, no1} – shows whether the students like to experiment;

@attribute *newthingdiscovery* with values {yes2, no2} – shows whether the students like to discover new and unknown for them things;

@attribute *curiosity* with values {yes3, no3} – indicates the students' self-evaluation of whether they are curious people;

@attribute *creativeperson* with values {yes4, no4, na} – indicates the students' self-evaluation of whether they are creative people.

The pruned tree is constructed by applying J48 algorithm and it consists of 8 nodes and 6 leaves. The top node shows the best predictor for classification the students as creative persons - *how often the idea is realized in practice.* Another predictor is *whether the students like to make experiments.* The model classifies a half of the students as creative persons, because they like experiments and often or very often transfer their ideas into some kind of new artifacts.

The full tree is also constructed through RandomThree classifier and it consists of 58 nodes as 46 of them are leaves. The top node used as the best predictor is the *frequency of new ideas coming to students.* Others found predictors are: *how often the students combine existing ideas to create something new, whether the students like to experiment* and *how often the new idea is realized in practice.*

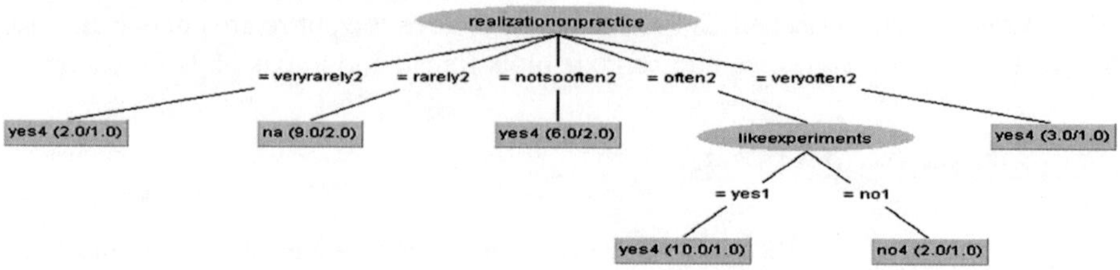

Fig. 1. Creativity classification model

The AdaBoost.M1 classifier reveals as the main predictor: *how often the idea is realized in practice* and as an additional predictor: whether the students *like to make experiments.*

The second model shows the students' learning performance during their course work preparation. The following attributes are taken into account:

@attribute *firstideaproject* with values {yes1, no1} – it contains the students' answers whether they create their course works according to their first idea;

@attribute *betterprojectperformance* with values {yes2, no2} – this attribute reflects on students' answers whether the course work is well accomplished;

@attribute *muchknowledgebetterproject* with values {yes3, no3} – whether the course work could be performed better if the students possess more knowledge;

@attribute *muchtimebetterproject* with values {yes4, no4} – whether the course work could be performed better if the students have more time for realization;

@attribute *projectimprovement* with values {yes5, no5} – it reflects on the students' answers according to their willing for course work improvement if they have this possibility;

@attribute *workingtemp*with values {veryslowly, slowly, average, fast, veryfast} – the students had to evaluate their pace at course work preparation;

@attribute *qualityfactors* with values {one, two, three, four, five, six} – how many factors influence on the course work quality (the factors are: necessity of new knowledge, more time for realization, specific computer skills, given interest to the course topics, motivation, other);

@attribute *selfassessmnt* with values {excellent, verygood, good, satisfactory} – the students had to self-assess the learning performance during their course works preparation.

The J48 algorithm is applied and the returned result contains just one leaf with label *very good* learning performance of students during their course work preparation: verygood (32.0/17.0). Predictors for learning performance are not identified at the usage of this algorithm.

The RandomThree classifier constructs full tree with 66 nodes as 52 of them are leaves. The best predictor for assessment of learning performance appears to be whether *the students create course works according to their first idea.* The next identified predictors with high impact are: *the temp at course work preparation* and *the students willing for course work improvement if they have this possibility.*

The AdaBoost.M1 algorithm identifies as the best predictor for assessment of learning performance: *the temp at course work preparation.*

The third classification model aims to identify the relationship between creativity skill and learning performance. It is built through the usage of all attributes from the first and second model. The pruned tree constructed after applying the J48 algorithm is shown on Fig. 2 and it points out the following:

IF the student possesses creative ability AND working pace is fast/very fast (average) (slowly/very slowly) THEN his/her learning performance is excellent (very good) (good).

IF the student does not possess creative ability THEN his/her learning performance is good.

IF the student cannot appraise whether he/she possesses creative ability AND rarely/not so often/often combines existing ideas to create something new THEN his/her learning performance is very good.

IF the student's learning performance is excellent THEN he/she is a creative person.

IF the student's learning performance is very good AND working pace is average AND he/she is willing to improve his/her course work if he/she has this possibility THEN the student is a creative person.

According to RandomTree algorithm, the best predictor for the relationship between creativity and learning performance is *the working pace at course work preparation.* The next predictors with high impact are: *the course works are created according to students' first idea, the factors that influence on the course work quality*, and *how often the new ideas are realized in practice.*

According to the applied AdaBoost.M1 algorithm, *the ability for a combination of existing ideas to create something new* is the best predictor for a relationship between creativity and learning performance.

The next models reveal the influence of students' personal characteristics like age on creativity (Fig. 3) and learning performance (Fig. 4). After applying J48 classifier and construction of the pruned tree it can be said that for the students from group 2 (26–30 years) is typical that they very often have new ideas, to the students from group 1 (21–25 years) often come new ideas and they often combine several ideas to prepare something new, the students from group 4 (36–40 years) often realize the ideas on practice, to the students from group 3 (31–35 years) and group 5 (41–45 years) not so often emerged new ideas, but they like to make experiments.

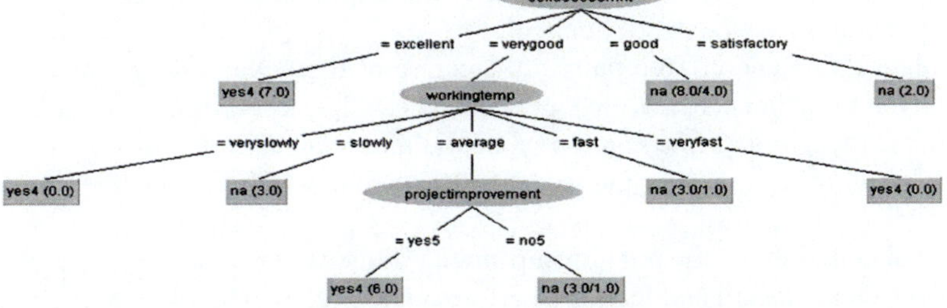

Fig. 2. The relationship between creative skill and learning performance

The full tree constructed through RandomTree algorithm points out that the root node is *how often the studentscome up with new ideas.* It splits to four nodes: *creative*

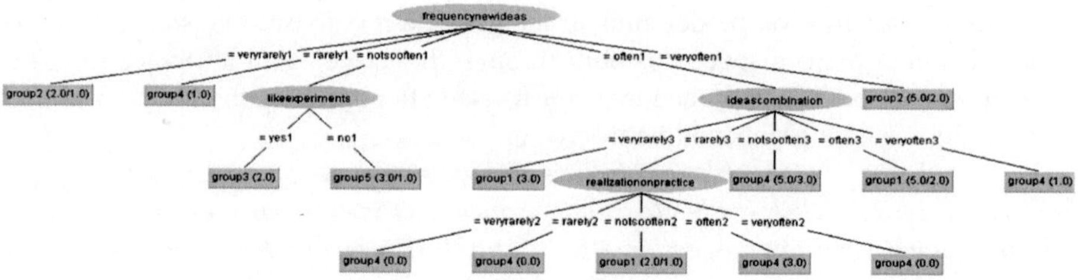

Fig. 3. The relationship between students' personal characteristics and creativity

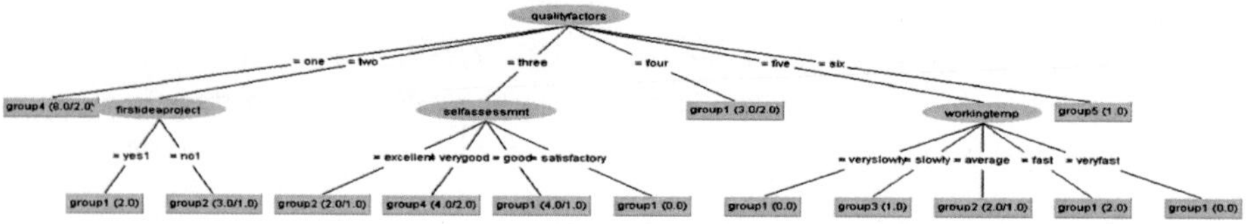

Fig. 4. The relationship between students' personal characteristics and learning performance

people, realization on practice the emerged new ideas, ability to combine ideas to receive something new and *like experimenting.* To the students from group 2 very often the new ideas are coming and they are creative persons and they often/very often combine ideas to obtain something new. To the students from group 1 very often the new ideas emerge and they are creative persons, but they rarely/very rarely combine ideas to obtain something new. To the students from group 4 often the new ideas emerge and they rarely combine ideas to receive something new, but often realize ideas on practice.

According to the AdaBoost.M1 algorithm to the students from group 3 and group 5 *not so often come new ideas.*

For description the relationship between students' age and learning performance according to the J48 algorithm the main predictor appears to be *the number of factors that influence on course work quality.* For the students from group 4 one or three factors influence on the course work quality, for the students from group 3 five factors and temp of working are important, and for students from group 5 six factors influence on the product quality. For students from group 1 and group 2, several factors (from two to five) are relevant for course work quality.

The algorithm RandomTree puts in the centre of the node: *working temp* as very important for the students' learning performance, while according to AdaBoost.M1 algorithm the main node is *the number of factors that influences on the product quality.* The last algorithm says that for the students from group 4 only one factor is important for the course work quality.

The final model shows the relationship among students' personal characteristics (age and gender), creative skill and learning performance (Fig. 5). The J48 algorithm allows to construct a pruned tree with the main node *the frequency for emerging new ideas.* The internal nodes are: *whether the students like experiments, the number of factors that influence on the course work quality* and *realization the idea on practice.*

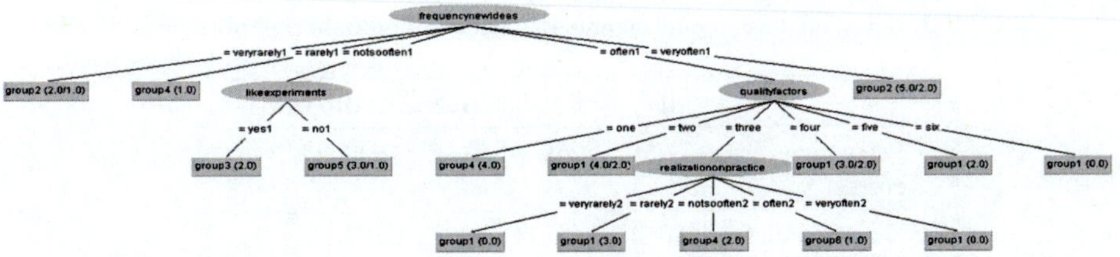

Fig. 5. The relationship among students' personal characteristics, creativity and learning performance

According to RandomTree algorithm, the main node is *whether the student is a creative person*. The sub-nodes are: *the number of factors that influence on the product quality* and *the frequency of new ideas emerging*. AdaBoost.M1 classifier identifies as the central node: *the frequency of new ideas emerging* and shows that it is typical for the students from group 3 and group 5.

The results point out that the students' age play somewhat role in creativity and learning performance, while the gender has influence neither on the creativity, nor on the learning performance.

4 Models Performance Comparison

Weka KnowledgeFlow environment is used to compare the performance of created models through the usage of J48, RandomTree and AdaBoost.M1 classifiers according to the arranged components and flows as they are shown on Fig. 6. For obtaining the training and test sets, a cross-validation technique is applied with 6 folds.

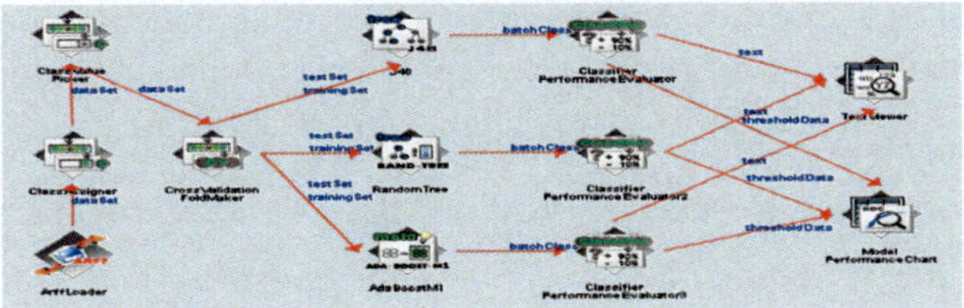

Fig. 6. Weka KnowledgeFlow environment for models performance comparison

After running the machine learning experiment, the received results, are summarized in Table 1. They point out that with the best parameters is the learning model of AdaBoost.M1 classifier, at the second place is J48 and the worst parameters are obtained from RandomTree algorithm. The comparison of visualized ROC areas and PRC areas of the examined classifiers for one class *excellent* of the attribute *selfassessmnt* is presented on Fig. 7, respectively on Fig. 8. The higher value of ROC and PRC indicates better model performance.

Table 1. Classification models comparison according to their performance

	J48	RandomTree	AdaBoostM1
Percentage correct	37.5%	25%	46.875%
Kappa_statistic	0.0643	−0.1179	0.0701
Precision	0.402	0.260	0.251
F-Measure	0.382	0.255	0.327
MCC	0.069	−0.137	0.116
ROC Area	0.558	0.477	0.566
PRC Area	0.417	0.334	0.372

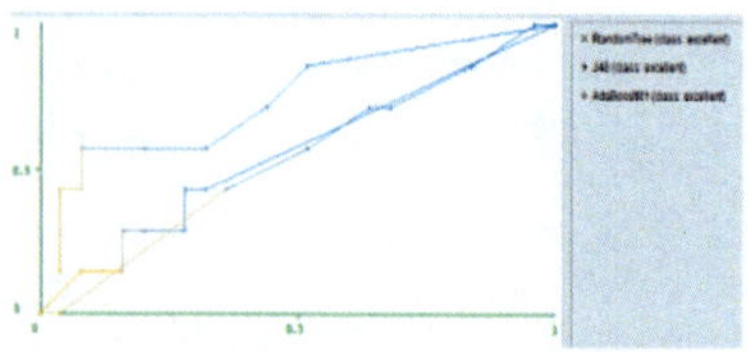

Fig. 7. The ROC areas

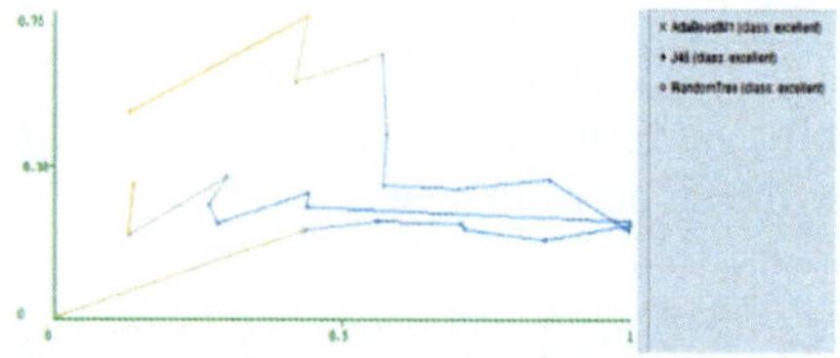

Fig. 8. The PRC areas

5 Conclusions

This exploration proves the mutual relationship between creative ability and learning performance. Creative persons, working on a given task, are characterized with optimized learning performance, while the learning performance of the students who self-described as non-creative persons is just good. After applying classification algorithms on the obtained dataset, several predictors for explaining this relationship are received: *the final product is created according to students' first idea, the importance of product quality factors, the frequency of realization in practice the new ideas, working temp* and *the ability for combination existing ideas to create something new*. Three machine learning algorithms are applied on "small data" for creation several classification models and their performance comparison shows that the AdaBoost.M1 parameters are better than J48 and RandomTree classifiers.

Acknowledgements. The authors would like to thank the Research and Development Sector at the Technical University of Sofia for the financial support.

References

1. Creativity. https://dictionary.cambridge.org/dictionary/english/creativity. Accessed 23 Feb 2020

2. Pusca, D., Northwood, D.O.: Curiosity, creativity and engineering education. Global J. Eng. Educ. **20**(3), 152–158 (2018)
3. Nakano, T., Wechsler, S.: Creativity and innovation: skills for the 21st century. Estudos Psicologia (Campinas), **35**(3), 237–246 (2018)
4. European Commission. Key Competences for LifeLong Learning (2018). https://eur-lex.europa.eu/legal-content/EN/TXT/PDF/?uri=CELEX:52018SC0014&from=EN. Accessed 23 Feb 2020
5. Zhang, Y., et al.: Review of creativity factors in final year design projects in China (2017). https://hdl.handle.net/2134/26039. Accessed 23 Feb 2020
6. Eishani, K.A., et al.: The relationship between learning styles and creativity. Procedia – Soc. Behav. Sci. **114**, 52–55 (2014)
7. Nakayama, M., et al.: Impact of learner's characteristics and learning behaviour on learning performance during a fully online course. Electron. J. e-Learn. **12**(4), 394–408 (2014)
8. Nakayama, M., et al.: Int. J. Educ. Technol. High. Educ. **14**, 6 (2017). https://doi.org/10.1186/s41239-017-0048-z
9. Shell, D., et al.: Associations of students' creativity, motivation, and self-regulation with learning and achievement in college computer science courses. In: CSE Conference and Workshop Papers, pp. 1637–1643 (2013)
10. Pretz, J.E., McColum, V.A.: Self-perceptions of creativity do not always reflect actual creative performance. Psychol. Aesthet. Creativity Arts **8**, 227–236 (2014)
11. Popescu, E., Leon, F.: Predicting academic performance based on learner traces in a social learning environment. IEEE Access **6**, 72774–72785 (2018). https://doi.org/10.1109/ACCESS.2018.2882297
12. Durga, V.S., Thangakumar, J.: A complete survey on predicting performance of engineering students. Int. J. Civ. Eng. Technol. **10**(2), 48–56 (2019)
13. Weka – open source software for machine learning. https://www.cs.waikato.ac.nz/ml/weka/. Accessed 23 Feb 2020
14. Freund, Y., Schapire, R.E.: Experiments with a New Boosting Algorithm, Machine Learning (1996). https://cseweb.ucsd.edu/~yfreund/papers/boostingexperiments.pdf. Accessed 23 Feb 2020
15. Salzberg, S.L.: C4.5: Programs for Machine Learning by J. Ross Quinlan. Morgan Kaufmann Publishers, Inc. (1993); Mach. Learn. **16**, 235–240 (1994)
16. Random Trees. https://docs.opencv.org/2.4/modules/ml/doc/random_trees.html. Accessed 23 Feb 2020
17. Bouckaert, R.R., et al.: WEKA Manual for Version 3-7-8, January 21, 2013
18. Mishra, A.: Metrics to Evaluate your Machine Learning Algorithm (2018). https://towardsdatascience.com/metrics-to-evaluate-your-machine-learning-algorithm-f10ba6e38234. Accessed 23 Feb 2020
19. The Data Scientist, Performance Measures: Cohen's Kappa statistic. https://thedatascientist.com/performance-measures-cohens-kappa-statistic/. Accessed 23 Feb 2020

Evaluating Statistical and Informatics Competencies in Medical Students in a Blended Learning Course

Vincenza Cofini(✉) and Pierpaolo Vittorini

Department of Life, Health and Environmental Sciences, University of L'Aquila, P.Le S. Tommasi, 1 - Coppito, 67100 L'Aquila, Italy
{vincenza.cofini,pierpaolo.vittorini}@univaq.it

Abstract. The paper reports on a case study aimed at investigating the use of technology in high-school in the subjects of informatics and statistics, as well as if the blended learning approach used in the same subjects - taught at the university level - was useful to introduce the students into scientific research. To this aim, a self-reported questionnaire was used, at the beginning and at the end of the courses of informatics and statistics. The results show limited use of technology in high school, at least on these subjects (15%), unless for spreadsheets (24%) and central tendency indicators (26%). Furthermore, the students reported that increased their ability to read scientific papers. As for the final outcomes, the students have small failing rates (3% and 9% for statistics and informatics, respectively) which are in-line with the self-reported knowledge on the subjects.

Keywords: Case study · TEL in high school · Blended learning · Statistics and informatics in medicine

1 Introduction

The curriculum of the degree course in Medicine and Surgery at the University of L'Aquila (Italy), includes - in the first year - the course of Health Informatics, Statistics and Scientific English. The course is made up of three modules, namely: "Health Informatics", "Medical Statistics" and "Scientific English". The course objectives are the comprehension of the base statistical methods and health informatics, the use of the English language to understand the international literature in the medical field and be able to discuss scientific topics. Both modules are offered with a blended learning approach [1, 2], using both traditional and state-of-the-art tools (i.e., Moodle [3] for the delivery of the course content and homework, a novel tool that provides both an automated evaluation and structured feedback regarding the homework [4, 5]).

In such a context, the present study has a manifold aim. First, if and which technology was used in high school to support the teaching of informatics and statistics. Second, if the same students - following the same subject at a University level - gained a deeper

Z. Kubincová et al. (Eds.): MIS4TEL 2020, AISC 1236, pp. 76–85, 2021.
https://doi.org/10.1007/978-3-030-52287-2_8

understanding of scientific research and, finally, if their self-reported knowledge in the subject was in line with the learning outcomes at the final exam.

Accordingly, the paper introduces in Sect. 2 the related work, describes in Sect. 3 the course structure and the two composing modules in detail. Section 4 then focuses on describing the study, whereas Sect. 5 reports the results. Finally, Sect. 6 ends the paper with a discussion about the most relevant findings.

2 Related Work

The impact of adopting (different) technologies in the classroom is a long and still discussed topic [6, 7]. It is commonly argued that the use of technology should excite students, prepare them for the future and encourage spontaneous learning [8]. However, results are still controversial. For instance, in [9] the authors found that overall technology use has no significant positive effect on the grades and attendance of at-risk students; in [10] the authors found that the primary benefit has been to increase information access and communication, but without leading to a significant improvement of the student performances on standardized tests.

Closer to our specific field, i.e., statistics and informatics in medicine, the available literature is not large. However, in [11] the authors report that blended learning appears to have a consistently positive effect (in comparison with no intervention) and to be more effective than (or at least as effective as) non-blended instruction for knowledge acquisition in health professions; in [12] the authors report on the adoption of computer-based activities to allow students to conduct data analysis and simulate statistical concepts, whose results suggest that the improved learning outcomes may be due to the increased attention, interest, and motivation that result from using the technology.

3 Course Structure

3.1 Medical Statistics Module

The module is organized in terms of 37 h of frontal lessons, two days a week, with homework exercises offered through the free Moodle learning platform available at https://moodle.univaq.it/ [3].

Prior to the students' beginning the course, the combination of traditional teaching and online learning was introduced, as suggested by in the literature. Blended basic operations such as registering, downloading learning materials, submitting homework, completing the test, and creating a new discussion post were introduced [13]. One hundred and thirty students registered on Moodle. During the course, from October to January, at the end of each week, the teacher uploaded on Moodle the files regarding the frontal lessons and the materials for the online exercises. During the holiday season (Christmas break), students were asked to fill ten tests online and to participate in the exam simulation on Moodle.

The lessons covered the following topics: Collecting and summarizing data, Common ways to describe data, Different ways to represent data, Frequency Tables, Cumulative Frequency, Probability, Simply linear Correlation and regression and Inferential Statistics [14].

To introduce the importance of medical statistics in scientific medical research, we started from published papers to analyse different study design and the statistical methods related to different primary objectives. We chose projects in which we participated because the databases were available and we were confident that the examples of the real-world help in the interpretation of statistical methods in medicine. We presented the cross-sectional design using available data from two studies to describe the questionnaire technique of data collection, to calculate the prevalence measure, the risk measure with 95% confidence intervals and to introduce statistical tests for categorical data [15, 16]. To demonstrate how to choose and to determine the sample size, to introduce the concept of randomization and the statistical tests for continuous data (paired or unpaired), we used data from two randomized controlled clinical trials [17, 18]. To explain the practical use of Bayes Theorem in medicine, sensibility and specificity concepts were based on data from a study conducted on infertile women [19]. Finally, to understand the use of probability, we used the results from a survival analysis [20].

3.2 Health Informatics Module

The module of Health informatics is organized in two parts. The first part covers the basics of informatics (representation of information, introduction to algorithms, computer architecture, operating systems, networking) with specific health-related topics (representation of medical information, examples of algorithms in computational epidemiology, imaging, telemedicine, PUBMED). The second part is instead practical and regards the statistical analysis of health data with R.

The course lectures were given both in-presence during normal classroom activities and online through the teacher's website. During the module, the students could take advantage of several formative assessment tools. For the theoretical topics, the students were invited - at the end of each topic - to use a computerized adaptive testing approach [21] to measure their current knowledge in the topic. For the practical topic, the students were invited to use a tool able to provide both an automated evaluation and structured feedback regarding the quality of the assigned homework [11, 12].

4 The Study

All the students admitted at the Medicine and Surgery degree in the Academic Year 2019/2020 were invited to participate in a project on "Teaching and learning experience in Health Informatics and Medical Statistics modules for Medical Students".

All participants were informed by the lecturers about the project objectives and were asked to fill an anonymous questionnaire, both pre and post-course.

We used a modified version of a questionnaire adopted in [22]. The questionnaire asked students to self-report about:

- in the pre-course, their knowledge about statistics and informatics, whether or not it was supported - during high school - by technology, as well as the ability to read and understand scientific papers;
- in the post-course, whether the learning outcomes were satisfactory or not and how the course may be improved.

5 Results

One hundred and four students participated in the survey, with mean age 20 years (± 3). Thirty-seven were males (36%) and forty-one were admitted for the first time. All questions used a rating scale from 0 to 10, where the higher the better. The self-reported knowledge for the pre-course, reported on average, 4 ± 2 in statistics and 4 ± 3 in informatics, respectively. Table 1 summarises the knowledge in statistics and informatics, by topic and the type of technology used to study it.

Table 1. Pre-course knowledge about Statistics and Informatics A = PC/Tablet, B = PC/Tablet/LIM, C = LIM, D = Other

Items	Type of technology					
	Yes (%)	A	B	C	D	%
Have you studied Statistics in the past?	69 (66%)					
Frequency Distribution	48 (74%)	7	2	5	6	19%
Central Tendency Measures	68 (99%)	14	4	2	7	26%
Measures of Dispersion	47 (73%)	10	3	4	4	20%
Probability	42 (67%)	8	2	5	4	18%
Linear Correlation	18 (28%)	5	1	3	3	12%
Linear Regression	17 (27%)	6	1	2	3	12%
Inferential Statistics for Mean	28 (43%)	9	2	2	2	14%
Inferential Statistics for Proportion	25 (38%)	6	2	1	3	12%
Confidence Intervals (CI)	18 (27%)	4	1	1	4	10%
Have you studied Informatics in the past?	54 (52%)					
Information and codification	30 (60%)	6	3	4	2	14%
Algorithms or programming	22 (43%)	10	2	1	3	15%
Computer architecture	37 (74%)	7	3	3	3	15%
Operating system	37 (74%)	9	3	3	1	15%
Digital/Analogic Conversion	16 (33%)	6	1	1	3	11%
Computer networking	16 (26%)	4	2	2	3	11%
Spreadsheet	24 (49%)	16	5	2	2	24%
Statistical Data Processing Software	11 (23%)	6	3	1	3	13%
Central tendency		7	2	2	3	15%

According to the students' answers, the use of technology in high school for informatics and statistics subjects is quite limited (on average, the 15%), mostly through PC or tablet (the median is 7 students). The topics in which technology is mostly used are central tendency indicators (26%), spreadsheet (24%) and dispersion (20%).

By comparing the final grades obtained by students that used (or not used) technology at school to support the didactic activities, the results are summarised in Table 2. With respect to the perceived knowledge in the two subjects, the use of technology at school makes the students more confident to better know the two subjects (both differences are statistically significant). When compared to course outcomes assigned by the teachers, the use of technology at school does not seem to make any difference. Actually, even if not statistically significant, the final grades in Informatics are lower in students that used technology at school. It is possible that the high-perceived knowledge at the beginning of the course could have caused an underestimation of the course difficulty.

Table 2. Differences in the perceived knowledge of the subjects (self) and the course outcomes (teacher) by the use of technology at school

Use of technology at school	Statistics		Informatics	
	Self [0,10]	Teacher [0,31]	Self [0,10]	Teacher [0,31]
Yes	5.4	29.3	5.9	25.8
No	2.9	28.1	3.0	27.6
p-value	0.002	n.s.	0.001	n.s.

Eighty-eight students read a scientific article, thirty-one students read the statistical analysis plan and most of them while surfing the internet (Fig. 1).

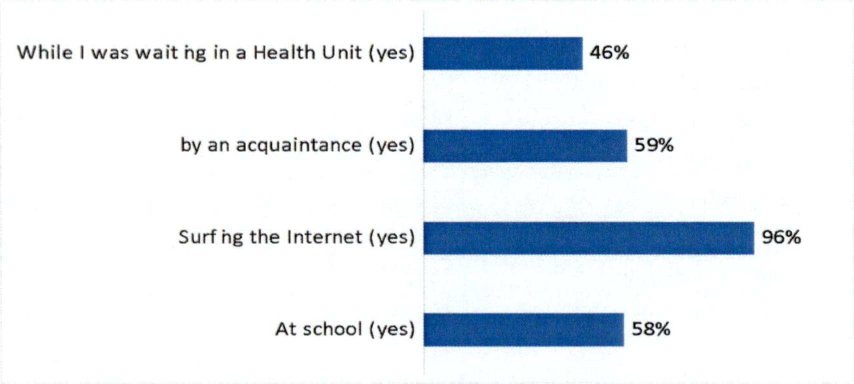

Fig. 1. Have you ever read a scientific article in the medical field?

As shown in Table 3, students perceived the importance of statistics and informatics to their career as a doctor, they preferred clinical study more than statistics and informatics competencies, as expected. They reported a high level of attitude towards lecture/lecturers and they understood the relationship between statistics, informatics and medicine, at this point of their education.

Table 3. Post-course Questionnaire Items (n = 60). SD = Strongly disagree; D = disagree; N = neutral; A = Agree; Strongly Agree = SA The total may be less than 60 in case of missing data

Medical Statistics Module			
I feel that	**SD/D**	**N**	**A/SA**
The course focuses on the concept of interpretation more than calculations	4 (6%)	19 (32%)	37 (62%)
The gained knowledge and experience is useful to my career as a doctor	3 (5%)	7 (12%)	50 (83%)
Sequencing of topics was logical	2 (3%)	8 (14%)	50 (83%)
Causes of no interest in the subject	**SD/D**	**N**	**A/SA**
I have to deal with numbers	40 (67%)	15 (25%)	5 (8%)
The subject needs creative thinking	33 (55%)	19 (32%)	8 (13%)
Lack of practising exercise for these topics	49 (82%)	6 (10%)	5 (8%)
I like clinical studies more than biomedical statistics	11 (19%)	21 (36%)	27 (45%)
Lectures were not interesting	48 (80%)	8 (13%)	4 (7%)
Lectures were lengthy	44 (74%)	11 (18%)	5 (8%)
Lectures were difficult to understand	46 (78%)	13 (21%)	1 (1%)
Too many lectures for one day	42 (71%)	14 (24%)	3 (5%)
There were no specific references	54 (90%)	4 (7%)	2 (3%)
I could not see the relationship between statistics and medicine at this level	45 (75%)	11 (18%)	4 (7%)
Simply am not interested in the subject	44 (74%)	14 (24%)	1 (2%)
Attitude towards lecture/lecturers	**SD/D**	**N**	**A/SA**
I was treated with respect	0 (0%)	7 (12%)	52 (88%)
Work and efforts were acknowledged	1 (1%)	9 (15%)	49 (84%)
The lecturer was supportive of the organization of the study	1 (2%)	15 (25%)	33 (73%)
The lecturer was the facilitator of understanding & guiding students	1 (2%)	6 (10%)	51 (88%)
The lecturer allowed to start debates/questions during the lessons	2 (4%)	16 (27%)	41 (69%)
Health Informatics Module			
I feel that	**SD/D**	**N**	**A/SA**
The course focuses on the concept of interpretation more than calculations	23 (38%)	23 (38%)	14 (24%)
The gained knowledge and experience is useful to my career as a doctor	8 (13%)	12 (20%)	40 (67%)

(*continued*)

The topic that was reported by students as the simpler to study in statistics was the central tendency indicators (18/45) while the more difficult was the hypothesis testing (25/43). With respect to informatics, there wasn't a topic that was clearly indicated as simpler to study, whereas the more difficult was the practical execution of statistical analyses with R (26/43).

At the end of the course, no student reported difficulties on using both the Moodle platform and the formative assessment tool, and most of them reported they realized the importance of the two modules to understand scientific research (see Table 4).

Eighty-five students participated in the final exam in Medical Statistics. Eighty-three passed the exam (97%) with a median grade of 29/30, two students failed: one with a final grade of 17/30 and the other one with the final grade of 11/30. This latter was

Table 3. (*continued*)

Medical Statistics Module			
I feel that	**SD/D**	**N**	**A/SA**
Sequencing of topics was logical	1 (2%)	2 (3%)	57 (95%)
Causes of no interest in the subject	**SD/D**	**N**	**A/SA**
I have to deal with technologies	44 (74%)	11 (18%)	5 (8%)
The subject needs creative thinking	37 (62%)	16 (27%)	7 (11%)
Lack of practising exercises for these topics.	47 (78%)	10 (17%)	3 (5%)
I like clinical studies more than informatics	19 (32%)	14 (23%)	27 (45%)
Lectures were not interesting	48 (81%)	6 (10%)	5 (9%)
Lectures were lengthy	50 (83%)	6 (10%)	4 (7%)
Lectures were difficult to understand	43 (72%)	12 (20%)	5 (8%)
Too many lectures for one day	49 (83%)	8 (14%)	2 (3%)
There were no specific references	54 (90%)	4 (7%)	2 (3%)
I could not see the relation between informatics and medicine at this level	37 (62%)	12 (20%)	11 (18%)
Simply I am not interested in the subject	45 (76%)	11 (19%)	3 (5%)
Attitude towards lecture/lecturers	**SD/D**	**N**	**A/SA**
I was treated with respect	1 (2%)	1 (2%)	58 (96%)
Work and efforts were acknowledged	0 (0%)	18 (30%)	42 (70%)
The lecturer was supportive of the organization of the study	0 (0%)	2 (3%)	58 (97%)
The lecturer was the facilitator of understanding & guiding students	1 (2%)	1 (2%)	58 (96%)
The lecturer allowed to start debates/questions during the lessons	2 (4%)	13 (21%)	45 (75%)

a student who had not attended the course with the new teaching methods because he came from another faculty. Fifty-eight students participated in the first round of exams for the Informatics subject. Fifty-three passed the exam (91%) and five failed (9%), with a median grade of 28/30. These results are in line with the self-reported levels of understanding of the two subjects: by referring to the number of students who disagreed on having acquired skills on the subjects, they were on average 3% for statistics and 6% for informatics, which nearly equates the 3% and 9% of failing students.

6 Discussion and Conclusions

The study investigated on a threefold objective, i.e., (i) the use of technology in high school to teach computer science and statistics, (ii) if in a blended learning method the students gained knowledge on scientific research and (iii) which were the pass/fail rate in the two subjects, with respect to the reported knowledge in the subjects.

Our findings showed that a bit more than half of the students, in high school, studied statistics (69%) and informatics (52%) and, as expected, the use of technology to study was low. We observed differences in the perceived knowledge on the two subjects between students that used (or not) technology at school, which - for Informatics,

Table 4. Skills acquired at the end of the Course. SD = Strongly disagree; D = disagree; N = neutral; A = Agree; Strongly Agree = SA

At the end of the Statistics module	**SD/D**	**N**	**SA/A**
I understood the main concepts of Medical Statistics	0 (0%)	6 (10%)	54 (90%)
I realized the relevance of statistics biostatistics to real health issues	1 (2%)	9 (15%)	50 (83%)
I gained confidence in my ability to do basic statistical & epidemiological analysis	2 (3%)	11 (18%)	47 (78%)
My skills improved in solving problems	3 (5%)	12 (20%)	45 (75%)
I gained skills to read scientific papers	3 (5%)	16 (27%)	41 (68%)
I gained skills to understand the importance of scientific research	1 (2%)	7 (12%)	51 (85%)
	3%	*17%*	*80%*
At the end of the Informatics module	**SD/D**	**N**	**SA/A**
I understood the main concepts of Informatics	3 (5%)	4 (7%)	53 (88%)
I realized the relevance of IT for health problems	4 (7%)	9 (15%)	47 (78%)
I gained confidence in my ability to make descriptive and inferential statistical analyses	3 (5%)	13 (22%)	44 (73%)
My skills improved in solving problems	4 (7%)	14 (23%)	42 (70%)
I gained skills to read scientific papers	6 (10%)	20 (33%)	34 (57%)
I gained skills to understand the importance of scientific research	1 (2%)	10 (17%)	49 (81%)
	6%	*21%*	*73%*

although not statistically different - resulted in opposite course outcomes, i.e., students that used technology at school had on average lower grades.

Furthermore, the results suggest that the two modules were well integrated, and the students reported improvement to understand the importance of scientific research and their methods. The blended learning methods used in the course did not produce problems in the students and seemed to help them, reinforcing a variety of statistical or informatics concepts and motivating the study of a new topic. Finally, the high success rates in the final exams suggest that the used methods helped the teachers to make the course an interesting subject.

Our findings can be summarised in terms of the following list of suggestions for the educators:

- medical students did not have problems in using both common and state-of-the-art tools, then we foster the adoption of technology in the classroom;
- for informatics: (i) particular attention should be placed to students that already used technology at school, which may underestimate the course difficulties and then result

in lower grades; (ii) strengthen the relationship of informatics with scientific research and the task of reading/writing a thesis or scientific papers;

- for statistics: (i) use technology in the classroom because it helps to motivate medical students to learn statistic methodology and data analysis; (ii) follow an applicative approach that seems to bridge the gap between medicine and statistics on medical course entry, introducing students to scientific research, from planning it, to the interpretation of the results.

To conclude, all these findings seem to be in line with the literature, and - even if our study does not include a control group - strengthen the evidence stating that the use of TEL in Medicine may help students to improve and better support their work in the subjects of informatics and statistics [23–25].

References

1. Garrison, D.R., Vaughan, N.D.: Blended Learning in Higher Education: Framework, Principles, and Guidelines. Wiley, San Francisco (2008)
2. Picciano, A.G., Dziuban, C.D., Graham, C.R. (eds.) Blended Learning: Research Perspectives, vol. 2. Routledge, New York (2013)
3. Al-Ajlan, A., Zedan, H.: Why moodle. In: 2008 12th IEEE International Workshop on Future Trends of Distributed Computing Systems. IEEE (2008)
4. Angelone, A.M., Vittorini, P.: The automated grading of R code snippets: preliminary results in a course of health informatics. In: Gennari, R., et al. (eds.) MIS4TEL 2019. AISC, vol. 1007, pp. 19–27. Springer, Cham (2020). https://doi.org/10.1007/978-3-030-23990-9_3
5. De Gasperis, G., Stefano, M., Sara, T., Pierpaolo, V.: Automated grading of short text answers: preliminary results in a course of health informatics. In: 18th International Conference on Web-Based Learning, ICWL 2019 (2019)
6. Means, B., Penuel, W.R., Padilla, C.: The Connected School: Technology and Learning in High School. Wiley, San Francisco (2001)
7. European Commission DG. Communications Networks, Content & Technology. 2nd Survey of Schools: ICT in Education Objective 2: Model for a Highly Equipped and Connected Classroom (2019)
8. Wenglinsky, H.: Using Technology Wisely: The Keys To Success In Schools. Teachers College Pr., New York (2005)
9. Muir-Herzig, R.G.: Technology and its impact in the classroom. Comput. Educ. **42**(2), 111–131 (2004)
10. Davies, R.S., West, R.E.: Technology integration in schools. In: Spector, J., Merrill, M., Elen, J., Bishop, M. (eds.) Handbook of Research on Educational Communications and Technology, pp. 841–853. Springer, New York (2014). https://doi.org/10.1007/978-1-4614-3185-5_68
11. Liu, Q., Peng, W., Zhang, F., Hu, R., Li, Y., Yan, W.: The effectiveness of blended learning in health professions: systematic review and meta-analysis. J. Med. Internet Res. **18**(1), e2 (2016)
12. Neumann, D.L., Neumann, M.M., Hood, M.: Evaluating computer-based simulations, multimedia and animations that help integrate blended learning with lectures in first-year statistics. Australas. J. Educ. Technol. **27**(2), 274–289 (2011)
13. Luo, L., et al.: Blended learning with Moodle in medical statistics: an assessment of knowledge, attitudes and practices relating to e-learning. Br. Med. Educ. **17**, 170 (2017)

14. Triola, M.M., Triola, M.F.: Fondamenti di statistica per le discipline biomediche. In: Giraudo, M.T., Sirovich, R. (eds.) Fondamenti di Statistica Per Le Discipline Biomediche. Pearson Italia (2017)
15. Bianchini, V., Cecilia, M.R., Roncone, R., Cofini, V.: Prevalence and factors associated with problematic internet use: an Italian survey among L'Aquila students. Riv. Psichiatr. **52**(2), 90–93 (2017). https://doi.org/10.1708/2679.27445
16. Cofini, V., Carbonelli, A., Cecilia, M.R., di Orio, F.: Quality of life, psychological wellbeing and resilience: a survey on the Italian population living in a new lodging after the earthquake of April 2009. Ann. Ig. **26**(1), 46–51 (2014). https://doi.org/10.7416/ai.2014.1957
17. Fusco, P., et al.: Unilateral paravertebral block compared with subarachnoid anaesthesia for the management of postoperative pain syndrome after inguinal herniorrhaphy: a randomized controlled clinical trial. Pain **157**(5), 1105–1113 (2016). https://doi.org/10.1097/j.pain.0000000000000487
18. Fusco, P., et al.: Transversus abdominis plane block in the management of acute postoperative pain syndrome after caesarean section: a randomized controlled clinical trial. Pain Phys. **19**(8), 583–591 (2016)
19. Carta, G., et al.: Office hysteroscopic-guided selective tubal chromopertubation: acceptability, feasibility and diagnostic accuracy of this new diagnostic non-invasive technique in infertile women. Hum. Fertil. (Camb) **21**(2), 106–111 (2018). https://doi.org/10.1080/14647273.2017.1384856
20. Casella, F., Sansonetti, A., Zanoni, A., Cofini, V., Capodacqua, A., Verzaro, R.: Radical surgery for gastric cancer in octogenarian patients. Updates Surg. **69**(3), 389–395 (2017)
21. Angelone, A.M., Vittorini, P.: A report on the application of adaptive testing in a first year university course. In: Proceedings of the 8th International Workshop on Learning Technology for Education Challenges (2019)
22. Daher, A.M., Amin, F.: Assessing the perceptions of a biostatistics and epidemiology module: views of year 2 medical students from a Malaysian university. A cross-sectional survey. BMC Med. Educ. **10**, 34 (2010)
23. Milic, N.M., et al.: Bridging the gap between informatics and medicine upon medical school entry: implementing a course on the Applicative Use of ICT. PLoS ONE **13**(4), e0194194 (2018). https://doi.org/10.1371/journal.pone.0194194
24. Andersson, C., Logofatu, D.: A blended learning module in statistics for computer science and engineering students revisited. iJEP **7**(4), 66–77 (2017)
25. Sánchez-Mendiola, M., et al.: Development and implementation of a biomedical informatics course for medical students: challenges of a large-scale blended-learning program. J. Am. Med. Inform. Assoc. **20**(2), 381–387 (2013)

Workshop on TEL in Nursing Education Programs (NURSING)

Workshop on Technology Enhanced Learning in Nursing Education, NURSING

The outbreak of COVID-19 required the adoption of national measures to tackle the spread of the virus all around the world. Consequently, a significant disruption in the traditional educational field occurred, placing both teachers and students in front of sudden challenges to ensure the continuity in education and training activities.

As in the basic educational field, healthcare workers also required an urgent and efficient educational approach to learning new skills essential for caring for this kind of patient, as well as protecting themselves.

In this regard, the online learning and web-based simulation activities suddenly become strategically invaluable ensuring the continuity of educational pathways, professional updates, and the promotion of health literacy for all citizens.

In this novel international framework, the content of the Workshop on Technology Enhanced Learning in Nursing Education strongly supports the international debate about the use of new technologies and learning outcomes.

In this regard, according to current evidence, the quality of learning outcomes in basic and post-basic nursing academic programs could be potentially improved through technology-based systems, that represent the basis for creating smart environments, where models like the high-fidelity simulation deserve great attention for the development prospects that they offer.

However, more robust confirmations are needed, as well as a discussion on the ethical and philosophical implications of technology-enhanced learning within the field of human caring.

Furthermore, little is known about the use of technology to enhance health literacy levels in the community.

For these reasons, this workshop aims to share the best and most timely knowledge about the application of technology-based systems into basic and post-basic nursing academic programs, as well as health educational programs aiming to enhance health literacy levels in the community.

In order to pursue this intent, workshop topics have been grouped into the following three main discussion points.

First, topics on education in nursing academic programs aim to discuss the effects of simulation and other technology-based systems on learning quality, including ethical, legal, and philosophical perspectives.

Second, topics on community health educational programs aim to discuss the impact of technology in improving health literacy levels in the community.

Finally, the workshop intends to provide a complete overview of technology-based methods as useful tools to improve the learning of the nursing process in clinical settings.

The 3rd edition of MIS4TEL Workshop on Technology Enhanced Learning in Nursing Education includes nine accepted papers discussing the use of high-fidelity simulation and other web- or computer-based learning technologies.

I would like to thank authors, reviewers, Professor Zuzana Kubincová, my co-chairman, and the PC members Professor Rosaria Alvaro and Prof. Celeste M. Alfes whose support made this work possible.

Organization

Organizing Committee

Loreto Lancia	University of L'Aquila, Italy
Rosaria Alvaro	University of Tor Vergata, Italy
Celeste M. Alfes	Case Western Reserve University, Cleveland USA

Program Committee

Angelo Dante	University of L'Aquila, Italy
Cristina Petrucci	University of L'Aquila, Italy
Pierpaolo Vittorini	University of L'Aquila, Italy

A Serious Game and Negotiation Skills in Nursing Students: A Pilot Study

Valentina Zeffiro[1(✉)], Raffaele Di Fuccio[2], Ercole Vellone[1], Rosaria Alvaro[1], and Fabio D'Agostino[3]

[1] University of Rome Tor Vergata, Rome, Italy
valentina.zeffiro@uniroma2.it
[2] Istituto di Scienze e Tecnologie della Cognizione – Consiglio Nazionale delle Ricerche, Rome, Italy
[3] Saint Camillus International University of Health Sciences, Rome, Italy

Abstract. Negotiation emerges in our lives whenever a decision needs to be made together with someone else. In the nursing field, negotiation is well represented in the relationship between the nurse and the patient. For this reason, it is important to help nurses to develop their communication skills, starting from the training period. The aims of this study were to identify the negotiation styles of nursing students with a serious game and to evaluate the effect of a tailored intervention on negotiation skills. The serious game was based on Rahim and Bonoma's theory of conflict management and included five scenarios of daily life situations. Descriptive and inferential statistics were performed to evaluate the students' negotiation styles before and after the intervention and the times spent to complete the game's scenarios. Results showed that nursing students preferred an integrating negotiation style and that they improved their compromising style after the intervention. The students were generally oriented towards using problem-solving reasoning and, after the intervention, renouncing something in order to reach a mutually acceptable decision in the included scenarios. To our knowledge, this is the first study to use a negotiation serious game in the nursing learning field and serves as an example of how technology can be accepted and integrated in education to help nursing students develop not only their technical but also relational skills.

Keywords: Serious game · Negotiation styles · Nursing students

1 Introduction

Negotiation is part of our life; people unconsciously negotiate with each other every day. Formally, negotiation is the act of conferring with others in order to reach a compromise. It is based on a bidirectional communication that takes place when there are different interests between the parties [1].

In nursing, the concept of negotiation is a key element and is present in different relationship situations: between nurses, between nurses and other healthcare practitioners,

Z. Kubincová et al. (Eds.): MIS4TEL 2020, AISC 1236, pp. 91–98, 2021.
https://doi.org/10.1007/978-3-030-52287-2_9

between nurses and patients and between nurses and patients' families [2]. Negotiation is particularly intrinsic in the nurse–patient relationship, a dynamic interaction in which various mutually negotiated relationships (e.g. grateful, coercive, manipulative) can exist depending on the patient's specific situation [3]. Research has shown that the patient can have a greater sense of control and power during the interaction when the nurse encourages an active negotiation relationship and allows shared decision-making [4, 5]. This active participation and sense of control lead to better patient self-care [6]. All nurses are expected to have negotiation and conflict management skills [7]. Therefore, it is important to develop these skills starting from the training period.

During their undergraduate years, nursing students develop relational skills that are essential for the caring process [8]. Moreover, they experiment with emotional challenges posed by conflicts [9], which could lead to negative organizational and caring outcomes [10]. Negotiation can help them manage these conflicts. To our knowledge, few studies have investigated negotiation in nursing students. Some authors emphasized that simulation can be helpful to students in "negotiating the role of the professional nurse" [11]; others showed that observing an expert nurse's behaviour patterns can improve students' negotiating style [12].

1.1 Serious Games

Serious games (SGs) are defined as "games in which education (in its various forms) is the primary goal, rather than entertainment" [13]. In digital game-based learning, teachers/educators "use digital games with serious goals (i.e. educational objectives) as tools that support learning processes in a significant way" [14].

An increasing number of SGs is being used in nursing education to develop students' clinical reasoning and decision-making skills, with positive results in specific clinical situations such as care of patients with chronic obstructive pulmonary disease [15], safe administration of blood transfusion [16], preterm newborn clinical assessment [17], inhaler techniques [18] and cardiopulmonary resuscitation [19].

To our knowledge, no previous study has used an SG to evaluate negotiation styles in nursing students.

1.2 Aims of the Study

The aims of this study were to evaluate negotiation styles in nursing students using an SG and to assess the effect of a tailored intervention on their negotiation skills.

2 Methods

2.1 Design, Setting, Sample and Data Collection

A pre–post-test intervention study was conducted at the Saint Camillus International University of Health Sciences in Rome, Italy, in November 2019. The pre-test involved students playing the SG to evaluate their negotiation styles. The subsequent intervention, consisting of two frontal lessons and two role-playing scenarios, was designed to improve students' knowledge about negotiation and help them develop their negotiation skills. The post-test involved repeating the SG to verify a possible effect of the intervention on students' negotiation styles. Finally, the students completed a questionnaire, the Player Experience of Need Satisfaction (PENS) scale, about their SG experience.

The participants were attending their second year of their basic nursing degree and came from several Asian, African and Central American countries. Students who did not attend both the pre- and the post-test were excluded from the study.

Data were collected before and after the intervention.

2.2 Intervention

The intervention was conducted on two different days by two experts: one was an expert in gamification and negotiation, and the other was an expert in nursing and relational skills. The intervention was divided into three sections: two frontal lessons and a role-playing component, which included two scenarios.

On the first day, after the SG, the students attended the first frontal lesson, which was centred around the main concepts of the role of soft skills with a focus on negotiation and its related theories, and with an additional input about the gamification approach.

The second day started with the second frontal lesson, which discussed the importance of negotiation in the nursing field, and continued with two role-playing scenarios that students performed in pairs. In the first scenario, one student played the role of a nurse negotiating his/her Christmas holidays with another student playing the role of the nurse coordinator. In the second scenario, one student played the role of a nurse negotiating a healthy behaviour with another student playing the role of a chronic patient.

2.3 Instruments

The following instruments were used to collect data:

a) A sociodemographic questionnaire was created to collect age, sex, marital status and education level data.
b) The ENACT Game, developed by a European Union project funded in the framework of the Lifelong Learning Programme[1] [20].

The ENACT Game is based on Rahim and Bonoma [21] theory of conflict management, which identifies five negotiation styles: (1) integrating, where the parties examine the different positions for an acceptable solution; (2) obliging, in which the parties underline the commonalities to satisfy each other's concerns; (3) dominating, when the behaviour of one party is strongly oriented towards winning; (4) avoiding, in which one party renounces something to satisfy the other party's concerns; and (5) compromising, in which both parties renounce something to reach a mutually acceptable solution.

[1] Enhancing Negotiation skills through on-line Assessment of Competencies and interactive mobile Training – GA 543301-LLP-1-2013-1-UK-KA3-KA3MP 2013 (http://enactgame.eu/site/).

The game includes five scenarios. In each of these, one student conversed with another to find a solution to an issue. For each scenario, the students had to answer five questions, selecting the preferred option from a list of five possible answers. According to these answers, one of the five negotiation styles was identified for each student.

The scenarios are as follows:

- Motorbike: Two young siblings need to use the same motorbike the same evening.
- CD shop: Two friends wait for two copies of their favourite band's CD for half an hour, but there is only one CD booked in their names.
- Restaurant: Two friends must decide where to go for dinner.
- Sport: The Plymouth team has decided to redesign their logo and elect the person in charge of the design among the team members. Two people receive the same number votes and meet to decide who should be in charge.
- TV programme: Two spouses live in a home with only one TV in the living room, and they must decide what to watch in the evening.

c) The PENS scale was used to evaluate the students' game experience through five dimensions: competence, autonomy, relatedness, presence/immersion and intuitive controls. It consists of 21 items and uses a 7-point Likert scale, where 1 represents "do not agree" and 7 represents "strongly agree". The higher the score in each dimension, the better the student's experience with the game [22]. The PENS scale is based on self-determination theory, which is focused on the intrinsic and extrinsic motivation that shapes individuals' identity and behaviours [23].

2.4 Data Analysis

A descriptive analysis including means with standard deviations, frequencies and percentages was performed to determine the students' sociodemographic characteristics and negotiation styles, the times spent to complete the ENACT SG scenarios and the students' game experience.

To evaluate the effect of the intervention on the style of negotiation, this variable was assessed before (Time 0) and after the intervention (Time 1). The possible significant difference between the two measurements was evaluated using a chi-squared test. A possible significant difference in the times spent to complete the scenarios between Time 0 and Time 1 was evaluated using a *t*-test. In all cases, a value of $p < 0.05$ was considered statistically significant.

IBM SPSS Statistics version 22.0 was used for the analysis.

2.5 Ethical Considerations

The study project was approved by the Research Committee of the Saint Camillus International University. All participants were informed in detail of the study's purpose, procedures, risks, benefits and data collection and signed informed consent forms.

As a project funded by the European Commission, the ENACT Game is free to use. The developers were informed about this study.

3 Results

Thirteen students (mean age: 24.6 ± 6.8 years) were included in this study. Most of them (92.3%) were female. All students were unmarried, and one of them was a nun. The majority (69.3%) had a previous degree.

No significant differences ($p > 0.05$) were observed in the frequencies of the five negotiation styles between Time 0 and Time 1. Nonetheless, after the intervention the use of the obliging style decreased by 5%, and that of the compromising style increased by 4.1%. The most used negotiation style both at Time 0 and at Time 1 was the integrating one (Fig. 1), which was prevalent especially in the motorbike and sport scenarios.

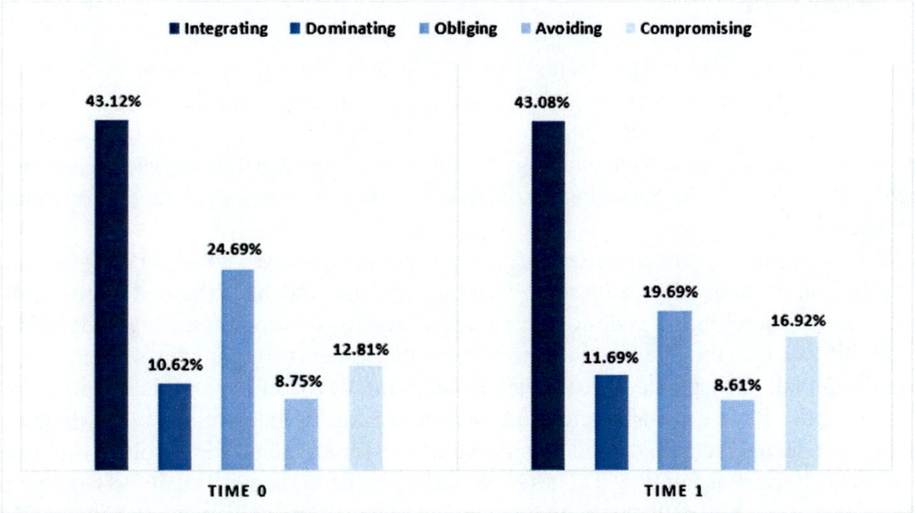

Fig. 1. Frequencies of the students' negotiation styles before (Time 0) and after (Time 1) the intervention among the five scenarios.

Students spent a mean of 5.1 ± 3.7 min to complete each scenario at Time 0 and a mean of 3.9 ± 2.6 min at Time 1 ($p = 0.073$).

The mean scores of the PENS dimensions are shown in Table 1.

Table 1. PENS dimensions' means scores according to the students' reported experiences.

PENS dimensions	Mean (standard deviation)
Competence	6 (0.5)
Autonomy	6 (0.9)
Relatedness	5.7 (1.1)
Presence/immersion	5.6 (0.8)
Intuitive controls	5.7 (0.8)

4 Discussion

This study evaluated nursing students' negotiation styles through the use of the ENACT SG before and after an intervention. To our knowledge, this is the first study to use a serious game to evaluate negotiation in the nursing field. Serious games have already been considered as a possible technology for future nursing education in order to achieve higher levels of clinical reasoning skills, evidence-based knowledge and professional autonomy [24].

Even though a significant difference in negotiation styles was not observed after the intervention, probably due to the small sample size, the students exhibited a reduction in the use of the obliging style to leave more space for the compromising style. This could indicate that the intervention encouraged the students to renounce something to reach a mutually acceptable decision rather satisfying the other person's concerns [21]. Moreover, the students mostly showed a preference for the integrating style, which is useful for reaching successful and creative solutions for all the parties involved through problem-solving reasoning [21]. These are skills that the nurse needs daily to negotiate with both patients and colleagues, and the students seemed to be able to implement them.

Regarding the time spent to complete each scenario, the students were faster the second time by about 1.2 min. Possible reasons could be that they remembered the scenarios in which they had already engaged or that their curiosity about the SG had diminished. The literature shows a very wide range of mean times spent to complete a scenario (e.g. as short as 2.1 min or as long as 32 min) [25, 26]. The time could be related to the specific situation, the setting and student involvement.

Finally, the students seemed to evaluate their ENACT SG experience positively. Specifically, they found the experience challenging but not difficult and felt free and interested in the activities. In general, it seems that SGs are embraced by nursing students, who find them useful, usable, satisfying and motivating for learning [16, 26, 27].

This study had certain limitations. One limitation was the small sample size. Another was that it evaluated students' negotiation styles not in clinical settings but in generic scenarios.

5 Conclusions

This pilot study is an example of how an SG can help to increasing negotiation knowledge and raising students' awareness about the process of negotiation in different scenarios of daily life.

The second-year students of Unicamillus University seemed to lean towards using the integrating negotiation style even though they had not received specific training in this before. The nursing practice itself pushes to find effective solutions in specific situations, and the students experienced this during the intervention.

Technology can make important contributions not only to technical but also to relational skills. For this reason, it should be considered a possible instrument that can be regularly incorporated in the various stages of learning for nurses. The students seemed to willingly accept the use of technology to test themselves in a simulation setting before engaging in a real clinical setting.

In the future, it would be interesting to repeat this study with a larger sample to confirm its results. Moreover, it would be useful to create an SG set with specific situations related to the nursing clinical practice in order to provide nursing students with scenarios more suitable for them.

References

1. Fisher, R., Ury, W., Patton, B.: L'arte del negoziato. Corbaccio, Milano (2007)
2. Keatinge, D.: Negotiated care-fundamental to nursing practice. Collegian **5**, 36–42 (1998)
3. Morse, J.M.: Negotiating commitment and involvement in the nurse-patient relationship. J. Adv. Nurs. **16**, 455–468 (1991)
4. Roberts, S.J., Krouse, H.J., Michaud, P.: Negotiated and nonnegotiated nurse-patient interactions: enhancing perceptions of empowerment. Clin. Nurs. Res. **4**, 67–77 (1995)
5. Krouse, H.J., Roberts, S.J.: Nurse-patient interactive styles power, control, and satisfaction. West. J. Nurs. Res. **11**, 717–725 (1989)
6. Roberts, S.J., Krouse, H.J.: Enhancing self care through active negotiation. Nurse Pract. **13**, 44, 47, 50–42 (1988)
7. Lee, C.J., Ahonen, K., Navarette, E., Frisch, K.: Successful student group projects: perspectives and strategies. Teach. Learn. Nurs. **10**, 186–191 (2015)
8. Zangao, M.O., Mendes, F.R.: Relational skills and preserving patient privacy in the caring process. Revista brasileira de enfermagem **68**(167–173), 191–197 (2015)
9. Arieli, D.: Emotional work and diversity in clinical placements of nursing students. J. Nurs. Sch. Off. Publ. Sigma Theta Tau Int. Honor Soc. Nurs. **45**, 192–201 (2013)
10. Labrague, L.J., McEnroe-Petitte, D.M.: An integrative review on conflict management styles among nursing students: implications for nurse education. Nurse Educ. Today **59**, 45–52 (2017)
11. Walton, J., Chute, E., Ball, L.: Negotiating the role of the professional nurse: the pedagogy of simulation: a grounded theory study. J. Prof. Nurs. **27**, 299–310 (2011)
12. Eddy, M.E., Schermer, J.: Shadowing: a strategy to strengthen the negotiating style of baccalaureate nursing students. J. Nurs. Educ. **38**, 364–367 (1999)
13. Michael, D.R., Chen, S.L.: Serious Games: Games that Educate, Train, and Inform. Muska & Lipman/Premier-Trade, Roseville (2005)
14. Sica, L.S., Veneri, A.D., Miglino, O.: Exploring New Technological Tools for Education: Some Prototypes and Their Pragmatical Classification. InTech, Croatia (2012)

15. Johnsen, H.M., Fossum, M., Vivekananda-Schmidt, P., Fruhling, A., Slettebo, A.: A serious game for teaching nursing students clinical reasoning and decision-making skills. Stud. Health Technol. Inform. **225**, 905–906 (2016)
16. Tan, A.J.Q., Lee, C.C.S., Lin, P.Y., Cooper, S., Lau, L.S.T., Chua, W.L., Liaw, S.Y.: Designing and evaluating the effectiveness of a serious game for safe administration of blood transfusion: a randomized controlled trial. Nurse Educ. Today **55**, 38–44 (2017)
17. Fonseca, L.M., Aredes, N.D., Dias, D.M., Scochi, C.G., Martins, J.C., Rodrigues, M.A.: Serious game e-Baby: nursing students' perception on learning about preterm newborn clinical assessment. Revista brasileira de enfermagem **68**(9–14), 13–19 (2015)
18. Chee, E.J.M., Prabhakaran, L., Neo, L.P., Carpio, G.A.C., Tan, A.J.Q., Lee, C.C.S., Liaw, S.Y.: Play and learn with patients—designing and evaluating a serious game to enhance nurses' inhaler teaching techniques: a randomized controlled trial. Games Health J. **8**, 187–194 (2019)
19. Boada, I., Rodriguez-Benitez, A., Garcia-Gonzalez, J.M., Olivet, J., Carreras, V., Sbert, M.: Using a serious game to complement CPR instruction in a nurse faculty. Comput. Methods Programs Biomed. **122**, 282–291 (2015)
20. Dell'Aquila, E., Marocco, D., Ponticorvo, M., Di Ferdinando, A., Schembri, M., Miglino, O.: ENACT: virtual experiences of negotiation. In: Educational Games for Soft-Skills Training in Digital Environments, pp. 89–103. Springer, Cham (2017)
21. Rahim, A., Bonoma, T.V.: Managing organizational conflict: a model for diagnosis and intervention. Psychol. Rep. **44**, 1323–1344 (1979)
22. Johnson, D., Gardner, M.J., Perry, R.: Validation of two game experience scales: the Player Experience of Need Satisfaction (PENS) and Game Experience Questionnaire (GEQ). Int. J. Hum Comput Stud. **118**, 38–46 (2018)
23. Ryan, R.M., Rigby, C.S., Przybylski, A.: The motivational pull of video games: a self-determination theory approach. Motiv. Emot. **30**, 344–360 (2006)
24. Johnsen, H.M., Fossum, M., Vivekananda-Schmidt, P., Fruhling, A., Slettebo, A.: Developing a serious game for nurse education. J. gerontol. Nurs. **44**, 15–19 (2018)
25. Nicolaidou, I., Antoniades, A., Constantinou, R., Marangos, C., Kyriacou, E., Bamidis, P., Dafli, E., Pattichis, C.S.: A virtual emergency telemedicine serious game in medical training: a quantitative, professional feedback-informed evaluation study. J. Med. Internet Res. **17**, e150 (2015)
26. Johnsen, H.M., Fossum, M., Vivekananda-Schmidt, P., Fruhling, A., Slettebo, A.: Teaching clinical reasoning and decision-making skills to nursing students: design, development, and usability evaluation of a serious game. Int. J. Med. Informatics **94**, 39–48 (2016)
27. Johnsen, H.M., Fossum, M., Vivekananda-Schmidt, P., Fruhling, A., Slettebo, A.: Nursing students' perceptions of a video-based serious game's educational value: a pilot study. Nurse Educ. Today **62**, 62–68 (2018)

Interprofessional High-Fidelity Simulation on Nursing Students' Collaborative Attitudes: A Quasi-experimental Study Using a Mixed-Methods Approach

Paola Ferri(✉), Sergio Rovesti, Alberto Barbieri, Enrico Giuliani, Chiara Vivarelli, Nunzio Panzera, Paola Volpi, and Rosaria Di Lorenzo

University of Modena and Reggio Emilia, Modena 41125, Italy
paola.ferri@unimore.it

Abstract. Background: interprofessional simulation appears to be effective training for nursing students, yet many questions remain about its feasibility, acceptability and efficacy in improving students' collaborative attitudes. Study design and participants: the aim of this quasi-experimental study, with a mixed-methods approach, was to evaluate changes in interprofessional collaborative attitudes after a training session based on an interprofessional high-fidelity patient simulation (IHFPS). The sample was composed of students attending the 2nd and 3rd year of the Nursing Degree Program and residents of the Anaesthesia Residency Program at University of Modena and Reggio Emilia in 2019. Methods: nursing students and residents were grouped into small interprofessional teams and participated in an IHFPS focused on acute care. To measure interprofessional collaboration attitude, the Jefferson Scale of Attitudes toward Physician-Nurse Collaboration (JSAPNC) and the Readiness for Interprofessional Learning Scale (RIPLS) were administered to nursing students. They completed a post-test to investigate their satisfaction with IHFPS and they replied to open-ended statements. Results and conclusions: 204 nursing students completed both the pre- and post-test surveys. Our results suggested that an IHFPS, with small teams of nursing students and residents, improved interprofessional collaborative attitudes of nursing students. We reported a statistically significant improvement in three factors of JSAPNC and in the RIPLS, which showed the positive effects of this experience on many collaborative skills. The students expressed high satisfaction with the training conducted in a realistic and safe setting, which improved their awareness of working in an effective multidisciplinary team.

Keywords: Interprofessional simulation · Nursing student · Mixed-methods approach

1 Introduction

Effective interprofessional collaboration (IPC) guaranties safe, high-quality and efficient nursing care and favors both patient and health professional satisfaction [1, 2]. In accordance with the World Health Organization (2010), "Learning together to work together

Z. Kubincová et al. (Eds.): MIS4TEL 2020, AISC 1236, pp. 99–110, 2021.
https://doi.org/10.1007/978-3-030-52287-2_10

for better health" is essential to the development of a collaborative practice [3]. The National League for Nursing believes that educational approaches must include opportunities for students to engage in interprofessional education (IPE), because health care professionals must have the competency to work in teams [4]. IPE occurs when students from two or more professions learn about, from and with each other to enable effective collaboration and improve health outcomes [3]. Despite the benefits and endorsements of IPC and IPE, many schools and health systems face numerous barriers to developing IPE initiatives [4]. Simulation is a teaching strategy that allows students to participate in the replication of clinical care in a safe environment, without the psychological pressure of a real care environment, in order to improve communication skills among nursing students and other healthcare professionals [5]. The scenarios planned for the simulated health activities create the conditions favoring teamwork and integration among professions in order to improve care programs and paths [6]. Many questions are still open about the feasibility, acceptability and efficacy of improving students' collaborative attitudes [7], since, in many cases, the effectiveness of these educational interventions has not yet been evaluated with the necessary psychometric assessment [8, 9]. Two recent reviews on the use of interprofessional simulation in undergraduate nursing programs concluded that simulation showed positive impact on the development of communication skills and collaborative attitude in nursing students, but additional research, utilizing a more robust method of research and reliable assessment methods, is needed [10, 11]. The aim of the research was to evaluate the attitude in interprofessional collaboration before (T0) and after (T1) a training session based on an interprofessional high-fidelity patient simulation (IHFPS) in undergraduate nursing students.

2 Method

2.1 Study Design and Participants

A pre-post study was implemented with an explanatory mixed-method approach. Participants were the students attending the 2nd and 3rd year of the Nursing Degree Program and residents of the Anesthesia Residency Program at University of Modena and Reggio Emilia in 2019.

2.2 IPE Training

All nursing students and residents participated in an experimental intervention based on an IHFPS at the Advanced Training and Medical Simulation Center, conducted by faculty facilitators who are experts in this methodology. Before the simulation, the topic of IPC was introduced and, afterwards, participants were divided into 34 groups composed of 6 nursing students and 1 resident. Each group was further divided into two subgroups, the first consisting of second year nursing students and resident, the second one consisting of third year nursing students and the same resident. All participants received an IPE training session in laboratory using IHFPS, which provided two different scenarios, one for each group. The first scenario, addressed to the first subgroup, simulated the conditions of a patient admitted to a surgery unit who, after surgery, presented a modification in his mental state, showing symptoms of acute cognitive deterioration. General clinical conditions, vital parameters and laboratory tests suggested the evolution of a septic state. Nursing students were able to recognize changes in mental status of the patient and in his vital signs. Therefore, they alerted the resident, on whose indication they took a blood sample. In the same simulation, the resident was able to formulate a first diagnostic suspicion. After having formulated the diagnosis, the resident prescribed a therapy and transferred the patient to the Intensive Care Unit (ICU). Following this first scenario, a faculty facilitator conducted a debriefing session in both subgroups in order to deal with clinical problems. The second subgroup, after having observed the first simulation described above through a video link, participated in the debriefing session and, afterwards, accepted the patient in ICU. Subsequently, they provided him with intensive monitoring and supportive therapies for vital functions, after having considered the necessity of mechanical ventilation. Nursing students and resident took into consideration the aforementioned procedures and treatments in accordance with their professional competency. The handover between the two subgroups had to take place in a standardized way, including a "patient" evaluation through the Confusion Assessment Method in ICU (CAMICU) and a passage of information through the Situation, Background, Assessment and Recommendation (SBAR) tool. At the end of the second simulation, a new debriefing session was conducted by the same faculty facilitator who conducted the first session, with all the participants of the two simulations.

2.3 Instruments

To measure the attitude in interprofessional collaboration, the following scales were administered to the nursing students before (T0) and after (T1) the simulation:

1. Jefferson Scale of Attitudes toward Physician-Nurse Collaboration (JSAPNC), a self-report questionnaire already validated in a sample of Italian physicians and nurses with a Cronbach's alpha of 0.70 [12]. Permission was obtained to use this tool by one of its developers (Dr. Mohammadreza Hojat). The most recent version of JSAPNC consists of 15 items with answers concerning participants' degree of agreement/disagreement on a 4-point Likert scale (from 1 = strongly disagree to 4 = strongly agree). The score ranges from 15 to 60 points; the higher the score, the greater the attitude of collaboration. JSAPNC items investigate four factors: "Shared education and collaboration"; "Caring versus curing"; "Nurse's autonomy"; "Physician's authority". In the present study, Cronbach's alpha was 0.69 at pre-test and 0.71 at post-test;
2. Readiness for Interprofessional Learning Scale (RIPLS) [13], validated and adapted to the Italian educational context by Sollami et al., with a Cronbach's alpha of 0.92 [14]. Permission to use the RIPLS was obtained from Dr. Alfonso Sollami, the author of Italian validation of the scale. This scale consists of 10 items with a Likert 5-point scale (from 1 = strongly disagree to 5 = strongly agree). In the present study, Cronbach's alpha was 0.91 at pre-test and 0.92 at post-test.

At the end of the IHFPS a short anonymous questionnaire was administered to investigate selected socio-demographic variables (gender, age) and level of satisfaction, with a Likert

4-point scale (from 1 = not at all satisfied to 4 = very satisfied). Data for qualitative analysis were collected by the following two open-ended statements:

- "Please describe in as much detail as possible the strengths of the interprofessional high-fidelity patient simulation in which you have just participated";
- "Please describe in as much detail your suggestions to improve the interprofessional high-fidelity patient simulation in which you have just participated".

2.4 Data Collection Procedure

This study was approved and authorized by the Nursing Degree Program Director and the Anesthesia Residency Program Director. All nursing students and residents were informed about the objectives and methods of this research and their participation was voluntary. This study was conducted in agreement with the Helsinki declaration. Nursing students were asked to anonymously complete the scales and investigative questionnaires before and after the simulation. Student anonymity was guaranteed by assigning a code to each participant. The faculty facilitators left the room during the survey administration.

2.5 Sample Size and Statistical Study Power

In accordance with the data collected in a pilot pre-post study focused on a pre- and post-intervention [15], assuming a minimum difference between the pre- and post-test of 2 points in the mean JSAPNC scores, the minimum sample to be enrolled in the present study was 42 students, based on the average total score of the JSAPNC and a standard deviation of 5, with an alpha error of 0.05 and a power of at least 0.80.

2.6 Data Analysis

Descriptive statistics such as mean and standard deviation were applied to summarize the socio-demographic characteristics of the participants and the JSAPNC and RIPLS scores. Statistical comparisons at and between T0 and T1 of both JSAPNC and RIPLS mean scores were conducted using the Student's paired and unpaired *t*-tests. A $p < 0.05$ value was defined as statistically significant. Data were analyzed using Stata 14 (StataCorp, College Station, TX, USA). Content analysis was used to analyze the qualitative data.

3 Results

204 nursing students agreed to participate in this study (response rate = 89%): 152 females and 52 males. Males were 22.62 years old (±0.49 SD) and females 21.97 years old (±0.22 SD) on average.

Table 1. Jefferson Scale of Attitudes toward Physician-Nurse Collaboration mean scores at pre- (T0) and post- (T1) training simulation.

JSAPNC factors and items	T0 Mean (SD)	T1 Mean (SD)	Statistical test *p* value
Factor 1. Shared education and collaboration	**24.62 (2.08)**	**26.20 (1.91)**	**$t = -12.61$ $p < 0.001$**
1. A nurse should be viewed as a collaborator and colleague with a physician rather than his/her assistant	3.90 (0.38)	3.89 (0.50)	$t = 0.16$ $p = 0.88$
3. During their education, medical and nursing students should be involved in teamwork in order to understand their respective roles	3.73 (0.49)	3.91 (0.31)	$t = -5.10$ $p < 0.0001$
6. There are many overlapping areas of responsibility between physicians and nurses	3.26 (0.68)	3.47 (0.68)	$t = -4.21$ $p < 0.0001$
9. Physicians and nurses should contribute to decisions regarding the hospital discharge of patients	3.83 (0.42)	3.90 (0.37)	$t = -1.96$ $p = 0.05$
12. Nurses should also have responsibility for monitoring the effects of medical treatment	3.34 (0.64)	3.69 (0.61)	$t = -7.64$ $p < 0.0001$
14. Physicians should be educated to establish collaborative relationships with nurse	3.32 (0.67)	3.68 (0.56)	$t = -8.21$ $p < 0.001$
15. Interprofessional relationships between physicians and nurses should be included in their educational programs	3.27 (0.71)	3.71 (0.56)	$t = -9.70$ $p < 0.001$
Factor 2. Caring versus curing	**10.67 (1.19)**	**11.30 (3.10)**	**$t = -2.79$ $p < 0.005$**
2. Nurses are qualified to assess and respond to psychological aspects of patient's needs	3.35 (0.62)	3.65 (0.52)	$t = -7.6$ $p < 0.0001$
4. Nurses should be involved in making policy decisions affecting their working conditions	3.90 (0.31)	3.90 (0.30)	$t = -0.19$ $p = 0.85$
7. Nurses have special expertise in patient education and psychological counseling	3.46 (0.59)	3.87 (2.92)	$t = -1.91$ $p = 0.06$
Factor 3. Nurse's autonomy	**10.49 (1.23)**	**11.30 (0.96)**	**$t = -11.35$ $p < 0.001$**
5. Nurses should be accountable to patients for the nursing care they provide	3.23 (0.72)	3.57 (0.65)	$t = -8.43$ $p < 0.0001$
11. Nurses should be involved in making policy decisions concerning the hospital support services upon which their work depends	3.52 (0.57)	3.81 (0.42)	$t = -7.51$ $p < 0.0001$

(*continued*)

Table 1. (*continued*)

JSAPNC factors and items	T0 Mean (SD)	T1 Mean (SD)	Statistical test *p* value
13. Nurses should clarify a physician's order when they feel that it might have the potential for detrimental effects on the patient	3.79 (0.45)	3.95 (0.25)	$t = -4.88$ $p < 0.0001$
Factor 4. Physician's authority	**6.83 (1.28)**	**6.49 (1.74)**	$t = 2.91$ $p < 0.004$
8. Doctors should be the dominant authority in all health care matters	3.19 (0.79)	3.04 (1.06)	$t = 2.03$ $p = < 0.05$
10. The primary function of the nurse is to carry out the physician's orders	3.67 (0.66)	3.45 (0.88)	$t = 3.54$ $p < 0.0005$
Total score	52.60 (3.67)	55.29 (4.64)	$t = -8.90$ $p < 0.0001$

3.1 Quantitative Analysis

The JSAPNC total mean score, at post-IPE survey (T1), showed a statistically significant difference when compared to that at IPE survey (T0), at paired t-test [55.29 ± 4.64 (SD) vs 52.60 ± 3.67 (SD), $t = -8.90$; $p < 0.0001$], indicating nursing students' greater attitude for interprofessional collaboration, as reported in Table 1.

Most of the JSAPNC items showed higher mean values at T1 in comparison with T0 with the exception of items grouped in factor 4. In factor 1, "Shared education and collaboration", all items except items 1 and 9 ("A nurse should be viewed as a collaborator and colleague with a physician rather than his/her assistant", "Physicians and nurses should contribute to decisions regarding the hospital discharge of patients") showed a statistically significant difference between T0 and T1; in factor 2, "Caring versus curing", at T1, only item 2 ("Nurses are qualified to assess and respond to psychological aspects of patient's needs") statistically significantly differed from T0; all 3 items grouped in factor 3, "Nurse's autonomy", showed a statistically significant difference between T0 and T1 as well as the two items of factor 4, "Physician's authority" (Table 1). The total mean score of JSAPNC did not show any statistically significant difference between males and females either at T0 [males: mean 51.87 ± 3.46 (SD) vs females: mean 52.90 ± 3.71 (SD), $t = -1.77$; $p = 0.08$] or T1 [males: mean 55.13 ± 7.20 (SD) vs females: mean 55.42 ± 3.22 (SD), $t = -0.39$; $p = 0.70$]. Paired *t*-test revealed significant changes from before to after the IPE training in the total mean score of RIPLS [43.97 ± 5.27 (SD) vs 47.75 ± 3.57 (SD), $t = -13.33$; $p < 0.0001$], showing a greater readiness for interprofessional learning among nursing students, as reported in Table 2. All 10 items of RIPLS, after the IHFPS, improved in a statistically significant way (Table 2).

Table 2. Readiness for Interprofessional Learning Scale mean scores at pre- (T0) and post- (T1) training simulation.

RIPLS Items	T0 Mean (SD)	T1 Mean (SD)	Statistical test *p* value
1. I would welcome the opportunity to work on small-group projects with other healthcare students	4.47 (0.71)	4.86 (0.40)	$t = -9.05$ $p < 0.0001$
2. Learning with students from other health care professions will help me to communicate better with patients and other professionals	4.36 (0.73)	4.83 (0.39)	$t = -10.20$ $p < 0.0001$
3. Learning with students from other health care professions will help to clarify the nature of patient problems	4.35 (0.75)	4.77 (0.48)	$t = -8.67$ $p < 0.0001$
4. Communication skills should be learned with other healthcare students	4.33 (0.71)	4.79 (0.48)	$t = -9.36$ $p < 0.0001$
5. Learning with students from other healthcare professions before graduation will help me to become a better team worker	4.54 (0.64)	4.80 (0.47)	$t = -5.60$ $p < 0.0001$
6. Learning with healthcare students before graduation would improve relationships after graduation	4.44 (0.68)	4.77 (9.48)	$t = -7.19$ $p < 0.0001$
7. Learning with students from other healthcare professions will help me to think positively about other professionals	4.19 (0.77)	4.65 (0.57)	$t = -8.68$ $p < 0.0001$
8. Learning with students from other healthcare professions will increase my ability to understand clinical problems	4.43 (0.66)	4.76 (0.47)	$t = -7.25$ $p < 0.0001$
9. Learning with other students will help me to become a more effective member of a healthcare team	4.46 (0.71)	4.79 (0.43)	$t = -6.96$ $p < 0.0001$
10. Learning with students from other healthcare professions will help me to understand my own limitations	4.44 (0.72)	4.74 (0.54)	$t = -6.47$ p < 0.0001
Total score	43.97 (5.27)	47.75 (3.57)	$t = -13.33$ $p < 0.0001$

The total mean score of RIPLS did not statistically significantly differ between the two genders at both T0 [males: mean 43.25 $\pm$ 5.04 (SD) vs females: mean 44.38 $\pm$ 5.03 (SD), $t = -1.41$; $p = 0.16$] and T1 [males: mean 47.11 $\pm$ 3.91 (SD) vs females: mean 48.12 $\pm$ 2.98 (SD), t = -1.9415; $p = 0.05$]. Post-test scores highlighted a high level of student satisfaction with the experience [mean: 3.87 $\pm$ 0.38 (SD)].

3.2 Qualitative Analysis

All 204 nursing students provided comments on the open-ended statement "Please describe in as much detail as possible the strengths of the interprofessional high-fidelity patient simulation in which you have just participated" and, using content analysis, the following 2 themes with 5 sub-themes emerged:

Interprofessional Collaboration

Awareness of Interprofessional Collaboration Efficacy in Patient Care

Most nursing students highlighted how IHFPS improved awareness of the importance and effects of interprofessional collaboration in patient care: "The positive aspects of this experience consist of greater awareness of the importance of collaboration and communication with other professionals for achieving a common goal" (Student80); "….it also makes clear the importance of team collaboration for ensuring patient well-being" (S95); "It highlighted how effective an equal collaboration between doctor and nurse can be and how much more positive the climate can be, all for the benefit of the patient" (S204).

Experimenting Collaboration and Interprofessional Communication

The participants agreed that the IHFPS had allowed them a true experimentation of interprofessional communication and collaboration: "This training allows the student to enter the role of the professional who has to deal with a multidisciplinary team. It therefore gives the student the opportunity to experiment his/her communicative and collaborative skills in the management of an emergency situation" (S4); "This training allows us to experience sharing and collaboration between peers and, at the same time, to maintain an open dialogue with the patient and colleagues. It is an experimentation in a realistic situation" (S16); "Very positive experience. I think it is useful because it permits us to deepen the interprofessional relationship and improve communication between doctors and nurses, positively impacting on clinical context and interprofessional collaboration" (S23).

Intra- and Inter-professional Learning

The IHFPS was perceived by nursing students as an effective and positive experience

of cooperation, communication, listening and intra- and interprofessional cooperation, without prejudicial misunderstanding: "This experience also provides a very interesting and inspiring context of cooperation with other students of our degree course" (S13); "It give us the opportunity for communication, listening and collaboration among group members (2nd and 3rd year nursing students/residents) without prejudice" (S177); "This opportunity allow us to see through the perspective of other professionals and to understand their way of interpreting health contexts and patients' needs in various situations. Also, it allows us to understand effective communication" (S48); "IHFPS helps us recognize and improve our own and others' relational and communicative limits to work better as a team" (S62).

Simulation as a Learning Mode

The Realism of Setting for Effective Learning

In the participants' opinion, the use of IHFPS permitted them to create a realistic training setting that favored learning: "A high-fidelity experimentation makes us live a real situation, making us feel real emotions" (S174); "High fidelity makes the situation almost real…seeing the collaboration among professionals is stimulating" (S175); "Very real setting allows us to fully immerse ourselves in the context" (S1); "Understanding the dynamics among the protagonists of an almost real situation, allows us to identify with them and gives us the opportunity to critically reflect on what we would have done or how we should have behaved" (S7).

Learning in a Protected Environment
IHFPS provides a context where students can get involved, making mistakes and learning from their limits and mistakes: "It is certainly useful to understand where I went wrong and to take these errors as a starting point to improve my skills. By making mistakes I learn. It is better to do it in a protected environment" (S19); "This experience is very useful because it allows us to get involved, to understand our limits and points to improve as well as our strengths" (S166).

127 nursing students provided comments on the open-ended statement "Please describe in as much detail as possible your suggestions to improve the interprofessional high-fidelity patient simulation in which you have just participated", and using content analysis, the following theme with 3 sub-themes emerged:
Further Application of IHFPS

More Frequent IHFPS
Many students hoped that this training will be conducted more frequently and with the involvement of other healthcare professionals, in addition to nurses and physicians: "To improve this experience it would be enough to carry it out more often, involving other professionals in addition to doctors and nurses" (S4); "More simulations could be done with more professional roles" (S72); "Different scenarios could be created for different interdisciplinary collaborations" (S47); "Nursing assistant could also be involved in these experiences" (S95).

Increase the IHFPS Duration
Only two students suggested increasing, even if only slightly, the duration of the IHFPS: "Increase the duration of the simulation even if only slightly" (S5).

Anxiety During IHFPS
Only two students highlighted that anxiety about training performance was an unpleasant element of the experience: "Less anxiety, due to being seen and judged during the simulation, should be felt by the protagonists of IHFPS" (S13); "Anxiety due to being observed" (S16).

4 Discussion

In our study, the students of the 2nd and 3rd year of the Nursing Degree Program who participated in the IHFPS showed an improved attitude towards interprofessional collaboration, highlighted by both validated scale scores. In our students' opinions, IHFPS increased their awareness of interprofessional collaboration efficacy on patient care and had a positive impact on the development of students' skills in communication and collaboration, similarly to what is highlighted in literature [10]. In fact, in our post-IHFPS survey, we reported an increase in students' ability to respond to specific patients' needs. This result is in line with a recent integrative review, which has highlighted that nursing students who participated in simulation demonstrated an increased levels of confidence in patient care [11]. Among our students, interprofessional simulation was appreciated as an effective and positive experience of cooperation, communication, listening and intra- and interprofessional teamwork, without prejudice, thus promoting a greater understanding of their own and other professionals' practices, as suggested by Granheim et al. [10]. In the opinion of our participants, consistent with other studies [16, 17], high-fidelity patient simulation created a realistic training setting that favored learning, allowing them to fully immerse themselves in experiencing real emotions. Moreover, the IHFPS stimulated the acquisition of technical competences, assessment, critical thinking as well as communicative-relational and collaboration skills [18]. In the current study we did not find a gender difference, contrary to the results observed in another study [19]. The perception of being in a context where it was possible to get involved, learning from one's limits and mistakes, without risking harming the assisted person, was greatly appreciated. Students expressed interest and desire for this type of training to be repeated, with the involvement of other health and technical professions as well. In particular, our participants expressed the opinion that nursing students and residents should be involved in more frequent activities aimed at developing skills of interprofessional relationships and teamwork during their education. A potential weakness with the current study is that participants were derived from just one university, which means that the sample may not be representative for nursing students in general. Another limitation is the quality of the evidence, depending on the pre- and post-test design. In future, larger studies which include students from different universities should be conducted. A strength of the current study is that validated questionnaires were used to assess attitudes toward IPC and readiness for interprofessional learning. Another point of strength is represented by the mixed-methods approach of this study.

5 Conclusions

This study suggests that an interprofessional high-fidelity patient simulation, with small teams of nursing students and residents, focused on emergency care, improves the interprofessional collaborative attitudes of students. We reported statistically significant improvement in three factors of JSAPNC and in the RIPLS, which showed the positive effects of this experience on many areas of collaborative skills. The students expressed high satisfaction with the IHFPS conducted in a realistic and protected setting, which improved their awareness of the importance and effectiveness of a multidisciplinary team. Changes in professional behavior should be further pursued to obtain a more appropriate cooperation in multidisciplinary teams thus improving health outcomes.

References

1. Reeves, S., Pelone, F., Harrison, R., Goldman, J., Zwarenstein, M.: Interprofessional collaboration to improve professional practice and healthcare outcomes. Cochrane Database Syst. Rev. **6**, CD000072 (2017)
2. Wei, H., Corbett, R.W., Ray, J., Wei, T.L.: A culture of caring: the essence of healthcare interprofessional collaboration. J Interprof. Care 1–8 (2019). [Epub ahead of print]
3. World Health Organization: Framework for action on interprofessional education and collaborative practice (2010). https://www.who.int/hrh/resources/framework_action/en/. Accessed 23 Jan 2020
4. National League for Nursing, Interprofessional education (IPE) (2015). http://www.nln.org/professional-development-programs/teaching-resources/interprofessional-education-(ipe). Accessed 23 Jan 2020
5. Jeffries, P.R.: A framework for designing, implementing, and evaluating simulations used as teaching strategies in nursing. Nurs. Educ. Perspect. **26**(2), 96–103 (2005)
6. Bray, B., Schwartz, C.R., Weeks, D.L., Kardong-Edgren, S.: Human patient simulation technology: perceptions from a multidisciplinary sample of health care educators. Clin. Simul. Nurs. **5**(4), e145–e150 (2009)
7. Liaw, S.Y., Siau, C., Zhou, W.T., Lau, T.C.: Interprofessional simulation-based education program: a promising approach for changing stereotypes and improving attitudes toward nurse-physician collaboration. Appl. Nurs. Res. **27**(4), 258–260 (2014)
8. Zhang, C., Thompson, S., Miller, C.: A review of simulation-based interprofessional education. Clin. Simul. Nurs. **7**(4), e117–e126 (2011)
9. Weller, J.M., Nestel, D., Marshall, S.D., Brooks, P.M., Conn, J.J.: Simulation in clinical teaching and learning. Med. J. Aust. **196**(9), 594 (2012)
10. Granheim, B.M., Shaw, J.M., Mansah, M.: The use of interprofessional learning and simulation in undergraduate nursing programs to address interprofessional communication and collaboration: an integrative review of the literature. Nurse Educ. Today **62**, 118–127 (2018)
11. Labrague, L.J., McEnroe-Petitte, D.M., Fronda, D.C., Obeidat, A.A.: Interprofessional simulation in undergraduate nursing program: an integrative review. Nurse Educ. Today **67**, 46–55 (2018)
12. Hojat, M., Gonnella, J.S., Nasca, T.J., Fields, S.K., Cicchetti, A., Lo Scalzo, A., Taroni, F., Amicosante, A.M., Macinati, M., Tangucci, M., Liva, C., Ricciardi, G., Eidelman, S., Admi, H., Geva, H., Mashiach, T., Alroy, G., Alcorta-Gonzalez, A., Ibarra, D., Torres-Ruiz, A.: Comparisons of American, Israeli, Italian and Mexican physicians and nurses on the total and factor scores of the Jefferson scale of attitudes toward physician-nurse collaborative relationships. Int. J. Nurs. Stud. **40**(4), 427–435 (2003)
13. Parsell, G., Bligh, J.: The development of a questionnaire to assess the readiness of health care students for interprofessional learning (RIPLS). Med. Educ. **33**(2), 95–100 (1999)
14. Sollami, A., Caricati, L., Mancini, T.: Attitudes towards interprofessional education among medical and nursing students: the role of professional identification and intergroup contact. Curr. Psychol. **37**(4), 905–912 (2018)
15. Ferri, P., Rovesti, S., Magnani, D., Barbieri, A., Bargellini, A., Mongelli, F., Bonetti, L., Vestri, A., Alunni Fegatelli, D., Di Lorenzo, R.: The efficacy of interprofessional simulation in improving collaborative attitude between nursing students and residents in medicine. A study protocol for a randomised controlled trial. Acta Biomed. **89**(7-S), 32–40 (2018)
16. Rubbi, I., Ferri, P., Andreina, G., Cremonini, V.: Learning in clinical simulation: observational study on satisfaction perceived by students of nursing. Prof. Inferm. **69**(2), 84–94 (2016)

17. La Cerra, C., Dante, A., Caponnetto, V., Franconi, I., Gaxhja, E., Petrucci, C., Alfes, C.M., Lancia, L.: Effects of high-fidelity simulation based on life-threatening clinical condition scenarios on learning outcomes of undergraduate and postgraduate nursing students: a systematic review and meta-analysis. BMJ Open **9**(2), e025306 (2019)
18. Buljac-Samardzic, M., Doekhie, K.D., van Wijngaarden, J.D.H.: Interventions to improve team effectiveness within health care: a systematic review of the past decade. Hum. Resour. Health **18**(1), 2 (2020)
19. Krueger, L., Ernstmeyer, K., Kirking, E.: Impact of interprofessional simulation on nursing students' attitudes toward teamwork and collaboration. J. Nurs. Educ. **56**(6), 321–327 (2017)

From High-Fidelity Patient Simulators to Robotics and Artificial Intelligence: A Discussion Paper on New Challenges to Enhance Learning in Nursing Education

Angelo Dante[1(✉)], Alessia Marcotullio[1], Vittorio Masotta[1], Valeria Caponnetto[1], Carmen La Cerra[1], Luca Bertocchi[1], Cristina Petrucci[1], and Celeste M. Alfes[2]

[1] Department of Health, Life and Environmental Sciences, University of L'Aquila, Edificio Rita Levi Montalcini - via G. Petrini, 67010 L'Aquila, Italy
{angelo.dante,cristina.petrucci}@univaq.it,
{alessia.marcotullio,vittorio.masotta1,valeria.caponnetto, carmen.lacerra,luca.bertocchi}@graduate.univaq.it

[2] Center for Nursing Education, Simulation, and Innovation, Frances Payne Bolton School of Nursing, Case Western Reserve University, Cleveland, OH, USA
cms11@case.edu

Abstract. High-fidelity simulation (HFS) is an educational method based on technological mannequins which faithfully reproduces both physiological or physiopathological human body responses to specific clinical conditions and nursing care. When the traditional education is integrated with HFS, improvements in nursing students' knowledge, performance, self-efficacy, self-confidence, problem solving ability, and critical thinking were reported, as well as relational and empathic skills. The level of realism reached in HFS sessions, defined as the 'degree to which a simulated experience approaches reality' demonstrated a positive association with students' learning outcomes. Most of high-fidelity patient simulators are computer-driven static mannequins which resemble adult or child human body dimensions. However, they show limits that should be overcome to provide a more realistic full-body experience in nursing education. In this regard, robotics and artificial intelligence have a key role for the technological evolution of nursing educational systems and their introduction in the simulation field is opening new perspectives that will produce unavoidably the redefinition of educational standards with beneficial implications for future nursing care. In this perspective, new challenges for nursing education has been discussed in this paper.

Keywords: High fidelity simulation training · Nursing students · Robotics · Artificial intelligence · Discussion paper

1 High-Fidelity Simulation in Nursing Education

High-fidelity simulation (HFS) is an educational method based on technological mannequins which faithfully reproduce both physiological or physiopathological human

Z. Kubincová et al. (Eds.): MIS4TEL 2020, AISC 1236, pp. 111–118, 2021.
https://doi.org/10.1007/978-3-030-52287-2_11

body responses to specific clinical conditions or nursing care [1–3]. With the aim of improving nursing students' learning outcomes, HFS is progressively integrating traditional nursing education all around the world. Considering that internships do not always allow students to reach adequate technical and non-technical skills useful to respond effectively to patients' health problems, especially those in life-threatening clinical conditions, HFS represents an opportunity for students to learn complex skills in a safe environment [4]. When traditional educational methods had been integrated with HFS, improvements in nursing students' knowledge, performance, self-efficacy, self-confidence, problem solving ability, and critical thinking were reported, as well as relational and empathic skills [5, 6].

1.1 Key Elements of High-Fidelity Simulation

Pre-briefing, scenario, and debriefing are key elements of an HFS session [7]. During the pre-briefing, faculty members provide students details about educational goals and the scenario they will face [8]. The debriefing following each HFS session is focused on students' critical reflection to allow them to identify their own weaknesses and orient future reinforcement strategies [9]. Debriefing should be facilitated by an expert faculty member and conducted in a dedicated environment suitable for learning, ensuring confidentiality, trust, effectiveness, and free communication. Finally, the scenario based on real clinical data allows students to repeatedly put their theoretical knowledge into practice [7]. The level of realism, which is the 'degree to which a simulated experience approaches reality', demonstrated a positive association with students learning outcomes [10].

1.2 Realism in High-Fidelity Simulation

The level of realism is also closely related to how learning environments, technologies, relationships, and emotions faithfully represent the reality besides the level of the simulator's technological advancement. In this regard, considering that nursing discipline expresses its contents mainly at the patient's bed through the relationship and caring [11], the simulator represents the core element of the educational experience based on HFS. In fact, students interact with the simulator making decisions and acting nursing interventions based on the amount and quality of the collected data. The level of the simulator technological advancement affects the level of realism of the HFS session and, consequently, the achievement of the learning objectives [10].

2 Characteristics and Limits of High-Fidelity Patient Simulators

As anticipated, most of high-fidelity patient simulators (HFPSs) are technologically advanced computer-driven static mannequins resembling full-scale adult or child human body dimensions. These simulators are able to reproduce autonomous breathing and respiratory, cardiac, and intestinal sounds. In addition, they can be cared for with several nursing activity as, for example, cardiopulmonary resuscitation manoeuvres, intubation, and manually ventilation, as well as intramuscular, intravenous, and subcutaneous injections. In order to enhance student's assessment skills, clinical conditions can be continuously checked through the monitoring system and they can be changed by faculty members through the software used to lead the simulation. Therefore, it is possible to increase learning opportunities for students through additional simulation modules, that allow to reproduce other signs and symptoms compared to the basic model. Typically, HFPSs have three operational modes. The first, i.e. the 'automated mode', is algorithm-based so that the simulated patient's clinical evolution is influenced by the students' interventions. The second one is the 'manual mode' and it is based on the direct operator control, while the last, i.e. the 'mixed mode', integrates the automated and manual operational modes allowing to join the preloaded simulator feedback with other manual-driven responses used by faculty members when students do not perform the expected nursing care or provide unnecessary care. However, the use of mixed mode requires clinical experienced operators able to interpret in real time the effects on an ideal patients' health derived from unexpected nursing care actions and able to quickly send a realistic simulator response. HFS systems are designed to enhance the acquisition of technical and non-technical skills. However, if not properly utilized, HFPSs only allow for partial reproduction of the patient's organic, relational and psychological responses to the health problem. Therefore, HFPSs risk to orient the student's learning to the treatment of organic problems, as typically happens in the biomedical approach, and results in distancing students from the holistic nursing care focus. In this regard, current HFPSs show limits that should be overcome to provide a more realistic full-body experience in nursing education. Examining the most used HFPSs, the head-feet model [12] makes it possible to systematically highlight most of their limits and grouping them into body systems (Table 1).

One of the limits underlying most simulator features is the feedback mechanism as mannequins do not reproduce patient's tactile and pain sensation, which are fundamentals to learn and accurately interpret the neurological assessment.

Further, since simulators are not 'intelligent', they provide only faculty-driven verbal answers and are not able to autonomously reproduce feelings or facial expressions related to clinical conditions. These features limit the possibility to perform a deep clinical assessment. When considering mouth, auricular, nasal, and ocular orifices, simulators only allow limited nursing assessment and interventions, as they reproduce the human anatomy, but they can faithfully simulate only few pathologic alterations, such as tongue oedema or throat sound. A further physical limit is represented by the absence of head mobility. The simulator head can only passively be moved by trainers or students during simulation, and the reproduction of the nape stiffness or other pathological conditions is not always possible, thus unavoidably conditioning the realism of events. In case of cardiorespiratory or abdominal examination, some aspects of nursing assessment cannot be fulfilled, due to the non-realistic feedback to nursing actions. The stiffness of both chest and abdomen does not allow to collect real data through the percussion or palpation, while auscultation is limited by the range of preloaded sounds. In addition, the intrathoracic respiratory pressures are reversed, as chest expansion is guaranteed by an air compressor. This allows to manually or mechanically support the ventilation only in those scenarios in which patients in a respiratory arrest do not have any type of ventilatory triggers. Conversely, a pressure contrast would occur between the ventilator and the lungs of the HFPS, making not feasible any ventilation mode different from Continuous Mandatory Ventilation. Furthermore, the peripheric vessels cannulation is generally possible for one arm only. At musculoskeletal level, static mannequins are not able to reproduce movements or express strength. Also, integumentary system could be improved since skin aspect (e.g. colour) does not vary following pathologic situations as e.g. hyperthermia, and it is not possible to evaluate skin hydration, dryness or elasticity. Furthermore, cyanosis can be reproduced only around the mouth but not in other body extremities, such as fingers, ears, nose or lips and skin appendages are not available. Finally, anthropometric evaluation can be performed only in supine position and body temperature can be displayed on the monitor, but it is not perceptible by touch.

Table 1. Limits of high-fidelity patient simulators

Body systems	Main limits
Sensorial and Neurological	Reflexes, tactile sensitivity and pain: absent Consciousness: partially reproducible Facial expression: not reproducible Language: only verbal
Vital signs	Body temperature: viewable only on the monitor Non-Invasive Arterial Body Pressure: only in one arm Pain: not reproducible at non-verbal level
Head-Neck	Head mobility: passive - ocular, auricular, nasal, buccal assessment: not fully effective due to the absence of physiopathological conditions
Chest	Percussion sounds: not executable due to rib cage stiffness providing not realistic feedback Morphology: generally, at physiopathological level it is only possible reproducing unilateral lung expansion (pneumothorax) Physiology: the intrathoracic respiratory pressures are reversed due to an air compressor useful to reproduce chest expansion
Abdomen	Palpation and percussion: not feasible due to the stiffness of abdominal wall
Peripheral vascular	Capillary refill time: assessment not possible
Integumentary	Skin colour: not variable - Skin hydration and turgor: assessment not possible - Skin appendages: absent
Muscle-skeletal	Amplitude and strength in movements: absent
Anthropometric measurements	Height and weight: only detectable supinely

3 From High-Fidelity Patient Simulators to Humanoid Patient Simulators

Robotics and Artificial Intelligence (AI) are key factors of the current technological evolution ongoing in nursing educational systems. Robots are operators not necessarily endowed with feelings or emotions created with the aim to systematically perform tasks [13]. Droids and humanoids are two distinct styles in robotics design and functionality available in the nursing field. Droids look like a box, while humanoids have a human-like design and also the ability to make their own decisions based on the AI which is an advanced technology used to reproduce intelligent behaviors and critical thinking like natural human ones [14, 15]. AI through 'machine learning' technology makes it possible for humanoids to learn from experience, adjust to new inputs and perform human-like tasks [13]. In addition, current developments in the 'ambient intelligence' that is an emerging discipline that enables environment to interact with and respond appropriately to the humans in that environment, allow to embed into robots the natural language processing, eye and face tracking, gesture and speech processing making them able to identify, face, and reproduce cognitive and emotional needs [16]. Since humanoids could be programmed to reproduce human emotional responses, an effective cognitive training and a strong relational engagement could be realized. The potential ability of humanoids to reproduce movements, thinking and interactions like humans could help overcome the current limits of HFPSs bringing learning experiences closer to the holistic nursing practice. Table 2 shows the characteristics of robots that are like HFPS and how humanoid robots currently overcome some of the main HFPS limits. Humanoids intelligent robots reproduce the 'whole' patient (not just a part) and allow students to learn nursing care based on a 'whole-patient' approach, as required in clinical training. In this regard, some pilot studies are reported in the recent nursing education literature about learning through experiences based on robotic [17–20]. Lin and colleagues [18] in a pre-post study revealed a significant improvement in the level of nursing students' skills related to patient transfer after students' participation in a learning experience performed through a human-like robot. Previously, also Zhifeng and colleagues [20] highlighted in a pilot experience the potential beneficial of robot technology in nursing education reporting that the robot-patient is effective for nursing students to self-study wheelchair transfer. Instead, Huang and colleagues [19], explored the potential effectiveness of a robot-patient able to reproduce paralysis, movements sensitive to pain, painful expression, and movements constrained by medical devices in enabling nurse to practice the skills required to transfer an actual patient. Involving nursing faculty, these authors demonstrated the potential benefit of robots not only for improving technical skills but also for stimulating empathy and relational abilities. In parallel with these research, following the current trend for improving the technological level of HFPSs, Gaumard® has released to market the S2225 Paediatric HAL® (https://www.gaumard.com/s2225), the world's most advanced paediatric patient simulator and the first able to simulate lifelike emotions through dynamic facial expressions, movement, and speech. HAL is designed to help educators to develop the specialized skills needed to effectively communicate, diagnose, and treat young patients in nearly all clinical areas. However, its effectiveness in improving nursing students' skills still under investigation as well as, considering the high cost of robot's patients, evidence on their cost-effectiveness are also expected.

Table 2. Robots' features and what they currently add to HFSS

Body systems	Robotic features similar to HFPS	What robotic add to HFPS
Sensorial and Neurological	Wireless streaming audio and store vocal responses in any language Pupillary response to light	Facial expression: present and dynamic Lifelike emotions: present and dynamic Reflexes, tactile sensitivity and pain: present Consciousness: total reproducible Language: reproduce thinking and interactions like human
Vital signs	Measure blood pressure: by palpation or auscultation Sounds Korotkoff: audible Pulse sites: carotid, radial, brachial, femoral, popliteal, and pedal Heart sounds: present and synchronized with ECG	Pain: reproducible at verbal and nonverbal level Non-Invasive Body Pressure: bilateral arms
Head-Neck	Airway sounds: present and synchronized with breathing	Ocular, auricular, nasal, buccal assessment: effective due to the presence of physiopathological conditions (e.g. tongue oedema) Head mobility: active
Chest	Morphology: possible unilateral chest rise simulates pneumothoraxes	Percussion sounds: present and synchronized with breathing patterns Care interventions: supports mechanical ventilation just like a real patient
Abdomen	Palpation and percussion: not possible Auscultation sounds: present at abdominal wall level	
Peripheral vascular	Capillary refill time: assessment not possible	
Integumentary	Central cyanosis: possible assessment	Skin colour: available in ethnic skin tones and variable
Muscle-skeletal		Articulation movement: realistic joint articulation, supports supine, prone, recumbent, and sitting positions Amplitude and strength in movements: present Movements: reproduce at 'whole' patient level (not just a part)
Anthropometric measurements		Detectable in standing position

4 Conclusion

We live in the era of technological and scientific development, in which the use of intelligent robot patient simulators as teaching and learning devices represents the start of a quiet revolution in the nursing students' educational field. For nurses, embracing this evolution means to actively contribute in knowledge development and decision-making processes regarding nursing education [14]. In this field, the introduction of robotics and AI is opening new perspectives that will produce unavoidably the redefinition of standards for simulation practices, with implications for future nursing care [21]. To improve nursing students' learning outcomes, nurses will be called to manage technologies and face moral and anthropological issues accompanying the use of intelligent robots. In this context, there will be a lot of issues that nurses will have to address to play an active role in the current technological era, some of which could be the following: a) Will new technologies be effective and calibrated for nursing student's education? b) What will be the role of nurses in technologies development? c) Will the interaction with intelligent robots improve non-technical skills, as empathy, or affect the relationship between nursing students and patients? d) What will be the moral implications for nursing? To address future challenges, nursing needs to have a strong presence in the technology-based educational field [22]. Specifically, nurses should have a key role in designing, testing, applying, and evaluating the impact of new technologies on learning outcomes to ensure patient safety and quality nursing care. At this regard, interdisciplinary collaborations among nurses, engineering, informatics and other experts in this fields are desirable for the future.

References

1. Gaba, D.M.: The future vision of simulation in health care. Qual. Saf. Health Care **1**, i2–10 (2004)
2. Maran, N.J., Glavin, R.J.: Low- to high-fidelity simulation - a continuum of medical education? Med. Educ. **37**(Suppl 1), 22–28 (2003)
3. Presado, M.H.C.V., Colaco, S., Rafael, H., Baixinho, C.L., Felix, I., Saraiva, C., Rebelo, I.: Learning with high fidelity simulation. Cien Saude Colet. **23**(1), 51–59 (2018)
4. Warren, J.N., Luctkar-Flude, M., Godfrey, C., Lukewich, J.: A systematic review of the effectiveness of simulation-based education on satisfaction and learning outcomes in nurse practitioner programs. Nurse Educ. Today **46**, 99–108 (2016)
5. La Cerra, C., Dante, A., Caponnetto, V., Franconi, I., Gaxhja, E., Petrucci, C., Alfes, C.M., Lancia, L.: Effects of high-fidelity simulation based on life-threatening clinical condition scenarios on learning outcomes of undergraduate and postgraduate nursing students: a systematic review and meta-analysis. BMJ Open **9**(2), e025306 (2019)
6. Smith, S.J., Roehrs, C.J.: High-fidelity simulation: Factors correlated with nursing student satisfaction and self-confidence. Nurs. Educ. Perspect. **30**(2), 74–78 (2009)
7. INACSL Standards Committee: INACSL standards of best practice: Simulation^{SM} debriefing. Clin. Simul Nurs. **12**(S), S21-S25 (2016)
8. Page-Cutrara, K.: Use of prebriefing in nursing simulation: a literature review. J. Nurs. Educ. **53**(3), 136–141 (2014)
9. Bae, J., Lee, J., Jang, Y., Lee, Y.: Development of simulation education debriefing protocol with faculty guide for enhancement clinical reasoning. BMC Med. Educ. **19**(1), 197 (2019)

10. Sherwood, R.J., Francis, G.: The effect of mannequin fidelity on the achievement of learning outcomes for nursing, midwifery and allied healthcare practitioners: systematic review and meta-analysis. Nurse Educ. Today **69**, 81–94 (2018)
11. Federspil, G.: Logica clinica - I principi del metodo in medicina. McGraw-Hill, Milano (2004)
12. Hogan-Quigley, B., Palm, M.L., Bickley, L.: Valutazione per l'assistenza infermieristica. Esame fisico e storia della persona assistita, 1st edn. Casa Editrice Ambrosiana, Milano (2017)
13. Maalouf, N., Sidaoui, A., Elhajj, I.H., Asmar, D.: Robotics in nursing: a scoping review. J. Nurs. Scholarsh. **50**(6), 590–600 (2018)
14. Erikson, H., Salzmann-Erikson, M.: Future challenges of robotics and artificial intelligence in nursing: what can we learn from monsters in popular culture? Perm. J. **20**(3), 15–243 (2016)
15. Amisha, P.M., Pathania, M., Rathaur, V.K.: Overview of artificial intelligence in medicine. J. Fam. Med. Prim. Care. **8**(7), 2328–2331 (2019)
16. Vogan, A.A., Alnajjar, F., Gochoo, M., Khalid, S.: Robots, AI, and cognitive training in an era of mass age-related cognitive decline: a systematic review. IEEE Access **8**, 18284–18304 (2020)
17. Miran, L., Murata, K., Ameyama, K., Yamazoe, H., Joo-Ho, L.: Development and quantitative assessment of an elbow joint robot for elderly care training. Intell. Serv. Robot. **12**(4), 277–287 (2019)
18. Lin, C., Huang, Z., Kanai-Pak, M., Maeda, J., Kitajima, Y., Nakamura, M., Kuwahara, N., Ogata, T., Ota, J.: Effect of practice on similar and dissimilar skills in patient transfer through training with a robot patient. Adv. Robot. **33**(6), 278–292 (2019)
19. Zhifeng, H., Takahiro, K., Masako, K.P., Yasuko, K., Mitsuhiro, N., Kyoko, A., Noriaki, K., Taiki, O., Jun, O.: Design and evaluation of robot patient for nursing skill training in patient transfer. Adv. Robot. **29**(19), 1269–1285 (2015)
20. Zhifeng, H., Chingszu, L., Masako, K.P., Jukai, M., Yasuko, K., Mitsuhiro, N., Noriaki, K., Taiki, O., Jun, O.: Robot patient design to simulate various patients for transfer training. IEEE ASME Trans. Mechatron. **22**(5), 2079–2090 (2017)
21. Robert, N.: How artificial intelligence is changing nursing. Nurs. Manage. **50**(9), 30–39 (2019)
22. Pepito, J.A., Locsin, R.: Can nurses remain relevant in a technologically advanced future? Int. J. Nurs. Sci. **6**(1), 106–110 (2019)

The Concept of High-Fidelity Simulation and Related Factors in Nursing Education: A Scoping Review

Vittorio Masotta[1(✉)], Angelo Dante[1], Alessia Marcotullio[1], Luca Bertocchi[1], Carmen La Cerra[1], Valeria Caponnetto[1], Cristina Petrucci[1], and Celeste Marie Alfes[2]

[1] Department of Health, Life and Environmental Sciences, University of L'Aquila, Edificio Rita Levi Montalcini - via G. Petrini, 67010 L'Aquila, Italy
{vittorio.masotta1,alessia.marcotullio,luca.bertocchi, carmen.lacerra,valeria.caponnetto}@graduate.univaq.it, {angelo.dante,cristina.petrucci}@univaq.it

[2] Center for Nursing Education, Simulation, and Innovation, Frances Payne Bolton School of Nursing, Case Western Reserve University, Cleveland, OH, US
cms11@case.edu

Abstract. High-Fidelity Simulation (HFS) is a tool used to expand nursing students' opportunities to care for critically ill patients while also overcoming limitations often faced during clinical placements. HFS provides students with opportunity to interact within a realistic clinical environment with a patient simulator able to reproduce a wide range of clinical conditions. Professional growth and improved learning outcomes are associated with the level of fidelity in simulation; however, heterogeneity of the adopted HFS definitions and studies reporting lead to a heterogeneity of research results. The aims of this scoping review, that analyzed 69 studies, were to clarify HFS concept, document its related main features, and provide an unambiguous definition of HFS. Thematic analysis of definitions and conceptual frameworks reported in the considered studies allowed for the identification of three main themes that represent the concept of HFS in nursing education: innovative educational strategy, realistic learning experience for skills improvement, and safe experience. Debriefing and pre-briefing were respectively the most common reported features related to HFS. The definition derived from thematic analysis considers HFS as a 'technology-based educational approach performed in a realistic and safe environment, that uses an interactive patient simulator able to reproduce life-like clinical conditions, allowing students to improve their technical and non-technical skills'.

Keywords: High-fidelity simulation · Scoping review · Nursing students · Concept development · Definition

1 Introduction

Healthcare facilities are often challenged to ensure nursing students are 'ready for practice', especially regarding situations not commonly faced during their standard clinical

Z. Kubincová et al. (Eds.): MIS4TEL 2020, AISC 1236, pp. 119–126, 2021.
https://doi.org/10.1007/978-3-030-52287-2_12

placements and the treatment of life-threatening conditions [1]. In this regard, High-Fidelity Simulation (HFS) is an effective educational strategy increasingly used in nursing education to develop skills and competencies through technology [2]. HFS improves abilities of students allowing them to overcome the limited learning opportunities offered by the standard clinical placements and the ethical issues related to the patients' care [3]. When performed in a realistic and safe learning environment with a human patient simulator capable of reproducing real clinical conditions, HFS significantly contributes to the development of essential nursing clinical practice skills [4]. The effects of HFS on nursing students' learning outcomes have been studied by several authors who documented its effectiveness both in undergraduate and postgraduate nursing courses [1, 5–7]. However, there is evidence that both the professional growth and learning outcomes improvement are associated with the level of fidelity reproduced during HFS sessions [7]. Moreover, the heterogeneous definitions of HFS and the variability of features and circumstances that occur during HFS sessions are correlated with the heterogeneity of the results [1]. In this regard, a clear definition of the HFS concept and its main features could improve the standardization and reproducibility of educational interventions in different international contexts, reduce heterogeneity in future research, and improve both the generalizability and comparability of results.

1.1 Aim

The aims were to clarify the HFS concept, documenting its related main features, and providing an unambiguous definition of HFS.

2 Methods

2.1 Study Design and Search Strategy

A scoping review was undertaken and reported against the relevant criteria of the Preferred Reporting Items for Systematic reviews and Meta-Analyses - extension for Scoping Review (PRISMA-ScR) checklist [8]. This methodology is used to determine the coverage of a topic in the literature and gives a clear overview of its focus, as well as clarifies concepts or examines emerging evidence [9]. PubMed, CINAHL, and Scopus databases were searched until November 2019 combining the following key words: 'Students, Nursing', 'High-Fidelity Simulation', 'High-Fidelity Simulation Training', and 'Realism'. No additional information sources were consulted.

2.2 Eligibility Criteria and Study Selection

To be considered, abstracts and titles of retrieved studies had to clearly refer to HFS as an educational strategy applied to nursing students. To be included in the review, after abstract screening, papers needed to: a) be focused on HFS as a learning method utilized exclusively in Bachelor's Degree in Nursing or Master of Science in Nursing; b) provide a description of HFS as applied educational methodology and its related characteristics; c) be published in English, Italian or French languages; d) be available in full-text version retrieved through the University online library system. Both abstract screening and study selection processes were independently conducted by two researchers and any disagreement was resolved discussing with a third author.

2.3 Data Charting Process and Variables

A data-charting form was developed and tested independently by two researchers to determine which variables had to be extracted. Two reviewers independently charted the data, discussed the results, and continuously updated the form through an iterative process [8]. Any disagreement was resolved through discussion with a third researcher. Data extracted was related to the article characteristics, definition of HFS, and main features of this method.

2.4 Critical Appraisal of Sources and Synthesis of Results

Since this review aimed at clarifying a concept instead of synthesizing the results of the included studies, their critical appraisal was not useful. In order to clarify the concept of HFS, a thematic analysis of definitions adopted by authors was performed and the occurrence of the main features of HFS was described.

3 Results

A total of 892 studies were retrieved from the explored databases (PubMed n = 154; CINAHL n = 95; Scopus n = 543). After duplicates were removed, 747 abstracts were screened and 146 of them met inclusion criteria (Fig. 1). Ninety-four full-text articles were retrieved through the University online library system. Of these, 69 were included in the scoping review. The remaining 25 studies were excluded from analysis due to: 1) focus on mid- or low-fidelity simulation, or e- or video-simulation, 2) inclusion of students different from Bachelor or Master's degree, and 3) enrollment of actors to play the patient role. In Table 1 the main characteristics of the included studies are reported, while the complete reference list of the included studies is available from authors, if required.

3.1 The Concept of High-Fidelity Simulation

The results of this scoping review confirm a heterogeneity in the definition of HFS adopted in the literature. Instead of clearly defining the concept of HFS, authors mainly described conceptual frameworks aimed at clarifying the broad meaning they attribute to HFS. In fact, authors reported the adopted educational strategies, employed technologies, and expected outcomes. Thematic analysis of these conceptual frameworks and definitions allowed to identify several descriptors' categories (Table 2) and three main themes related to the HFS.

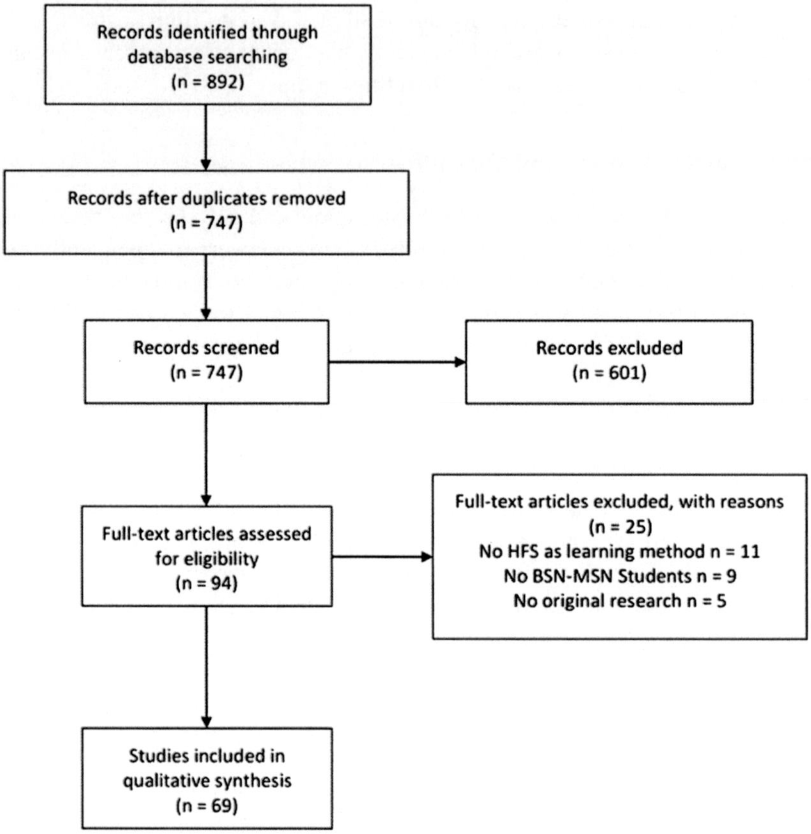

Fig. 1. Flow-chart of included studies

Table 1. Characteristics of the included studies (n = 69)

Characteristics	Description
Continent (%)	America (47.8%), Europe (24.6%), Other (27.5%)
Students (%)	Undergraduate (98.6%), Postgraduate (1.4%)
Scenarios (%)	Life-threatening conditions (36.2%), Other/Unclear (63.8%)
Settings (%)	Simulation Laboratory or Centre (44.9%), Other/Unclear (55.1%)
Students in each session (%)	≤4 (48.5%), >4 (13.3%), Unclear (38.2%)

3.2 HFS as an Innovative Educational Strategy

Most of authors considered HFS as an innovative teaching and learning strategy that allows students to integrate theory and practice. Therefore, HFS is considered as a structured pedagogical approach based on technology, in which the interactivity of learning experience is realized using a computer-driven manikin allowing students to perform tasks needed to solve health problems in a virtual patient.

Table 2. Descriptive categories and themes related to HFS

Categories	Themes
Effective pedagogical approach Innovative learning strategy Integrate theory and practice Interactive learning experience Structured learning experience Teaching and learning strategy Technology for education	Innovative educational strategy
Feedback Interactivity Patient simulator Realism Scenario Skill improvement	Realistic learning experience for skills improvement
Patient safety Safe environment	Safe experience

3.3 HFS as a Realistic Learning Experience for Skills Improvement

Authors considered HFS useful for the improvement of students' technical and non-technical skills in managing several clinical conditions ranging from home care to life-threatening scenarios. In this regard, the following factors have a key role in the achievement of nursing students' skills: 1) participation in a realistic scenario, 2) interaction with the patient simulator, and 3) feedback provided to students both during and after their performance. The relationship with the patient simulator seems to be the core part of the simulation sessions as the simulator is the main training tool used to develop students' skills. However, authors argued that other elements contribute to achieving realistic learning. In fact, students live an immersive experience in which instruments, environment, relationships, and technologies can also have a positive impact on their skills improvement.

3.4 Main Features Related to HFS

In the analyzed literature, pre-briefing and debriefing moments were the most frequent feature related to HFS. In 47.1% of the considered papers, students received pre-briefing before entering in an HFS session. It should be noted that students frequently received pre-briefing already in the simulation environment, to get them comfortable with its tools and technologies. Whereas, when considered alone, debriefing is treated in the 81.2% of the included studies, highlighting its importance (in 29.0% of the studies debriefing was guided through videos of students' performances). Debriefing seems to be a cornerstone of HFS in nursing educational settings. Instead, the presence of a control room to manage the simulation was only described in 21.7% of the studies.

3.5 How to Define High-Fidelity Simulation

Considering the main themes and descriptors' characteristics reported above, it is possible to give an unambiguous definition of HFS that could embrace the plurality of information that the literature gives on this educational methodology. In this regard, HFS can be considered as a 'technology-based educational approach performed in a realistic and safe environment through an interactive patient simulator able to reproduce life-like clinical conditions, that allows students to improve their technical and non-technical skills' (Fig. 2).

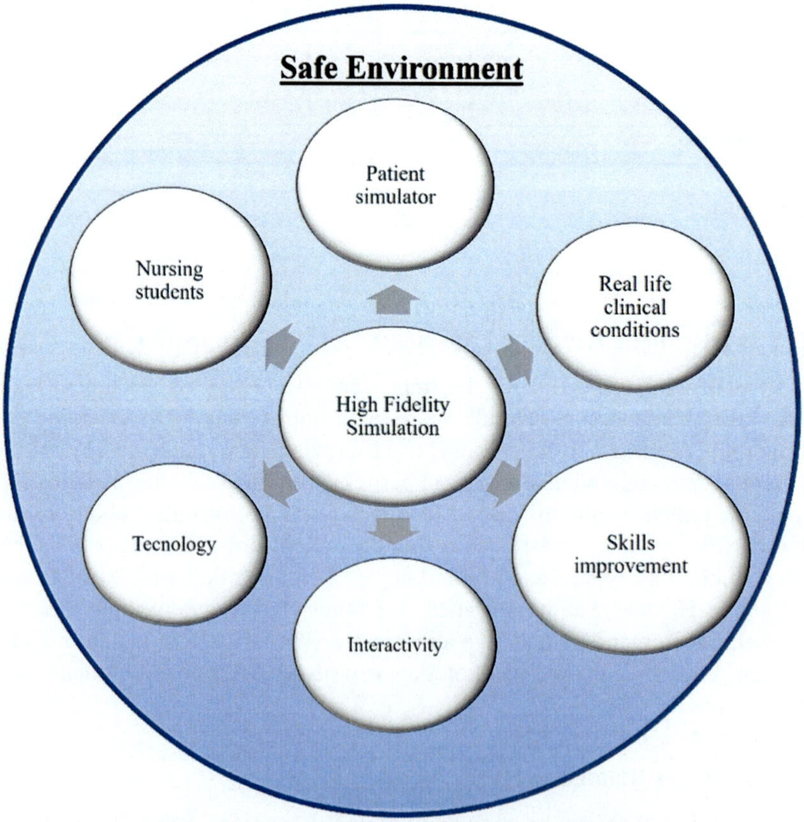

Fig. 2. The concept of High-fidelity simulation

4 Discussion and Conclusions

In the last decade, studies analyzing HFS implementation and effectiveness have been constantly increasing, likely due to the technological improvement and research interest on how to obtain highly trained nurses. Studies were predominately published in America, thus limiting the generalizability to the European context due to differences in the academic programs. The Bologna Process [10], that created a homogeneous European high-education context, should act as a driving force to produce evidence useful not only in improving nursing students technical and non-technical skills, but also in enhancing researchers' and educators' experience on HFS. In the literature, HFS is reported as an innovative educational strategy in which realistic experiences are provided in a safe context. Next Gen technologies, e.g. simulators, are expected to implement HFS as an educational approach for students' skills growth, providing a broad range of scenarios in which learners could train while receiving feedback of actions they perform. However, the interaction with the simulator is not the only fundamental feature of HFS, but also the environment where the simulation is performed needs to be considered. In this regard, tools and equipment, as well as rooms where students act, are integral part of the simulation technique and contribute to enhance realism, adding power to this educational method. Such power is further increased by the safety and forgiveness of the environment towards students, as well as by the briefing and debriefing activities. In fact, students are introduced in HFS scenario during the briefing and experience a concrete educational activity by a guided reflection on their performance during the debriefing. Hence, students experience the first two stages of the 'experiential learning' cycle that they are expected to complete individually and during the clinical practice. This process allows to translate the experience into concepts, which guide learners in their future as professionals [11]. In this regard, the INACSL guidelines on simulation highlight the importance of debriefing, which is particularly useful if guided through video playback of students' performances [12]. The main strength of this work is the achievement of an unambiguous definition of HFS and its main related descriptors that could be taken as a framework for future research. Instead, the principal study weakness is rather represented by the exploration of HFS only in the nursing educational field. In conclusion, the more realistic HFS it is, the higher students' abilities will be. Therefore, providing a framework, educators and researchers should consider that implementing simulation could bring to a broader amount of homogeneous results that may definitively clarify HFS contribution in nursing education.

References

1. La Cerra, C., Dante, A., Caponnetto, V., Franconi, I., Gaxhja, E., Petrucci, C., Alfes, C.M., Lancia, L.: Effects of high-fidelity simulation based on life-threatening clinical condition scenarios on learning outcomes of undergraduate and postgraduate nursing students: a systematic review and meta-analysis. BMJ Open **9**(2), e025306 (2019)
2. Hallin, K., Bäckström, B., Häggström, M., Kristiansen, L.: High-fidelity simulation: assessment of student nurses' team achievements of clinical judgment. Nurse Educ. Pract. **19**, 12–18 (2016)

3. Lewis, R., Strachan, A., Smith, M.M.: Is high fidelity simulation the most effective method for the development of non-technical skills in nursing? a review of the current evidence. Open Nurs. J. **6**, 82–89 (2012)
4. Abram, M.D., Forbes, M.O.: High-fidelity simulation: an application to psychopharmacological training for the psychiatric nurse practitioner student. Issues Ment. Health Nurs. **40**(3), 260–267 (2019)
5. Dante, A., La Cerra, C., Masotta, V., Caponnetto, V., Gaxhja, E., Petrucci, C., Lancia, L.: Efficacy of high-fidelity simulation on learning outcomes: immediate results for a postgraduate intensive care nursing course. In: Popescu, E., Gil, A.B., Lancia, L., Sica, L.S., Mavroudi, A. (eds.) Advances in Intelligent Systems and Computing, vol. 1008, pp. 32–39. Springer, Cham (2020)
6. Tuzer, H., Dinc, L., Elcin, M.: The effects of using high-fidelity simulators and standardized patients on the thorax, lung, and cardiac examination skills of undergraduate nursing students. Nurse Educ. Today **45**, 120–125 (2016)
7. Sherwood, R.J., Francis, G.: The effect of mannequin fidelity on the achievement of learning outcomes for nursing, midwifery and allied healthcare practitioners: systematic review and meta-analysis. Nurse Educ. Today **69**, 81–94 (2018)
8. Tricco, A.C., Lillie, E., Zarin, W., O'Brien, K.K., Colquhoun, H., Levac, D., Moher, D., Peters, M.D., Horsley, T., Weeks, L.: PRISMA extension for scoping reviews (PRISMA-ScR): checklist and explanation. Ann. Intern. Med. **169**(7), 467–473 (2018)
9. Munn, Z., Peters, M.D., Stern, C., Tufanaru, C., McArthur, A., Aromataris, E.: Systematic review or scoping review? Guidance for authors when choosing between a systematic or scoping review approach. BMC Med. Res. Methodol. **18**(1), 143 (2018)
10. European Ministers in charge of Higher Education: The Bologna Declaration of 19 June 1999, Joint Declaration of the European Ministers of Education (1999)
11. Kolb, D.A.: Experiential Learning: Experience as the Source of Learning and Development, 2nd edn. FT Press, Upper Saddle River (2014)
12. INACSL Standards Committee: INACSL standards of best practice: Simulation^{SM} simulation design. Clin. Simul. Nurs. **12**, S5–S12 (2016)

The Use of Simulation for Teaching Therapy Management: An Observational Descriptive Study on 2nd and 3rd Year Students of the Nursing Degree Course of Reggio Emilia

Mecugni Daniela[1,2](✉), Turroni Elena Casadei[1], Doro Lucia[1], Franceschini Lorenza[1], Lusetti Simona[1], Gradellini Cinzia[1], and Amaducci Giovanna[1]

[1] Nursing Degree Course, University of Modena and Reggio Emilia, Reggio Emilia, Italy
daniela.mecugni@unimore.it,
{elena.casadeiturroni,luciamariagrazia.doro,
franceschini.lorenza,lusetti.simona,gradellini.cinzia,
amaducci.giovanna}@ausl.re.it

[2] Department of Surgery, Medicine, Dentistry and Morphological Sciences, University of Modena and Reggio Emilia, Reggio Emilia, Italy

Abstract. The patient safety is considered a strategic and ethical objective of health, also one of the factors that determine the care quality. To ensure the patient safety, the prevention and the reduction of medication errors is a crucial aspect. Therapy errors are identified as the most frequent committed in the healthcare context. Nurses plays an extremely important role in the patient safety, as the professional responsible for the correct application of the diagnostic/therapeutic prescriptions. Nursing students must have the opportunity to experiment the management of the therapy to acquire skills that can prevent the related risks. This study describe the perception of nursing students, regarding the acquisition of skills, in the therapy management, through a didactic module in where they can try the simulation using a therapy software. The study involved 503 students attending the second and the third year of the Nursing Course, considering two different cohorts. The data collection tool was used is a questionnaire with 10 closed-ended questions, using a Likert scale, and one open question. The students has perceived as an effective training, considering the achievement of the didactic objectives. The simulation it allowed the students to consolidate skills and/or to acquire new and more complex ones, through the experience and the learning from errors.

Keywords: Clinical simulation · Medication error · Electronic Medication Administration Record (eMAR) · Baccalaureate nursing education

1 Introduction

1.1 Background

Patient safety is considered a strategic goal and an ethical imperative of all current health and care settings, as well as one of the factors which determine the quality of care. To

Z. Kubincová et al. (Eds.): MIS4TEL 2020, AISC 1236, pp. 127–137, 2021.
https://doi.org/10.1007/978-3-030-52287-2_13

ensure patient safety, the prevention and the reduction of Medication Error (ME) is a crucial aspect [1–4, 10].

Therapy errors are identified as the most frequent type of error committed in the healthcare sector, and they represent the 12–20% of the total errors in healthcare. In the United States it has been estimated that, approximately, 1.5 million people have been damaged by treatment errors [2–4]. In the United Kingdom, it has been calculated that administration errors increase the 2–7% of the hospitalizations of more than 2 days with related additional costs [2–4]. The percentage of therapy errors in the Netherlands is between 12% and 20% of the total errors [5], and in Italy, therapy errors affect about 4% of the hospitalizations [6].

The Italian Ministry of Health (2008) defines ME as any preventable event that can cause or lead to an inappropriate use of a drug or an harm to the patient. This event may be related to errors of prescription, transcription, labelling, preparation, packaging, distribution, dispensing, administration, monitoring, patient education and self-prescription [6, 7].

Considering the therapy, the nurse plays an extremely important role in patient safety, as responsible for the correct application of the diagnostic and therapeutic prescriptions [8]. The importance of this role comes from the fact that, the nurse is directly involved in the entire process, from the prescription till the patient intake [1, 4, 10].

Several authors [9] estimated that up to 40% of a nurse's work shift is used in therapy management; in fact, the increased acuity of hospitalized patients, the concomitance of multiple pathologies, associated with an increased volume of prescribed drugs, make the management process of therapeutic prescriptions extremely complex, that's why it could arrive to an errors' risk [10].

The therapy management is an important skill for nursing students which involves the different learning areas related to: 1) knowledge of pharmacology; 2) dosage calculation skills; 3) gestural skills in the administration of drugs through the different ways; 4) critical thinking in the choice of whether or not to administer the prescribed drug; 5) communication-relational skills in education for the management of long-term therapy; 6) application of the "8 rights", which means to check the Correct patient, drug, time, dosage, administration route, drug's storage, effects' monitoring, and registration [11].

To prepare students who are able to manage the complexity of the current prescriptions, the multitude of the contextual factors that can increase the risk of error, the use of the computerized therapy system, is a responsibility of the trainers. To acquire these skills, the student must have the opportunity to experiment theme, in the management of scenarios adequately representative of this complexity, during the training [4, 11].

High fidelity (e. g, simulation, computerized mannequins or virtual reality) and Low fidelity simulation (e.g. human simulator, role playing) are both recommended for the training of healthcare professionals and they can be used for teaching the therapy management to students [2, 10, 12–15]. In fact, through the artificial creation of a work environment or a specific care scenario, the simulation allows the students to: 1) to put into practice the skills that the situation requires (cognitive, gestural, communicative-relational, critical thought and decision making), in a protected setting; 2) to learn from error, through a constant stimulus to the reflection before, during and at the end of the action, without the fear of causing damage and/or feeling judged [2, 13–15]. In addition to this, the introduction, in the simulation, of disturbing contextual elements (e.g. interruptions, multiplicity of concurrent requests, background noises, telephone ringing), and/or the use of instruments/devices specific to the real clinical contexts (e.g. computerized therapy, computerized systems for patient recognition), represent further possibilities to guarantee the needed realism and the complexity to the situation [2, 4]. These peculiarities make the simulation a teaching methodology which is particularly suitable for the teaching of the therapy management, because they let to construct diversified training courses, with progressively increasing complexity considering the 1st, 2nd and 3rd year students; in this way the students have the opportunity to progress in the level of the training objective in the three-year period [11].

1.2 Context of the Study

The Nursing Degree Course of the University of Modena and Reggio Emilia (Reggio Emilia campus), decided to analyse the typology and the causes of the nursing students' therapy errors, during their last year of the course, through a pilot study. The errors emerged from this study resulted similar to those reported by Harding & Patrick (2008) [16], and by Wolf, Hicks, Serembus (2006) [17]. Specifically, the errors that students make most frequently are: 1) wrong dosage of the drug; 2) wrong patient, 3) wrong drug; 4) wrong time of administration. The main causes of these errors are: 1) non-compliance with procedures/protocols; 2) performance deficits (in the use of computerized therapy or in the drug management); 3) lack of knowledge (about pharmacology and/or dosage calculation); 4) inadequate transmission of the information (in the prescription or between the different team members). In addition to these, the risk of error related to the inexperience and to some contextual factors, such as chaotic situation, frequent interruptions or high patient turnover, are contributing factors.

The pilot study have allowed the teachers to:

1. Rethink the teaching offer of the laboratories, in terms of therapy management, by proposing an additional two-year laboratory using the simulation, in addition to the traditional laboratory practice, The previous laboratory was dedicated to: 1) dosage calculation exercises; 2) training in administration techniques 3) dilution of drugs; 4) preparation of a syringe pump; 5) setting up a volumetric pump; 6) application of the critical thinking, 7) application of the decision-making process in the solution of cases report available in a paper document.
2. Strengthen and evaluate the students' ability to calculate the dosage of drugs, using a dosage calculation test in the 2nd and 3rd practical training exam.

From the Academic Year (AY) 2015-2016, in the 2nd and 3rd year of the Course (2nd semester), a laboratory became an integral part of the curriculum, with the use of the simulation and the application of a computerized therapy software. The teaching outcomes of this module area: 1) letting students to identify the critical issues related to reading a prescription, preparing a therapy, 2) choosing the most appropriate therapy in relation to the symptom reported by the patient; 3) making the student acquire the ability to act self-control and effective strategies to prevent possible errors. These main objectives are divided into specific ones, considering a progressively increasing complexity (e.g, from the 2nd to the 3rd year), thanks to the possibility to propose different contents in the simulations proposed and in the complexity of the created settings.

2 Scope

This observational descriptive study aimed at investigating the perception by nursing students about: a) the laboratory experience with the use of simulation and computerized therapy software, b) the achievement of their educational objectives, and c) the impact of the acquired learning on their training and future professional role'.

3 Method

The sample consisted of students attending the 2nd and 3rd year of the nursing degree course during the AY 2016/17, 2017/18, and 2018/19. An observational study was conducted on these students who attended the laboratory planned with the use of a simulation and computerized therapy software.

In each of the aforementioned AYs, the students have been divided into groups of about 20 students each. The proposed laboratories, named 'Medication Error 2 and 3, lasted 5 h, in co-management by a pedagogical tutor and a pharmacist of the Internal Pharmacy Service of the hospital Santa Maria Nuova - AUSL/IRCCS of Reggio Emilia. During the laboratory, three role-playing situations have been proposed to the students, simulating set a 3-bed hospital room of a surgery and medicine units. In these contexts, a therapy trolley was set up, with a laptop having the computerized therapy software in use at the operating units of the hospital. The software used in the simulation fully reproduces its real functionality, including the presence of a terminal used to register patient and drug recognition through a bar-code.

The actors on the scene, dressed and trained for the situation, are: one student, in the role of nurse, intent in the administration of the therapy, three students in the role of patients, other students in the role of a family member, a doctor or a supporting operator. The other students act as simulation observers.

In relation to the scenario of the simulation, the student who is acting as the nurse is called to effectively and safely manage a specific phase of the therapy management process, considering the recognition, the prescription, the administration, the caregiver's education in the administration of therapy. These have been requested considering different scenario:

1. the administration by a gastrostomy tube;
2. the adjusting of the care responses considering to the available data (e.g. diabetic patient complaining of symptoms of hypoglycaemia to whom the nurse is preparing to administer insulin therapy);
3. the critical evaluation of the collected data (e.g. pain management starting from the prescription of different types of pain reliever drugs);

4. the implementation of effective strategies for the management of some context disturbing factors (e.g. defining priorities, assigning activities to the support professionals, collaborating with the doctor);
5. the correct use of the computerized therapy system (e.g. to recognize incompleteness of a prescription).

In all these scenarios, the possibility to take inappropriate and unsafe decisions represent further elements of complexity.

At the end of the simulation, having collected the contribution of the actors and of the observers, the university tutor and the pharmacist retrace what happened and guide an interdisciplinary analysis about the subject of the risk management in the administration of the therapy, using emerged errors as a source of learning.

At the end of the laboratory course, students' satisfaction was collected by administering an anonymous questionnaire consisting of 10 closed-ended questions using the Likert scale from 1) 'Not at all' to 5) 'Completely' or the dichotomous' answer Yes/No. At the end of the questionnaire, an open question allows the student to report suggestions, personal notes, expectations, or expectations not got (If the laboratory were re-proposed, would you like).

Before the administration of the questionnaire, the students were informed about the purpose of the study, authorized by the President and the Council of the Nursing Degree Course. They were also informed about the guarantee of the anonymity and that handing the questionnaire would have been as a consent to the study.

4 Results

A total of 503 students, 250 from the 2nd year and 253 from the 3rd year, participated in the laboratory session with the use of simulation and computerized therapy (Table 1).

Table 1. Data of students who participated in the laboratory course during the three AY

	2nd year			3rd year		
	F	M	Average age	F	M	Average age
2016–2017	51	21	22 years	79	18	24 years
2017–2018	79	15	23 years	46	16	23 years
2018–2019	69	15	22 years	77	17	24 years
Total	199	51		202	51	

The questionnaire was filled by a total of 404 (80.3%) students, 192 (76.8%) from the 2nd year, and 212 (83.8%) from the 3rd year (Table 2).

Table 2. Students who completed the questionnaire/course year/AY

	Number of students	
	2nd year	3rd year
AY 2016/17	57	72
AY 2017/18	80	51
AY 2018/19	55	89
Total	192	212

4.1 The Perception of the Laboratory Objectives Achievement

The total of the 2nd and 3rd year students believes that they have achieved the educational objectives of the laboratory, almost unequivocally: 182/192 students of the 2nd year (95%), and students of the 3rd year 207/212 (98%). Only 7 students of the 2nd year on 192 (4%) and 4 of the 3rd on 212 (2%) believe the objectives have been not achieved (Fig. 1).

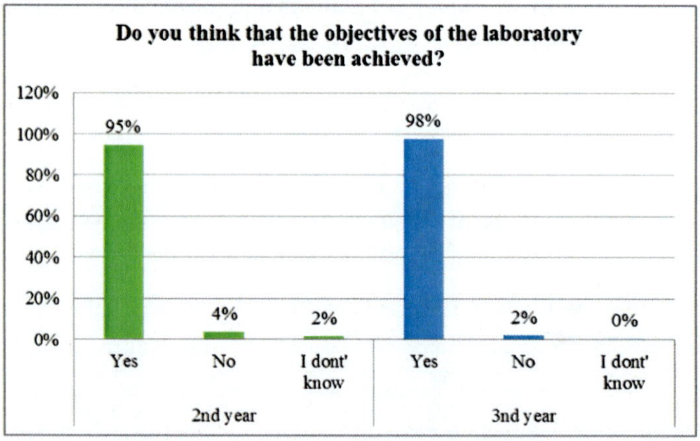

Fig. 1. Perception of achievement of objectives out of the total number of students divided by course year

The following Table 3 shows the detail of the perception of the achievement of the laboratory objectives divided by each AY and by course year. From this, it is possible to highlight that, in the 2016/17 AY, 100% of the 3rd year students (72/72) and 98% of the 2nd year (56/57) believe they have achieved the laboratory objectives. In the AY 2017/18 this percentage becomes 100% (80/80) if we consider the students of the 2nd year, and it becomes 94% (48/51) for those of the 3rd year: Finally, in the AY 2018/19, only 84% (46/55) of the 2nd year students and 98% (87/89) of 3rd one believe they have been achieved the expected educational objectives.

Table 3. Perception of achievement of objectives for each AY and course year

Objectives achieved	2nd year						3nd year					
	Yes		No		Don't know		Yes		No		Don't know	
AY 2016/17	56	98%	1	2%			72	100%	0	0%		
AY 2017/18	80	100%	0	0%			48	94%	3	6%		
AY 2018/19	46	84%	6	11%	5%	3	87	98%	1	1%	1	1%
Total	182	94.8%	7	3.6%	1.6%	3	207	97.6%	4	1.9%	1	0.5%

4.2 The Perception of the Usefulness of the Learning Outcomes in the Internship Activities

Most of the 2nd and 3rd year students, of the three examined AYs, consider the laboratory learning outcomes useful for the internship, in particular: 187/192 (97%) students of 2nd year, and 207/212 (98%) students of the 3rd; only 3 students (1%) of the 3rd year do not consider them useful (Fig. 2).

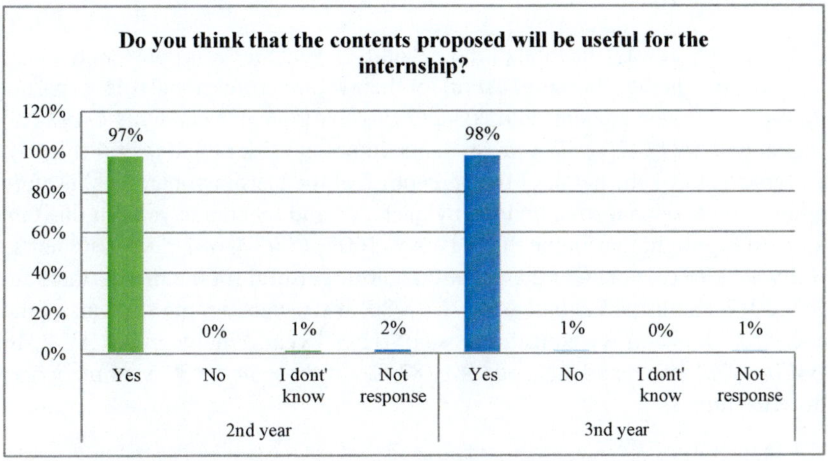

Fig. 2. Perception of usefulness of the learning acquired in the laboratory for the internship, out of the total number of students divided by course year

In the Table 4 the detail of the perception of the outcomes' usefulness related to the internship is presented, divided by each AY and by course year. From this table it is possible to highlight that while in the AY 2016/17, 100% (57/57) of the 2nd year students and all the ones of the 3rd year (72/72) consider the acquired learning outcomes useful for the internship.

Table 4. Perception of usefulness for the internship, learning acquired, for each AY and course year

Usefulness for internship	2nd year								3rd year							
	Yes		No		Don't know		No answer		Yes		No		Don't know		No answer	
AY 2016/17	57	100%	0	0%			0	0%	72	100%	0	0%			0	0%
AY 2017/18	77	96%	0	0%			3	4%	48	94%	2	4%			1	2%
AY 2018/19	53	96%	0	0%	2	4%	0	0%	87	98%	1	1%			1	1%
Total	187	97%	0	1%	2	1%	3	2%	207	98%	3	1%			2	1%

In the 2017/18 AY this percentage is 96% (77/80) if we consider the students of the 2nd year and 94% (48/51) for those of the 3rd year.

Finally, in the AY 2018/19, the percentage remains at 96% (53/55) for the 2nd year students and arrive to the 98% (87/89) for the 3rd year students.

4.3 The Perception of the Usefulness of the Learning Outcomes for the Future Professional Role

In an almost univocal way, the total of the 2nd and 3rd year students consider the learning outcomes acquired in the laboratory useful for their future professional role, in particular, 188/192 (98%) 2nd year students and 203/212 (96%) of the 3rd year. Only 1/192 (%) 2nd year student and 5/212 (2%) 3rd year students do not consider them useful (Fig. 3).

The Table 5 shows the detail of the perception of the learning outcomes' usefulness, for the future professional role, divided by each AY and by course year. In this table, it is possible to highlight that in the 2016/17 AY, 100% (57/57) of 2nd year students, and 97% of thy 3rd (70/72) consider the acquired learning useful for the professional future. In the AY 2017/18 this percentage is 98% (78/80) if we consider the students of the 2nd year, and 90% (46/51) if we consider those of the 3rd year. Finally, in the AY 2018/19, 96% (53/55) of 2nd year students and 98% (87/89) of those in the 3rd year recognize the laboratory usefulness.

5 Discussion

The growing complexity of the therapy management process, the associated unpredictability of the current clinical-care situations, the stressful dynamics of the clinical settings, together with the limited experience and skills expose the students, during the training, to the risk of therapy errors [1, 4, 10, 11].

The simulation, aimed at the acquisition of skills in the safe management of therapy, lets possible to create a training context, with a progressively increasing complexity,

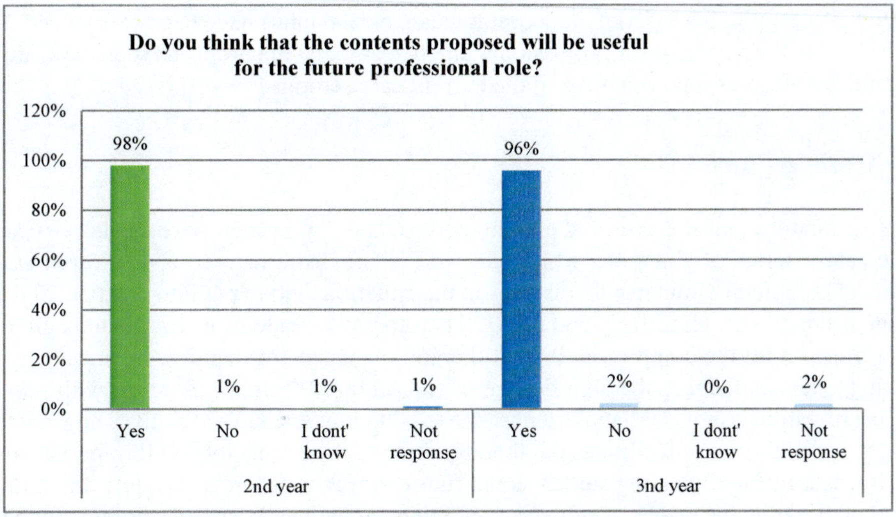

Fig. 3. Perception of usefulness of learning acquired in the laboratory for the future professional role, out of the total number of students divided by course year.

Table 5. Perception of the usefulness of the acquired learning for the future professional role, for each AY and course year.

Usefulness for future role	2nd year								3rd year							
	Yes		No		Don't know		No answer		Yes		No		Don't know		No answer	
AY 2016/17	57	100%	0	0%			0	0%	70	97%	1	1%			1	1%
AY 2017/18	78	98%	0	0%			2	2,5%	46	90%	3	6%			2	4%
AY 2018/19	53	96%	1	2%	1	2%	0	0%	87	98%	1	1%			1	1%
Total	188	98%	1	1%	1	1%	2	1%	203	96%	5	2%			4	2%

considering the 2nd and 3rd years students, and it enriches and complete the students learning in this area. The collected results showed that the laboratory was perceived by all the students, in the three academic years, as an effective training offer in terms of achieving the training objectives. The students state that the use of the simulation, associated with the computerized therapy software, allowed them to consolidate skills and/or to acquire new and more complex ones, through the experience and the learning from mistakes [2, 3, 11, 12].

The results of the study also showed that all the students, in the three academic years, perceive the usefulness of the acquired learning for both internship and future professional role, In specific, it has been underline that, in the laboratory, the skills (cognitive, technical-gestural, managerial and communicative-relational) have been recognized as specific and fundamental for professional practice and that the proposed scenarios have been recognized as representative of the current care settings [11–13].

6 Conclusions

The simulation and the use of the computerized therapy system have made possible to create a series of situations which, due to their realistic nature, have allowed students to completely immerse themselves in the situation, and to put into practice all the skills needed to act effectively and safely. Therefore, the students were able to begin to experiment with the complex skills that therapy management requires, in current care settings, acknowledging the significance of the acquired learning. Studies, with more robust research designs, will make it possible to better understand and analyse how much this type of laboratory facilitates the therapy management capacity and it improves the performance of both nursing students, and future professional nurse. Despite this limit, it is believed that the results of students' perception about the described laboratory experience can contribute to develop further researcher with a larger sample and can to open up the possibility of a comparison with other Nursing Degree Courses in the Italian context.

References

1. Alvaro, R., Bagnasco, A., Del Negro, L., Lancia, L., Rubbi, I., Sasso, L., Vellone, E., Venturini, G., Banacaro, A., Delpidio, G., Fierro, A., Lentini, K., Maricchio, R., Menoni, S., Scialò, G., Sili, A., Sperlinga, A., Tibaldi, L.: La sicurezza nella somministrazione della terapia farmacologica: una revisione narrativa della letteratura. L'Infermiere **3**, 22–26 (2009)
2. Sarfati, L., Ranchon, F., Vantard, N., Schwiertz, V., Larbre, V., Parat, S., Faudel, A., Rioufol, C.: Human-simulation-based learning to prevent medication error: a systematic review. J. Eval. Clin. Pract. **25**(1), 11–20 (2019). https://doi.org/10.1111/jep.12883. Epub 2018 Jan 31 PubMed PMID: 29383867
3. Berdot, S., Roudot, M., Schramm, C., Katsahian, S., Durieux, P., Sabatier, B.: Interventions to reduce nurses' medication administration errors in inpatient settings: a systematic review and meta-analysis. Int. J. Nurs. Stud. **53**, 342–350 (2016). https://doi.org/10.1016/j.ijnurstu.2015.08.012. Epub 2015 Sep 7. Review. PubMed PMID: 26365701
4. Chan, R., Booth, R., Strudwick, G., Sinclair, B.: Nursing students' perceived self-efficacy and the generation of medication errors with the use of an electronic medication administration record (eMAR) in clinical simulation. Int. J. Nurs. Educ. Scholarsh. **16**(1), 1–10 (2019). pii: /j/ijnes.2019.16.issue-1/ijnes-2019-0014/ijnes-2019-0014.xml. https://doi.org/10.1515/ijnes-2019-0014. PubMed PMID: 31539361
5. Hudson, P.T., Guchelaar, H.J.: Risk assessment in clinical pharmacy. Pharm. World Sci. **25**(3), 98–103 (2003). Review. PubMed PMID: 12840962
6. Anonymous: Dossier Cineas e Makno: Errori in ospedale: 35 mila vittime l'anno e un costo di 10 miliardi di euro. ASI 2002 17–25 Aprile, pp. 16–19 (2002)
7. Ministero della Salute: Raccomandazione per la prevenzione della morte, coma o grave danno derivati da errori in terapia farmacologica. Raccomandazione no 7 (Marzo 2008). Scaricabil all'indirizzo internet: http://www.salute.gov.it/imgs/C_17_pubblicazioni_675_allegato.pdf. Data ultima consultazione 31 Jan 2020

8. Profilo professionale dell'infermiere: DM Sanità 14/09/1994 no 739
9. Armitage, G., Knapman, H.: Adverse events in drug administration: a literature review. J. Nurs. Manag. **11**(2), 130–140 (2003). Review. PubMed PMID: 12581401
10. Lee, S.E., Quinn, B.L.: Incorporating medication administration safety in undergraduate nursing education: a literature review. Nurs. Educ. Today **72**, 77–83 (2019). https://doi.org/10.1016/j.nedt.2018.11.004. Epub 2018 Nov 14. Review. PubMed PMID: 30453203
11. Fothergill Bourbonnais, F., Caswell, W.: Teaching successful medication administration today: more than just knowing your 'rights'. Nurs. Educ. Pract. **14**(4), 391–395 (2014). https://doi.org/10.1016/j.nepr.2014.03.003. Epub 2014 Mar 15 PubMed PMID: 24857050
12. Sears, K., Goldsworthy, S., Goodman, W.M.: The relationship between simulation in nursing education and medication safety. J. Nurs. Educ. **49**(1), 52–55 (2010). https://doi.org/10.3928/01484834-20090918-12. Epub 2010 Jan 4 PubMed PMID: 19810664
13. Shearer, J.E.: High-fidelity simulation and safety: an integrative review. J. Nurs. Educ. **52**(1), 39–45 (2013). https://doi.org/10.3928/01484834-20121121-01. Epub 2012 Nov 21. Review. PubMed PMID: 23181458
14. Sponton, A., Iadeluca, A.: La simulazione nell'infermieristica. Metodologie, tecniche e strategie per la didattica, 1st edn. Casa Editrice Ambrosiana, Milano (2014)
15. Tosterud, R., Hedelin, B., Hall-Lord, M.L.: Nursing students' perceptions of high-and low-fidelity simulation used as learning methods. Nurs. Educ. Pract. **13**(4), 262–270 (2013). https://doi.org/10.1016/j.nepr.2013.02.002. Epub 2013 Feb 28 PubMed PMID: 23454066
16. Harding, L., Petrick, T.: Nursing student medication errors: a retrospective review. J. Nurs. Educ. **47**(1), 43–47 (2008). PubMed PMID: 18232615
17. Wolf, Z.R., Hicks, R., Serembus, J.F.: Characteristics of medication errors made by students during the administration phase: a descriptive study. J. Prof. Nurs. **22**(1), 39–51 (2006). PubMed PMID: 16459288

Computer Laboratory: The Key to Access the Electronic Databases in Learning Evidence-Based Practice

Stefano Finotto[1], Marika Carpanoni[1], Patrizia Copelli[1], Chiara Marmiroli[1], and Daniela Mecugni[1,2](✉)

[1] Nursing Degree Course, University of Modena and Reggio Emilia, Reggio Emilia, Italy
{stefano.finotto,patrizia.copelli,daniela.mecugni}@unimore.it, {marika.carpanoni,chiara.marmiroli}@ausl.re.it

[2] Department of Surgery, Medicine, Dentistry and Morphological Sciences, University of Modena and Reggio Emilia, Reggio Emilia, Italy

Abstract. The Evidence-Based Practice can be defined as the integration of the best available research evidence with information related to patient preferences, the level of competence of clinicians, and the resources available to make decisions related to patient care. Skills in using computers and interrogating electronic resources are among one the most important barriers to using the Evidence-Based Practice process. The Nursing Degree Course of the University of Modena and Reggio Emilia, seat of Reggio Emilia, included in the study plan a three-year Evidence-Based Practice laboratory for to develop skills related to the ability to use electronic health resources available on the web. This study describe the nursing students' perception relative to their skills about the use of computer lab for research of scientific evidence in electronic databases learned during the laboratory's three-year Evidence-Based Practice. The sample is composed of 164 students attending the third year of the Nursing Degree Course in Reggio Emilia. The data collection tool was used is the Italian version of Evidence Based Practice Competence Questionnaire. A high percentage of students believe they are able to formulate structured clinical questions using the PICO methodology and more than half of the sample feels they are able to perform research on scientific evidence in a structured and systematic way on the main databases. The nursing students' perception relative to their skills about the use of the computer lab for the research of scientific evidence in electronic databases learned during the laboratory's three-year EBP, are positive.

Keywords: Evidence-Based Practice · Nursing students · Computer laboratory · Electronic databases · Learning

Z. Kubincová et al. (Eds.): MIS4TEL 2020, AISC 1236, pp. 138–147, 2021.
https://doi.org/10.1007/978-3-030-52287-2_14

1 Introduction

1.1 Background

The Evidence Based Medicine (EBM) movement has developed considerably since the last decades of the last century so that today clinical professionals strive to make their decisions based on the best available evidence [1, 2].

During this time, EBM has been declined in the specifications of the health disciplines, including nursing care through Evidence Based Nursing (EBN). In the last few years the parochial phase linked to the different health disciplines has been overcome to arrive at Evidence Based Practice (EBP), a mental representation that perhaps translates better the knowledge of the EBM movement [3].

The EBP has become one of the most important principles on which modern health-care is based since it is able to increase the quality of care through the standardization of the care process [4]. Despite the impetus given by the EBM movement, still today the use into clinical practice of knowledge derived from scientific research is often hindered [5] and this is a cause of variability in clinical practice with the consequence that not all patients have an equal access to the most appropriate treatments [6].

The EBP is defined by Sackett et al. [7] as the use of the best available evidence complemented by the clinical experience and the patient's values to make decisions about individual patient care, DiCenso et al. [8] integrated this model with the consideration of resources. Therefore, EBP can be defined as the integration of the best available research evidence with information related to patient preferences, the level of competence of clinicians, and the resources available to make decisions related to patient care [9]. As claimed by McInerney & Suleman [10], except for nurses to whom elements of EBP were introduced during basic training, nurses may not consider the concept of best practice and continue to base their practice based on their own experience or that of their colleagues.

The data reported by Pravikoff et al. [11] in a survey conducted in the United States indicate a lack of understanding and appreciation of the value of research in practice. In this study it is noted that nurses are not eligible to integrate the practice with scientific evidence; in fact, almost half of the sample was not familiar with the term EBP, more than half do not believe colleagues who use evidence in practice, the majority (58%) of the sample do not seek evidence to support the practice, 82% do not have never used hospital libraries, 76% never researched CINAHL and 58% Medline.

In their conclusions, the authors believe US nurses unprepared for EBP and among the causes of this situation indicate the failure of nurses' educationprograms that prepare students by maintaining a practice based mostly on tradition, intuition and experience. Melnyk et al. [12] focused on EBP competencies for nurses and recognized that nurses estimate that they were not yet competent in meeting any of the 24 Evidence-Based Practice Competencies for Practicing Registered Nurses and Advanced Practice Nurses. Determining whether education has an effect on practice is hardly measurable although studies have shown positive correlations between education and changes in practice [1–13]. The link between education and EBP seems to be the keystone in the implementation of the EBP process in clinical practice.

The World Health Organization in his technical guide for Member States [14], European strategic directions for strengthening nursing and midwifery towards Health 2020 goals, sets as the fourth priority area "Promoting evidence-based practice and innovation", where it declares that "Evidence-based practice is every nurse's and midwife's concern. It should be enabled by means of education, research, leadership and access to evidence sources". However, nursing education, both basic and subsequent, has historically focused on preparing nurses to become evidence generators through research methodology courses, rather than evidence users, that is, professionals who can effectively translate the results of research into clinical practice [15].

The European Higher Education Area (EHEA) framework specifies expected learning outcomes for candidates with a Bachelor's degree, including skills in finding, evaluating, referring and applying scientific information [16]. Skills in using computers and interrogating electronic resources are among the most important barriers to using the EBP process.

As stated by Ivanitskaya et al. [17], the students have no competence in online research, in the ability to critically evaluate the evidence and about 90% of them overestimate their ability to locate the online health literature and to evaluate its reliability. In the study by Ruzafa-Martinez et al. [18], which aimed to evaluate the effectiveness of an EBP for undergraduate nursing students course, it is highlighted that among the teaching strategies that increase the attitudes, knowledge and skills there is the practical classes with access to computers.

Also Young et al. [19] in their review highlight how combined teaching methods, including computer lab sessions, are more likely to be successful in increasing knowledge, skills and attitudes in EBP than individual interventions. In the study of Fernandez et al. [20] four teaching methods, including computer lab sessions and distance learning sessions using an EBP-DVD, were compared to measure the skills related to the search for evidence; the Authors conclude that significant improvement in cognitive and technical EBP skills can be achieved for postgraduate nursing students by integrating an EBP-DVD teaching resource at computer lab sessions.

1.2 Context of the Study

Given the lack of evidence relating to the presence of EBP courses in nursing education curricula [3] and, in view of its importance in order to train future nurses capable of making decisions based also on scientific evidence, the Nursing Degree Course of the University of Modena and Reggio Emilia, seat of Reggio Emilia, included in the study plan a three-year EBP laboratory and a teaching of Scientific Evidence for Nursing.

The three-year laboratory develops skills related to the use of electronic tools, specifically the computer, and the ability to use electronic health resources available on the web. The laboratory is developed over the three years of the degree course with objectives that recall the first three steps of the EBP in nursing by Melnyk et al. [21]. The other five steps of EBP are discussed in the teaching of Scientific Evidence for Nursing.

The objectives of the laboratories for the first year of the course are: the student is able to identify a clinical problem; the student is able to translate the clinical problem using the P.I.C.O. methodology into a clinical question; the student is able to recognize the structure of a scientific article; the student is able to cite scientific evidence; the student is able to search for evidence in the Medline-PubMed database.

Those in the second year are: the student is able to search for scientific evidence in the CINAHL and Cochrane databases, and for the third year: the student is able to search for scientific evidence in the Guidelines databases (SIGN, National Guideline Claringhouse, New Zealand Guidelines Group, RNAO, NICE, SNLG).

The laboratory is organized in groups of students not exceeding 20 and is conducted by an expert EBP academic tutor. The didactic organization of the first year laboratory includes two 5 and 3 h meetings. The first meeting begins with an introduction by the expert academic tutor to the clinical problem and the P.I.C.O. methodology, and then the tutor introduces the Medline-PubMed database. Each student is in a computer station and performs all the steps illustrated by the tutor in order to understand the strategies and resources available in the database. Next, the tutor assigns two prepackaged clinical problems to students who need to set up the clinical question and interrogate Pubmed to find evidence that can answer the question. The second meeting takes place during the clinical training period; students have the mandate to identify a clinical question related to their clinical experience and, using Medline-PubMed with the supervision of the tutor, to find scientific evidence that allows to answer the clinical question.

The didactic organization of the second year laboratory foresees two meetings of 3 h each. The first lab begins with an introduction to the CINAHL and Cochrane databases. Each student is in a computer station and performs all the steps illustrated by the tutor in order to understand the strategies and resources available in the databases. Thereafter, the tutor assigns two pre-clinical issues to students who have set the clinical question and interrogate CINAHL, and Cochrane to find evidence that could answer the question. The second meeting takes place during the clinical training period; students have the mandate to identify a clinical question related to their clinical experience and, using CINAHL and Cochrane databases with the supervision of the tutor, to find scientific evidence that allows to answer the clinical question.

The didactic organization of the third year laboratory is the same as that of the second year but the databases that are introduced by the tutor and used by the students are Guidelines databases (SIGN, National Guideline Claringhouse, New Zealand Guidelines Group, RNAO, NICE, SNLG). Also in this laboratory, each student works on a computer station in order to understand the strategies and resources available in the databases.

2 Scope

The purpose of this study was to describe the nursing students' perception relative to their skills about the use of computer lab for research of scientific evidence in electronic databases learned during the laboratory's three-year EBP and to their competencies on EBP process.

3 Method

A convenience sample consisted of 164 students attending the third year of the Nursing degree course in Reggio Emilia, 99 of the 2017–2018 AY and 65 of the 2018–2019 AY.

The data collection tool used is the Italian version of Evidence Based Practice Competence Questionnaire (EBP-COQ) developed by Ruzafa-Martinez et al. [22]; EBP-COQ is an instrument specifically created for nursing students. It analyses the following dimensions: attitudes, skills and knowledge to EBP. The Italian version of the EBP-COQ questionnaire consists of 33 questions divided into 3 sections: student master data (8), attitudes (13) and skills - knowledge (12). The response choice for sections attitudes and skills - knowledge is expressed by a Likert scale at 5 with extreme strongly agree (1) - strongly disagree (5). For data analysis the Likert scale was considered in a trichotomic sense and the results were divided into positive (strongly agree - agree), neutral (neither agree nor disagree) and negative (disagree – strongly disagree). The linguistic-cultural validation of the Italian version of the EBP-COQ has been carried out, the study relating to it has been accepted by a nursing journal and is awaiting its publication.

Descriptive data analysis and frequency calculation was performed.

The Degree Course Nursing Board approved the study. The identity of respondents was protected at all times. Using a box in which to deposit the questionnaire, we could not match the responses to respondents. This maintained total anonymity. Respondents were provided with a complete instruction sheet, which included: an explanation of the research and its purpose, expectations of the respondent, a statement outlining anonymity issues and an indication that return of completed questionnaire would constitute informed consent to participate.

4 Results

The sample is composed of 83% of females and 17% of males, has an average age of 24 (SD 4) years and 3% has another degree. The 75% of students have done more than 4 scientific evidence searches using electronic databases (EDBs) in the past 3 years and the 48% have read more than 4 scientific journals in the past month. The data relating to the student master data are shown in Table 1.

Regarding the attitudes for EBP, the 96% of the sample believe that EBP helps to make decisions in clinical practice, the 66% believe that applying EBP would better define the role of the nurse, the 97% think that the use of EBP improves patient care and the 77% are interested in reading scientific articles.

With regard to EBP knowledge, the 76% of the students in the sample feel able to formulate a clinical question to search for best scientific evidence, the 56% feel able to perform research of scientific evidence in a structured and systematic way on the main databases and the 64% feel able to search for scientific evidence in the guideline and systematic reviews databases.

Regarding EBP skills, the 88% of the sample believe is able to formulate structured clinical questions using the PICO methodology and the 71% know the main sources that offer controlled and catalogued information from the point of view of evidence.

The attitudes, skills and knowledge to EBP sections are shown in Table 2.

5 Discussion

Our study showed students' perception of the effectiveness of the computer laboratory for the research of scientific evidence in electronic databases and for their skills in applying

Table 1. The student master data

Gender	N (%)
Female	106 (83)
Male	22 (17)
Age	*Average (SD)*
	24 (4)
Education	*N (%)*
Anyone	158 (96)
Other degrees	4 (3)
Master	0 (0)
Other	2 (1)
EDB search last 3 years	*N (%)*
Anyone	0 (0)
1–2	6 (5)
3–4	26 (20)
>4	96 (75)
Reading scientific journal last month	*N (%)*
Anyone	14 (11)
1–2	32 (25)
3–4	20 (16)
>4	62 (48)

the EBP process are positive. The teaching methodology that includes computer lab sessions led by an expert EBP academic tutor seems to be effective in making students acquire the skills necessary to search for evidence in the EDBs. The results of our study are in line with that of Fernandez et al. [19] which demonstrates how computer lab sessions are more effective than distance learning for the acquisition of search skills.

Furthermore, the results of our study confirm those of Ruzafa-Martinez et al. [17] and Young et al. [18], which demonstrate how teaching methods that include computer lab sessions are more likely to be successful in increasing knowledge, skills and attitudes for EBP. The limitation of this study is related to the description of the skills of the student only at the end of the educational path and its monocentric approach.

Table 2. Data relating to the attitudes, skills and knowledge to EBP (items with * have an inverted score).

Item	Strongly agree	Agree	Neither agree nor disagree	Disagree	Strongly disagree
Section attitudes	*N (%)*	*N (%)*	*N (%)*	*N (%)*	*N (%)*
EBP helps make decisions in clinical practice	78 (49)	76 (47)	6 (4)	0 (0)	0 (0)
I think I can critically evaluate the quality of a scientific article	17 (11)	77 (48)	33 (21)	23 (14)	10 (6)
Applying EBP would better define the role of the nurse	51 (32)	58 (36)	20 (13)	21 (13)	9 (6)
The nurse employment contract should include time to reading and critical analysis of literature	50 (31)	83 (52)	22 (14)	5 (3)	0 (0)
The application of EBP would allow to increase autonomy compared to other professions	61 (38)	62 (39)	14 (9)	15 (9)	8 (5)
I would like that in my work as a nurse apply EBP	57 (36)	82 (51)	20 (12)	1 (1)	0 (0)
The use of EBP improves patient care	90 (56)	65 (41)	4 (2)	1 (1)	0 (0)
In the future I would like to be able to contribute to the practical application of EBP	54 (34)	66 (41)	33 (20)	6 (4)	1 (1)
I am not interested in reading scientific articles*	8 (5)	14 (9)	15 (9)	56 (35)	67 (42)
With the application of EBP, the changes that would occur in patients would be minimal*	8 (5)	25 (15)	22 (14)	48 (30)	57 (36)
I am pleased that the 'EBP is only a theoretical current that is not put into practice*	2 (1)	8 (5)	14 (9)	61 (38)	75 (47)
If I had the opportunity, I would attend an EBP course	34 (21)	74 (46)	43 (27)	8 (5)	1 (1)
I wish I had better access to the evidence scientific papers published in nursing	86 (53)	72 (44)	4 (3)	0 (0)	0 (0)
Section skills	*N (%)*	*N (%)*	*N (%)*	*N (%)*	*N (%)*
I feel able to formulate a clinical question to begin the search for the best scientific evidence	33 (20)	91 (56)	38 (23)	1 (1)	0 (0)
I don't feel able to carry out scientific research in structured and systematic way on the main bibliographic databases*	14 (9)	33 (20)	25 (15)	69 (43)	21 (13)
I do not feel able to search for scientific evidence in the guidelines and systematic reviews databases*	12 (7)	27 (17)	20 (12)	77 (47)	27 (17)
I feel able to critically evaluate the quality of a scientific article	26 (16)	88 (54)	39 (24)	8 (5)	2 (1)

(continued)

Table 2. (*continued*)

Item	Strongly agree	Agree	Neither agree nor disagree	Disagree	Strongly disagree
I don't feel able to evaluate whether the results obtained in a scientific study are valid*	19 (12)	24 (15)	35 (21)	66 (40)	19 (12)
I feel able to verify the practical usefulness of a scientific study	35 (21)	95 (59)	26 (16)	6 (4)	0 (0)
Section knowledge	*N (%)*	*N (%)*	*N (%)*	*N (%)*	*N (%)*
I am able to formulate clinical questions structured according to the PICO methodology	44 (27)	100 (61)	17 (10)	1 (1)	1 (1)
I know the main sources that offer controlled and catalogued information from the point of view of evidence	33 (20)	83 (51)	21 (13)	13 (8)	13 (8)
I do not know the most important characteristics of the main research designs*	2 (1)	9 (6)	25 (15)	83 (51)	44 (27)
I know the different levels of evidence of research studies	21 (13)	60 (49)	26 (16)	19 (12)	16 (10)
I don't know the different degrees of recommendation with respect to adopting a specific health intervention	9 (5)	27 (17)	38 (23)	57 (35)	32 (20)
I know the main association measures (RR, OR, etc.) and the measures of potential impact (NNT, NND, relative risk, etc.) for evaluating the effect size analyzed in the research designs	39 (24)	80 (49)	36 (22)	8 (5)	0 (0)

6 Conclusions

The laboratory's three-year EBP, which provides computer lab sessions conducted by an expert EBP academic tutor, in which the student learns to use the computer to interrogate EDBs, represents an effective teaching method. The laboratory's three-year EBP allows nursing students to acquire skills, attitudes and knowledge related to the application of the EBP process and, in particular, to increase their skills in finding evidence through the use of the computer.

The nursing students' perception of their skills about the use of the computer lab for the research of scientific evidence in EDBs learned during the laboratory's three-year EBP and their competencies on EBP process is positive.

More robust study design will be necessary to analyze deeper and with more details the results of this, just descriptive, study. In particular, it would be interesting to use a pre-post study design to understand more effectively the effectiveness of the computer lab sessions for the acquisition of skills related to the use of the computer in the search for evidence in electronic databases by nursing students; in addition, a pre-post study could better highlight the increase in skills, attitudes and knowledge in the application of the EBP process.

References

1. Morris, J., Maynard, V.: The value of an evidence based practice module to skill development. Nurse Educ. Today **27**(6), 534–541 (2007)
2. Gibbs, L., Gambrill, E.: Evidence-based practice: counterarguments to objections. Res. Soc. Work Pract. **12**(3), 452–476 (2002)
3. Finotto, S., Carpanoni, M., Turroni, E.C., Camellini, R., Mecugni, D.: Teaching evidence-based practice: developing a curriculum model to foster evidence-based practice in undergraduate student nurses. Nurse Educ. Pract. **13**(5), 459–465 (2013)
4. Florin, J., Ehrenberg, A., Wallin, L., Gustavsson, P.: Educational support for research utilization and capability beliefs regarding evidence-based practice skills: a national survey of senior nursing students. J. Adv. Nurs. **68**(4), 888–897 (2012)
5. Thompson, D.S., Estabrooks, C.A., Scott-Findlay, S., Moore, K., Wallin, L.: Interventions aimed at increasing research use in nursing: a systematic review. Implement Sci. **2**(1), 1–16 (2007)
6. Grol, R., Grimshaw, J.: From best evidence to best practice: effective implementation of change in patients' care. Lancet **362**(9391), 1225–1230 (2003)
7. Sackett, D.L., Rosenberg, W.M.C., Gray, J.A.M., Haynes, R.B., Richardson, W.S.: Editorials evidence based medicine: what it is and what it isn't. BMJ **312**, 71 (1996)
8. DiCenso, A., Cullum, N., Ciliska, D.: Implementing evidence-based nursing: some misconceptions. Evid.-Based Nurs. **1**(2), 38–40 (1998)
9. Ciliska, D.K., Pinelli, J., DiCenso, A., Cullum, N.: Resources to enhance evidence-based nursing practice. AACN Clin. Issues **12**(4), 520–528 (2001)
10. McInerney, P., Suleman, F.: Exploring knowledge, attitudes, and barriers toward the use of evidence-based practice amongst academic health care practitioners in their teaching in a South African University: a pilot study. Worldviews Evid.-Based Nurs. **7**(2), 90–97 (2010)
11. Pravikoff, D.S., Tanner, A.B., Pierce, S.T.: Readiness of U.S. nurses for evidence-based practice. Am. J. Nurs. **105**(9), 40–52 (2005)
12. Melnyk, B.M., Gallagher-Ford, L., Zellefrow, C., Tucker, S., Thomas, B., Sinnott, L.T., et al.: The first U.S. study on nurses' evidence-based practice competencies indicates major deficits that threaten healthcare quality, safety, and patient outcomes. Worldviews Evid.-Based Nurs. **15**(1), 16–25 (2018)
13. Adamsen, L., Larsen, K., Bjerregaard, L., Madsen, J.K.: Moving forward in a role as a researcher: the effect of a research method course on nurses' research activity. J. Clin. Nurs. **12**(3), 442–450 (2003)
14. WHO: European strategic directions for strengthening nursing and midwifery towards Health 2020 goals. http://www.euro.who.int/__data/assets/pdf_file/0004/274306/European-strategic-directions-strengthening-nursing-midwifery-Health2020_en-REV1.pdf?ua=1
15. Fineout-Overholt, E., Melnyk, B.M., Schultz, A.: Transforming health care from the inside out: advancing evidence-based practice in the 21st century. J. Prof. Nurs. **21**(6), 335–344 (2005)
16. Lahtinen, P., Leino-Kilpi, H., Salminen, L.: Nursing education in the European higher education area - variations in implementation. Nurse Educ. Today **34**(6), 1040–1047 (2014)
17. Ivanitskaya, L.V., Hanisko, K.A., Garrison, J.A., Janson, S.J., Vibbert, D.: Developing health information literacy: a needs analysis from the perspective of preprofessional health students. J. Med. Libr. Assoc. **100**(4), 277–283 (2012)
18. Ruzafa-Martínez, M., López-Iborra, L., Armero-Barranco, D., Ramos-Morcillo, A.J.: Effectiveness of an evidence-based practice (EBP) course on the EBP competence of undergraduate nursing students: a quasi-experimental study. Nurse Educ. Today **38**, 82–87 (2016)

19. Young, T., Rohwer, A., Volmink, J., Clarke, M.: What are the effects of teaching evidence-based health care (EBHC)? Overview of systematic reviews. PLoS ONE **9**(1), e86706 (2014)
20. Fernandez, R.S., Tran, D.T., Ramjan, L., Ho, C., Gill, B.: Comparison of four teaching methods on evidence-based practice skills of postgraduate nursing students. Nurse Educ. Today **34**(1), 61–66 (2014)
21. Melnyk, B.M., Fineout-Overholt, E., Stillwell, S.B., Williamson, K.M.: Evidence-based practice: step by step: the seven steps of evidence-based practice. Am. J. Nurs. **110**(1), 51–53 (2010)
22. Ruzafa-Martinez, M., Lopez-Iborra, L., Moreno-Casbas, T., Madrigal-Torres, M.: Development and validation of the competence in evidence based practice questionnaire (EBP-COQ) among nursing students. BMC Med. Educ. **13**(1), 19 (2013)

Perspectives in Nursing Education: From Paper Standardized Taxonomies to Electronic Records Applied in Nursing Practice

Luca Bertocchi[1](✉), Annamaria Ferraresi[2], Vianella Agostinelli[3], Giuliana Morsiani[3], Federica Sabato[4], Luisa Anna Rigon[5], Gianfranco Sanson[6], and Loreto Lancia[1]

[1] Department of Health, Life and Environment Sciences, University of L'Aquila Edificio Rita Levi Montalcini, Via G. Petrini, 67100 L'Aquila, Italy
luca.bertocchi@graduate.univaq.it
[2] Local Health Unit of Ferrara, Ferrara, Italy
[3] Local Health Unit of Modena, Modena, Italy
[4] Maggiore Dialysis Unit, University Hospital of Trieste, Trieste, Italy
[5] Formazione in Agorà (Scuola di Formazione alla Salute), Via Svezia, 9, 35127 Padua, Italy
[6] University Hospital of Trieste, Piazzale Di Valmaura 9, 34148 Trieste, Italy

Abstract. Nursing process has a relevant impact on overall quality of healthcare services and patient outcomes, however nursing contribution appears to be almost 'invisible'. Use of an international Standardized Nursing Language (SNL), as NANDA-I NOC NIC (NNN) taxonomy, allows making nursing visible in education, practice and research with final better outcomes on patients. Electronic Health Record (EHR) could facilitate learning of standardized languages in nursing undergraduate students, but only few studies analyzed it. This paper aims to report the experience of the 'Caring project' utilizing a SNL in an Italian healthcare setting and to discuss potentialities of EHR in improving nursing education and practice. 'Caring project' was an Italian pilot observational study, conducted on 231 patients of 8 Units retrospectively (January 2018–March 2019), using a 'paper' assessment form with Gordon's Functional Health Patterns (FHPs) and an Individual Care Plan (ICP), an 'Excel® based health record prototype'. In the study a total of 484 Nursing Diagnosis (NDs) has been identified, corresponding to a mean of 2.0 ± 1.5 NDs/patient. Integration of SNL with EHRs could help nurses to use a holistic approach and, in nursing education, it could improve undergraduate students clinical reasoning skills. Further researches are necessary in this subject, in particular regarding the nursing diagnostic accuracy.

Keywords: Nursing process · Electronic health records · NANDA-I taxonomy · Nursing education · Standardized nursing terminology · Nursing diagnosis

1 Introduction

Nursing Process (NP) is defined as a systematic approach to care using the fundamental principles of critical thinking, client-centered approaches to treatment, goal-oriented

Z. Kubincová et al. (Eds.): MIS4TEL 2020, AISC 1236, pp. 148–153, 2021.
https://doi.org/10.1007/978-3-030-52287-2_15

tasks, evidence-based practice recommendations, and nursing intuition. Holistic and scientific postulates have been integrated to provide the basis for compassionate, quality-based care [1]. NP includes stages of patient's assessment and data collection, making the nursing diagnoses, determining the expected outcomes, planning and implementing the appropriate interventions and evaluating the level of outcomes' achievement [2]. NDs, as part of NP, has a relevant impact in the improvement of the quality of care predicting patients' and organizational outcomes [3]. Unfortunately, at an international level, nursing documentation appears to be heterogeneous, incomplete and inaccurate, making it difficult to measure nursing outcomes [4]. Nursing contribution on patients' health and organizational costs risks to remain mostly 'invisible' [3, 5]. One of the strategies to avoid these critical issues is the adoption of an international SNL, able to improve the health care quality [6] enhancing nurses and patients' outcomes [7, 8]. All healthcare agencies are expected to use EHRs in the near future to increase effectiveness and patient safety, reducing medical errors and costs [9, 10]. For this reason, it may be helpful to track the NP using a SNL embedded in the EHR and to measure the final outcomes relevant to nursing care [11]. In addition, several studies found that NDs are an independent predictor of patient hospital outcomes (e.g. patient mortality) and should be included in the hospital discharge abstract as a complement to the medical Diagnosis Related Groups [12–14]. Education, together with adequate nurse staffing and an enhanced care environment, contributes to better patient outcomes [7]. Consequently, nursing students education on accurate clinical reasoning using SNL represents the first step to allowing future nurses to make the difference on clinical practice outcomes [15]. In this regard, the use of SNLs, such as International Classification of Nursing Practice (ICNP®) and NANDA International (NANDA-I), are recommended in nursing undergraduate education as useful tools facilitating the learning of NP [16, 17]. Hong and colleagues (2015) highlighted in Korean nursing students that using a SNL improves competency of education and helps to ratify the steps of the NP. For this reason, using educational strategies, as EHR, could be useful to enhance diagnostic accuracy. This paper aims to report the experience of the 'Caring project' utilizing a SNL in an Italian healthcare setting and to discuss the implications of shifting from 'paper' to 'electronic' records in nursing education and practice.

2 The Caring Project Experience

In 2018, a revolutionary multi-phase project for Italian nursing profession called 'Caring project' started in a North Italian Local Health Unit with the aim of providing a personalized care, creating a culture on NNN model, utilizing a unique care documentation and implementing the primary nursing model. The project involved 34 pilot Inpatients Units (IUs), in which the Switzerland philosophy of Relationship-Based Care model was introduced [18]. A retrospective observational study had been carried out on eight IUs (six medical wards and 2 Intermediate Care Facilities). Data were collected between January 2018 and March 2019 using the Gordon's FHPs paper form and the ICP in an Excel® form. ICP ('Individual Care Plan', in Italian 'Piano Assistenziale Individuale') is a clinical nursing documentation system for data collection related to NP according to the Gordon's FHP and NNN taxonomy. It is constituted by NDs, nursing outcomes, and nursing interventions that are classified in NANDA-I, Nursing Outcomes Classification (NOC), and Nursing Interventions Classification (NIC), respectively [19–22], and includes a set of 64 NDs selected from the existing 244 of the total NDs by expert facilitators (n = 34) intensively trained on FHPs and NNN taxonomy (nurses followed a six months period of training and received a manual to support an appropriate ICP compilation). Training was carried out at the beginning with a 'paper' care plan, then with data inserted directly in Excel®. However, nurses had the freedom to add to the ICP other NDs, which were originally not included. To avoid patient's identification, all data were anonymized by assigning a progressive numeric code to each patient included in the database. Patients admitted in the 15 months period in the selected IUs were included, while the exclusion criteria concerned nursing assessment not filled out. Data analysis was performed using the Statistical Package for Social Science, Version 24.0 software (IBM Corp., Armonk, NY, USA). A two-tailed *p*-values $\leq .05$ was set for statistical significance.

2.1 Results of the Caring Project

Of 255 patients: 231 were enrolled (90.6%); 131 (56.7%) were females. Their mean age was 77.5 ± 13.8 years (range 32–102), without a significant gender-based difference (males: 77.5 ± 14,3 vs. females: 77.5 ± 13.7, $p = 0.995$). Patients were distributed for 78% in Medicine Units and for the remaining 22% in the Intermediate Care Facilities. The mean number of Gordon's FHPs for patient was 1.9 (on overall 432 FHPs selected). Five main FHPs detected were *Health Perception Health Management* (45.5%), *Nutritional Metabolic* (41%), *Activity Exercise* (33.8%), *Elimination* (23.4%), *Cognitive-Perceptual* (22.1%). FHPs less selected were *Self-Perception-Self-Concept* (2.8%), *Role-Relationship* (1.4%), *Value-Belief* (0.5%) and the *Sexuality-Reproductive* (0%). A total of 484 NDs were reported corresponding to a mean of 2.0 ± 1.5 NDs/patient. Of the 64 NDs available in the ICP, 47 (76.6%) were identified as priorities and 2 NDs (*Complicated grieving* and *Availability to improve family coping*) have been added. The five most frequent NDs were *Risk for falls* (31.6%), *Imbalanced nutrition - less than body requirements* (10.4%), *Impaired physical mobility* (10.4%), *Acute pain* (10.4%) and *Dysfunctional Gastrointestinal motility* (10%). The number of NDs categorized by gender was similar (females: 2.2 ± 1.4; males 2.0 ± 1.6; $p = 0.261$).

2.2 Implication for Nursing Practice and Education

Analyzing the findings, the average number of Gordon's FHPs detected in this study (1.9), is not comparable with the literature due to the lack of supporting data. Nurses assigned 2.0 NDs/patient; this value is lower than that reported in the literature, that ranged from 4.5 to 9.4 [23–27]. In the study, it has been observed a scarce use of FHPs belonging to *Self-Perception-Self-Concept*, *Role-Relationship*, *Sexuality-Reproductive*, and *Value-Belief*. Therefore, there is a need to increase awareness among nurses on the correct form's compilation and evaluation of the psycho-social-spiritual-sexual needs (frequently neglected). *Sexuality-Reproductive* FHP was never selected by nurses and no NDs were assigned; we believe that, besides to the typology of analyzed IUs, this result is probably due to patients and nurses' tendency to raise barriers on sexual needs, as previous studies have already shown [28, 29]. The findings of 'Caring project' should be considered in the light of some limitations: retrospective data collection; convenience sample; 64 NDs in the ICP on 244 NDs available (26%); lack of nurse's familiarity with ICP; low sample size. These biases could have had led to a probable underestimation of NDs number in comparison to the literature data [24–26, 30, 31]. In regard to clinical practice, the transition from a 'paper and Excel® based health record' to an Electronic Health Records Standardized Nursing Language (EHRSNL) could allow to promote a cultural evolution from a 'biomedical' to a 'holistic' approach, interiorize clinical reasoning, simplify data collection and enhance nurses' communication [32]. In addition, using EHRSNL could allow to have useful data for research in order to highlight nursing contribution and enhance nursing outcomes using the NP. In education, using EHRSNL could allow undergraduate nursing students to facilitate interiorizing NP and improving their clinical reasoning skills and, finally, enhance diagnostic accuracy and clinical decision making [15, 33]. Therefore, in the nursing field the translation from 'paper' to 'electronic' records may have important implications both in clinical and educational settings.

3 Perspectives and Frontiers of Nursing EHR in Education

Integrating a EHR with a SNL represents an opportunity that may simplify the learning of NP; we think it could make it easier for learners to assess patients' needs using a conceptual model (e.g. Gordon's FHPs), to apply the decision making process and to link better nursing care to beneficial outcomes. For this reason, it could have a positive impact on educational and clinical outcomes. ICP used in 'Caring project' was a prototype tool and should be implemented with more sophisticated software. In Italy, a couple of software based on Gordon's FHPs and NNN taxonomy are currently available for nursing education (i.e. Florence®) and for nursing practice (i.e. Kairos®). Florence® is a virtual environment based on the updated standardized nursing record that allow students to learn actively the NP using simulated cases. In Italy, the advanced age and comorbidity of population requires a cultural and organizational nursing transformation in the health care that overcomes the biomedical approach and empowers nurses to assume more visible roles, through the improvement of their professional competences and the measurement of clinical practice. Future research needs to validate new EHRSNLs and to investigate how these could improve NDs accuracy of nursing students and nurses in their daily clinical practice.

References

1. Toney-Butler, T.J., Thayer, J.M.: Nursing process. In: StatPearls. StatPearls Publishing, Treasure Island (FL) (2020)
2. Aydin, N., Akansel, N.: Determination of accuracy of nursing diagnoses used by nursing students in their nursing care plans. Int. J. Caring Sci. **6**(2), 252–257 (2013)
3. Sanson, G., Vellone, E., Kangasniemi, M., Alvaro, R., D'Agostino, F.: Impact of nursing diagnoses on patient and organisational outcomes: a systematic literature review. J. Clin. Nurs. **26**(23–24), 3764–3783 (2017)

4. Paans, W., Sermeus, W., Nieweg, R.M.B., Van der Schans, C.P.: Prevalence of accurate nursing documentation in patient records. J. Adv. Nurs. **66**(11), 2481–2489 (2010)
5. Welton, J.M., Dismuke, C.E.: Testing an inpatient nursing intensity billing model. Policy Polit. Nurs. Pract. **9**(2), 103–111 (2008)
6. Müller-Staub, M., Lavin, M.A., Needham, I., Van Achterberg, T.: Nursing diagnoses, interventions and outcomes – application and impact on nursing practice: systematic review. J. Adv. Nurs. **56**(5), 514–531 (2006)
7. Aiken, L.H., Clarke, S.P., Sloane, D.M., Lake, E.T., Cheney, T.: Effects of hospital care environment on patient mortality and nurse outcomes. J. Nurs. Adm. **39**(7–8 SUPPL.), 223–229 (2009)
8. Petrucci, C., Calandro, M.T., Tresulti, F., Baldacchini, A., Lancia, L.: Relationship between nurse staffing and nursing outcomes: a narrative review of literature. Prof. Inferm. **68**(4), 195–202 (2015)
9. Rosenbloom, S.T., Miller, R.A., Johnson, K.B., Elkin, P.L., Brown, S.H.: Interface terminologies: facilitating direct entry of clinical data into electronic health record systems. J. Am. Med. Inform. Assoc. **13**(3), 277–288 (2006)
10. Moody, L.E., Slocumb, E., Berg, B., Jackson, D.: Electronic health records documentation in nursing: nurses' perceptions, attitudes, and preferences. Comput. Inform. Nurs. **22**(6), 337–344 (2004)
11. Flanagan, J., Weir-Hughes, D.: NANDA-I NIC and NOC, the EHR, and meaningful use. Int. J. Nurs. Knowl. **27**(4), 183 (2016)
12. Welton, J.M., Halloran, E.J.: Nursing diagnoses, diagnosis-related group, and hospital outcomes. J. Nurs. Adm. **35**(12), 541–549 (2005)
13. Sanson, G., Welton, J., Vellone, E., Cocchieri, A., Maurici, M., Zega, M., Alvaro, R., D'Agostino, F.: Enhancing the performance of predictive models for hospital mortality by adding nursing data. Int. J. Med. Inform. **125**, 79–85 (2019)
14. D'Agostino, F., Vellone, E., Cocchieri, A., Welton, J., Maurici, M., Polistena, B., Spandonaro, F., Zega, M., Alvaro, R., Sanson, G.: Nursing diagnoses as predictors of hospital length of stay: a prospective observational study. J. Nurs. Scholarsh. **51**(1), 96–105 (2019)
15. Hong, H., Park, J., Park, W.: Usage patterns of nursing diagnoses among student nurses in psychiatric unit: relation with NANDA and SNOMED CT. J. Korean Acad. Psychiatr. Ment. Health Nurs. **24**(1), 1–11 (2015)
16. Macchi, B., Sironi, C., Di Mauro, S., Ausili, D.: Using ICNP® in undergraduate nursing education with nursing theoretical models: a literature review. Prof. Inferm. **69**(3), 159–166 (2016)
17. Palese, A., De Silvestre, D., Valoppi, G., Tomietto, M.: A 10-year retrospective study of teaching nursing diagnosis to baccalaureate students in Italy. Int. J. Nurs. Terminol. Classif. **20**(2), 64–75 (2009)
18. Koloroutis, M.: Relationship-Based Care: A Model for Transforming Practice, 1st edn. Creative Health Care Management, Minneapolis (2004)
19. Gordon, M.: Nursing Diagnosis: Process and Application, 3rd edn. Mosby Inc., St. Louis (1994)
20. Herdman, H.T., Kamitsuru, S.: NANDA International Nursing Diagnoses: Definitions & Classification 2018–2020, 11th edn. Thieme, New York (2017)
21. Moorhead, S., Johnson, M., Maas, M.L., Swanson, E.: Nursing Outcomes Classification (NOC): Measurement of Health Outcomes, 6th edn. Elsevier Inc., St. Louis (2018)
22. Butcher, H.K., Bulechek, G.M., Dochterman, J.M.M., Wagner, C.M.: Nursing Interventions Classification (NIC), 7th edn. Elsevier Inc., St. Louis (2018)
23. Rosenthal, G.E., Halloran, E.J., Kiley, M., Pinkley, C., Landefeld, C.S.: Development and validation of the nursing severity index. A new method for measuring severity of illness

using nursing diagnoses. Nurses of University Hospitals of Cleveland. Med. Care **30**(12), 1127–1141 (1992)

24. D'Agostino, F., Sanson, G., Cocchieri, A., Vellone, E., Welton, J., Maurici, M., Alvaro, R., Zega, M.: Prevalence of nursing diagnoses as a measure of nursing complexity in a hospital setting. J. Adv. Nurs. **73**(9), 2129–2142 (2017)
25. Sanson, G., Alvaro, R., Cocchieri, A., Vellone, E., Welton, J., Maurici, M., Zega, M., D'Agostino, F.: Nursing diagnoses, interventions, and activities as described by a nursing minimum data set: prospective study in an oncology hospital setting. Cancer Nurs. **42**, E39–E47 (2019)
26. Castellan, C., Sluga, S., Spina, E., Sanson, G.: Nursing diagnoses, outcomes and interventions as measures of patient complexity and nursing care requirement in Intensive Care Unit. J. Adv. Nurs. **72**(6), 1273–1286 (2016)
27. Cabral, V.D.H., Andrade, Í.R.C., Melo, E.M., Cavalcante, T.D.M.C.: Prevalence of nursing diagnoses in an intensive care unit. Rev Da Rede Enferm. Do Nord. **18**(1), 84–90 (2017)
28. Zampieron, A., Aldo, S., Corso, M.: A retrospective study of nursing diagnoses, outcomes, and interventions for patients admitted to a cardiology rehabilitation unit. Int. J. Nurs. Terminol. Classif. **22**(4), 148–156 (2011)
29. Iannicelli, A.M., De Matteo, P., Vito, D., Pellecchia, E., Dodaro, C., Giallauria, F., Vigorito, C.: Use of the North American nursing diagnosis association taxonomies, nursing intervention classification, nursing outcomes classification and NANDA-NIC-NOC linkage in cardiac rehabilitation. Monaldi Arch. Chest Dis. **89**(2), 137–146 (2019)
30. Dias de Araujo, D., Rodrigues de Carvalho, R.L., Machado Chianca, T.C.: Nursing diagnoses identified in records of hospitalized elderly. Invest. Educ. Enferm. **32**(2), 225–235 (2014)
31. Jomar, R.T., de Souza Bispo, V.R.: The most common nursing diagnosis among adults/seniors hospitalised with cancer: integrative review. Ecancermedicalscience **8**, 462 (2014)
32. Carrington, J.M.: The usefulness of nursing languages to communicate a clinical event. Comput. Inform. Nurs. **30**(2), 82–88 (2012)
33. Peres, H.H., de Almeida Lopes Monteiro da Cruz, D., Lima, A.F., Gaidzinski, R.R., Ortiz, D.C., Mendes e Trindade, M., Tsukamoto, R., Batista de Oliveira, N.: Conceptualization of an electronic system for documentation of nursing diagnosis, outcomes, and intervention. Stud. Health Technol. Inform. **160**(Pt 1), 279–283 (2010)

The Perceived Usefulness of a Problem-Solving Incorporated into Blended Learning in Nursing Education: A Descriptive Study

Loredana Pasquot(✉), Letteria Consolo, and Maura Lusignani

University of Milan, Carlo Pascal 36, 20133 Milan, Italy
{loredana.pasquot,letteria.consolo,maura.lusignani}@unimi.it

Abstract. Second edition of problem-solving incorporated into blended learning was implemented in a Masters Nursing Degree. The new redesign took into account the critical issues recorded in the previous edition and made the necessary changes. The changes should facilitate students' online activities and allow them to become more involved in learning how to solve health problems in communities. It was considered interesting to investigate if the students perceived the problem-solving' usefulness in a blended learning environment. The descriptive study is based on a questionnarie completed by 24 students at the end of problem-solving. The students' perceptions of different elements were investigated both including their attitude in using the Moodle forum and worshop and the Mindomo software upload on Moodle, in learning through the sharing questions and ideas in an blended environment. The empirical results show the students' agreement on the usability and applicability of blended problem-solving, with the exception of Mindomo. The answers given by students lead to reflection on the need for a better their blended learning adaptability. In fact, students' responses are neutral, both in terms of their greater involvement in the construction of their learning, and in terms of the extension of problem-solving incorporated into blended learning to other courses

Keywords: Student's perceptions · Problem-solving · Blended learning · Moodle forum · Moodle workshop

1 Introduction

Innovations in advanced digital technology can work well with teachers, allowing them to redesign their course and have more resources to engage students in active and constructive learning.

Blended learning, combination of offline and online learning in a way that one complimenting the other is the most promising trend in higher education [1]. Technology is a core component of blended course and support online teaching and learning. The configuration of the proportion of the online compared to offline are recommended with different ranges. Mcgee [2] highlights these different proportion, by reporting the opinions of Allen, Seaman, and Garrett that suggest a range from 30 to 79% in either

Z. Kubincová et al. (Eds.): MIS4TEL 2020, AISC 1236, pp. 154–163, 2021.
https://doi.org/10.1007/978-3-030-52287-2_16

online or face-to-face; while Brown recommend a range from between 90–10 and 10–90 distributions of face-to-face and online sessions.

In the Health and Prevention Health Care in the Community course of the Masters Nursing Degree at University of Milan, five modules have been redesigned and implemented in a blended learning manner. From a pedagogical point of view, the aim was: a students' learning experience on a series of separate but synchronously taught modules [3]. The activity chosen for the students was problem- solving, articulated in an innovative and original architecture and contents. In fact, in this problem - solving students are required to work in a group and in an autonomous way, in analysing and answering to problems that are related to the contents of the five modules. The activities and interactions have been integrated between online (10%) and face-to-face (90%) and for this integration, the learning experience has been defined Problem-Solving incorporated into Blended Learning (referred to below as PSIBL).

The new redesign (second edition, a.y. 2019–2020) has taken into account the critical issues recorded in the previous edition (a.y. 2018–2019) and has made the necessary changes. The changes should facilitate students' online activities and allow them to become more involved, in learning how to solve health problems in communities.

In relation to the changes made, it was considered interesting to investigate the perceptions of students on the Problem-solving incorporated into Blended learning (PSIBL), to see if the redesign has actually solved the critical issues identified in the last edition, and their satisfaction with the teaching and learning experienced. Particularly we wanted to know:

What is the students' perceived usefulness of the PSIBL?

2 Blended Learning: Theoretical Elements

Graham [4] says that most common use of the term blended learning or hybrid Learning denotes a combination of traditional face-to-face and online instruction. Online and face-to-face are planned in a strategic manner, to engage students in a truly active learning way. In other words, this method uses online technology to not just supplement but transform and improve learning of students with varying learning styles [5].

So, it is possible to state that blended learning is the integration of three components: digital tools (e-learning), face-to-face activities (tradizional teaching and learning) and autonomous study (on student's own time).

A blended learning environment occurs through online Learning Management System (LMS), like for example Moodle and Blackboard. In these digital platforms, students can interact with teachers and other students and teachers can upload learning materials for student to study at home before the class. So, in-classroom students can be more active and share their thoughts on the topic which they have just studied at home. In this way, teachers are no longer the main source of information in class but they change role as facilitators. Main facilitators' roles are: preparing in-class and online learning materials, ensuring the connection between online and in-class students' activities, following up learning process to evaluate students' understanding and give constructive feedback.

In the blended learning, educators can focus on student understanding, rather than the delivery method itself. On the students' side, the blended learning provides students with more efficient environment, where they have more learning support by the availability of online learning facilities and more quality in-class meetings because they benefit from the online activities and materials. Multi-format resources, more time for discussion and reflection, teachers' changing role as facilitators have been obtained thanks to technology [5], which has allowed students to become more involved in the construction of their learning.

From another point of view in the blended learning approach, the online teaching and learning substitutes for class-time. The reduction in class-time and flexible online options are facilitations for many students who work [4] and can schedule a time and place to access activities and resources when they can.

An important question regards blended learning's effectiveness for helping students' learning. Generally, the researchs have found small to moderate effect size in favor of blended learning when compared with face-to-face or full-online learning [5, 6].

A meta-analysis study on effect sizes of blended learning in higher education setting by using disciplines and method of end-of course evaluation as categorital moderators, found a small summery effect size of blended learning on student achievement compared to that of tradizional teaching in-class instruction. The authors have also found no significant different effect sizes regarding the end-of course evaluation [5]. Dziuban et al. analyzed the meta-analyses conducted by Means and her colleagues in 2013, state that there is an additional concern with blended learning effect size on students' achievement [6]. The manner on which the blended learning is configurated in the studies included, for example: online instructions, e-mail, computer laboratory, mapping and scaffolding tools, electronic portfolio etc. This by no means invalidates these studies, but effect size associated with blended learning should be interpreted with caution where the impact is evaluated within a particular learning context [6].

As online learning is growing, institutions and instructors have become more interested in knowing what factors influence students' learning and satisfaction in online learning environment [7]. According to literature the elements that affect the student' satisfaction on e-learning course concern the course design, the activities of the course, the ease of using the online activities [8]. If the course material and activities are well prepared and easy to use, students feel comfortable and have greater motivation to study. A positive learning climate depends on the efficient interaction of the student and the teacher, especially by way of quick feedback and frequent interaction in the e-course [8]. More analytical factors are included in a model of evaluating student satisfaction in blended learning environment: e-learning adaptability, perceived usefulness, in-time of teacher's response, perceived ease of use and course applicability [9]. The findings of student's satisfaction are important, because if the students feel the blended learning is useful for the study, they are motivated to use it. Yet, it is evident that some students could be reluctant to use a e-learning system [8].

3 Architecture and Components of Problem-Solving Incorporated into Blended Learning (PSIBL)

In a problem-solving method the students learn by working on the problems. Problem-solving methods provide the architecture and components for implementing the reasoning part of knowledge-based systems [10]. The PSIBL has been developed in an innovative architecture of problem-solving, to seek the best way to marry the online and face-to-face activities with the need to grow capacity of analyzing problems and generating solutions from the students. Online environments and classrooms influenced problem-solving performance through tutorials, exercises, assignments, and different examples [11]. In the PSIBL, the following asynchronous online activities have been chosen to improve problem solving performance: ill problem analysis, mind map construction, peer assessment of the solutions proposed by the students. It is also stated that the problem-solving performance might be improved by examining and generating ideas through knowledge sharing, which could be organized by online tools such as a discussion board, chat room, voting, or electronic mail [11]. In the PSIBL this sharing is related to analyzing the three ill problems in an online forum and using knowledge, learned in the teachers' lecture sessions in-class, to solve them and assess among peers the solutions. Between the online forum that ends with the construction of mind maps on the three ill problems and peer assessment, students read the material uploaded online by teachers to prepare them for the classroom, where they attend face-to-face lessons. The knowledge learned prepares each student to give the answers autonomously to the three ill problems. For its architecture, PSIBL can be defined as a transformative blended learning design, i.e. as Mecgee states, "a model where learners actively construct knowledge through dynamic interactions" [2]. In fact, in the PSIBL, the students, divided in small groups, are required to demonstrate the ability to construct the mind maps in using questions and ideas raised from ill problems' analyzing. The ill problems, in contrast to being well-structured, generate perplexity, confusion or doubt [12] and open to different solutions. In this online activity, the importance of graphic organizers is significant, not only to see the relationships between ideas but also to organize and structure thoughts in a more readable and understandable manner [13, 14]. Working together allows the students to compare alternative interpretations of the ill problems and increase involvement in the learning process [15]. In the PSIBL, the learning process has also been fostered by the peer assessment for a deeper understanding of the course material [16] and a viewing of the ill problems' solution prospected by their peers.

4 Objective

In this study, the focus is on the students' subjective opinions on some aspects of the PSIBL's, in order to determine how the students, perceive the usability of the online activities and the usefulness for their learning with the online and face-to-face integrated activities.

5 Method

Design. A descriptive study was carried out by collecting data with a questionnaire, after the implementation of the PSIBL, which was redesigned in order to correct the critical issues recorded in the previous first edition (a.y. 2018–2019). The PSIBL was redesigned in September 2019 and implemented from December 2019 to January 2020, at the Master Nursing Degree at University of Milan.

All the units of meaning listed in Table 1 were used to formulate a questionnaire. The three categories of units of meaning chosen show the intention of measuring of all three impact on students' perceptions. The first unit of meaning is intended to measure the impact of the asynchronous Moodle forum, which replaced the Chat that in the first edition was evaluated as ineffective. In the Moodle forum the students worked in collaborative small groups in constructing two mind maps for each of the ill problem. In the mind maps, the students had to associate the questions and ideas, raised and shared in the forum, by Mindomo software. Mindomo was delivered on Moodle and each student of every small group, as secretary, take turns in the construction of the required six mind maps. In the activities of the asynchronous online forum, the students were divided into five groups, with five students in each. The second unit of meaning is linked to the content provided online and taught face-to-face by the teachers of the five modules involved in the PSIBL. The third unit of meaning is related to the online peer assessment in the asynchronous Moodle workshop. The peer assessment wasn't in the first edition, and it was planned to give students a further opportunity to learn by evaluating the answers of their group peers. The students uploaded their answers to Moodle workshop and each student of the small group evaluated the quality of the other's.

Table 1. Categories and units of meaning inside PSIBL

Collaborative learning in asynchronous Moodle forum Ease use of technology	Using online and face-to-face delivered contents	Learning from the peer assessment in asynchronous Moodle workshop
• Analyze ill problems • Share questions and ideas raised from ill problem • Construct mind maps with questions and ideas • Use online activities and interactions	• Materials upload online by teacher • Face-to-face teacher's sessions	• Collect each of the students' written answers to the ill problems • Evaluation criteria and scale for peer assessment

Participants. First year students of the Masters Nursing Degree. The eligibility criterion was the participation in all online and face-to-face activities of PSIBL. All twenty-five students were admitted to the study. Twenty-four students answered the questionnarie (20 females, 4 males, average age 28 years, 20 already worked as nurses or midwifes).

Students were recruited by e-mail at the end of PSIBL and they were asked to answer a questionnaire in the Moodle feedback.

Ethical Issues. The questionnaires were filled in anonymously. The data have been collected in compliance with Legislative Decree 30/06/2003, n.196. It has been declared that the data will be used exclusively for scientific research purposes.

6 Results

The research question, what is the students' perceived usefulness of the PSIBL, requires the analysis of the data collected with the questionnaire on the subjective perception of learning interactions and activities of the students. Each Likert-type question asked the student to select one of five responses that are ranked in order of strength, to measure the satisfaction rate. The students who answered the questionnaire are 24 of the 25 attending the course. Table 2 shows the percentage, the median and inter-quarter range findings of each of the questions.

Most students indicated agreement with the idea that Moodle forum is efficacy in collecting peers' questions and ideas (Question 2. Mdn = 4, IQR = 0) and in sharing these questions and ideas on three ill problems, between the peers' in the small group (Question 1. Mdn = 4, IQR = 1). Infact, most students agree that the sharing in the Moodle forum is useful to deepen the analysis of the ill problems (Question 4. Mdn = 4, IQR = 1) and the questions and ideas, raised in the small groups are effectively summarized in the mind maps (Question 5. Mdn = 4, IQR = 1). Most students, however, agree with the idea that itsn't easy to use Mindomo Software on Moodle for building mind maps (Question 3. Mdn = 2.5, IQR = 1). The peer assessment on Moodle workshop finds agreement in most student that it is an occasion in expanding learning about other answers to ill problems (Question 7. Mdn = 4, IQR = 1); but it is not known whether the evaluation criteria provided were a clear guide in judging and in scoring the peers' work, because most students have a neutral position (Question 9. Mdn = 3, IQR = 1). Regarding the online and face-to-face activities managed by the teachers, most student agree that the knowledge provided face-to-face was useful in giving answers to the three ill problems (Question 8. Mdn = 4, IQR = 0), but they are neutral on the usefulness of the material provided online to be better prepared for the class (Question 6. Mdn = 3, IQR = 1). Opinion seems to be divided with regard to sufficient information on the Moodle platform in applying the interactions and activities (Question 13. Mdn = 3, IQR = 2). Many respondents (N = 11, 46%) expressed strong agreement or agreement, but a different number (N = 7, 29%) indicated that they disagreed or strongly disagreed.

Instead, the opinion suggests consensus regard with the support in solving the difficulties of using the Moodle platform (Question 14. Md = 4, IQR = 1) and the feedback received in a timely manner on the Moodle platform (Question 16. Md = 4, IQR = 0). In two questions on the Moodle platform, opinions are neutral, both on the opportunity in improving digital literacy (Question 12. Mdn = 3, IQR = 2) and as a learning system in carrying out the interaction and activities required by blended learning (Question 15. Mdn = 3, IQR = 1). Finally, in the two general questions about PSIBL, most students seem uncertain about the opportunity to extent it to other courses (Question11. Mdn = 3, IQR = 1). Instead, there is a dissonance of opinion about this method, as an opportunity to involve students more in the construction of their own learning than traditional face-to-face lessons (Question 10. Mdn = 3, IQR = 2). Many respondents (N = 11, 46%) expressed strong agreement or agreement, but a different number (N = 7, 29%) indicated that they disagreed or strongly disagreed.

Table 2. Findings of the items of the questionnaire

Questions	% of Students					Mnd	IQR
	SD1	D2	N3	A4	SA5		
1-Asynchronous Moodle forum is adapt in sharing questions and ideas	0	0	29	59	12	**4**	**1**
2- Asynchronous Moodle forum is effective in collecting all questions and ideas	4	0	8	79	**9**	**4**	**0**
3- Mindomo software on Moodle is easy to use in building mind maps	4	42	33	17	4	**2.5**	**1**
4- Asynchronous Moodle forum is useful to deepen the analysis of ill problem	0	0	17	54	29	**4**	**1**
5- Mind maps built with Mindomo summarize well the questions and ideas	0	0	17	54	29	**4**	**1**
6- It is useful to read the online material before the class.	4	17	50	29	0	**3**	**1**
7- Peer evaluation in Moodle workshop is useful in expanding learning about other answers to ill problems	0	8	21	58	13	**4**	**1**
8- Teachers provide knowledge useful in writing ill problem answers (in-class)	0	0	20	63	17	**4**	**0**
9- Evaluation criteria provide clear guidance in assigning judgement and score	4	17	33	46	0	**3**	**1**
10- PBLI involves the student more in constructing their own learning than traditional lesson	8	21	25	42	4	**3**	**2**
11- It is useful to extend PSIBL to other courses	13	33	29	21	4	**3**	**1**
12- Use of the Moodle was useful in improving my "digital literacy" skills	8	21	29	34	8	**3**	**2**
13- I received sufficient information on Moodle	13	16	25	37	9	**3**	**2**
14- I received adequate support in using the Moodle platform	4	4	17	54	21	**4**	**1**
15-Moodle makes the interactions and activities in blended learning easier	0	8	50	42	0	**3**	**1**
16- I received the required feedback in a timely manner on the Moodle platform	0	0	8	71	21	**4**	**0**

SD = Strongly Disagree; D = Disagree; N = Neutral; A = Agree; SA = Strongly Agree; Mdn = Median; IQR = Inter Quarter Range

7 Discussion

Reflecting on the findings that were generated from a single Likert-type items it is necessary to discuss them with caution.

The two answers (questions 1 and 2) on the Moodle forum (Mdn = 4, IGR = 1; Mdn = 4, IQR = 0) seem to indicate that the online environment created a sense of community, where students could express and share their opinion. The online forum seems to be further confirmed effective for collaborative learning with the answers to question 4, indicating that students feel the forum is useful to deepening the analysis of the ill problems (Mdn = 4, IQR = 1). The findings of the question 7 show students' agreement with the idea that the Moodle workshop is useful in expanding their learning about other answers to ill problems (Mdn = 4, IQR = 1). The agreement of many of the students on the answers to the questions 1, 2, 4 e 7 seem, therefore, to indicate a substantial usefulness of the online environment for both collaborative learning and individual learning through the evaluation of peers' work. By contrast, there is no substantial convergence of opinion, as would have been expected, in perceiving that problem-solving in blended involves the student more in constructing their own learning than traditional lesson (Question 10. Mdn = 3, IQR = 2) and only the 46% of respondents agreed or strongly agreed. Much more the contrast emerges from the answers to question 11, where the students converge in neutral opinion on extending PSIBL to other courses (Mdn = 3, IQR = 1), and the 46% (11 students) expressed strong disagreement or disagreement, and only 21% (6 students) indicated that they agreed or strongly agreed. These findings (question 10 e 11) could be related to factors affecting students' blended learning adaptability or PSIBL applicability. The Mindomo software could have affected the student perception of the PSIBL applicability. The students infact seem disagree about the ease in using it to build the mind maps (Mdn = 2.5, IQR = 1). It cannot be excluded the influence on the student perception of the second factor, blended learning adaptability, as the problem-solving is the first blended learning experience for the students, and all the other courses of the Nursing Master Degree are carried out with the traditional in-class lessons. Thus, many of students' neutral opinion to questions (6, 12, 13, 15) could depend on the student's internal contrast between understanding the importance of digital innovation, commitment to learning to use digital tools and changing one's habits. In contrast to these neutral responses most students agree with the support to overcome the difficulties in using Moodle platform (Question 14. Mdn = 4, IQR = 1), in receiving feedback in a timely manner on the Moodle platform (Question 16, Mdn 4, IQR = 0), and on the usefulness of the content transmitted by the teachers in giving the answers to the problems (Question 8, Mdn = 4, IQR = 0). These findings (Question 14, 16, 8) seems provide positive perceptions with regard PSIBL applicability.

8 Conclusion

Measuring student perceptions was important to highlight the strong and weak points of the course for further improvement. The interpretation of students' perceptions seems to indicate that the critical issues of the first edition of the problem-solving in BL have been solved, with regard to perceived of PSIBL applicability. The questions asked to the students concerned the online and the face-to-face interactions and activities. The students found the asynchronous Moodle forum and workshop to be effective and useful in collaborative learning and autonomous learning. But, the analysis of the students' opinions, measured with Likert scale, showed their disagreement on Mindomo Software, element of problem-solving applicability, thus it is necessary a greater training in the use of the Mindomo software to build mind maps on the ill problems. Some neutral answers given by students lead to reflection on the need for their better blended learning adaptability. The improvements of the students blended learning adaptability requires greater and more engaging information action in using of Moodle platform for problem-solving. Satisfaction with the technical support given positively influence students' acceptance of e-learning. On the other hand, the blended experience is a novel approach in this Nursing Master Degree, and because of this the students may have some resistance or difficulty adapting to a blended learning.

These findings provide the perceptions of a small number of students (24) in a local setting. Adding the correlation between perceived usefulness and the success of students in the final exam could give a better insight into the effectiveness of e-learning. Despite this limitation, the findings could provide elements of reflection and management for those who wish to apply a PSIBL in other contexts.

References

1. Castro, R.: Blended learning in higher education: trends and capabilities. Educ. Inf. Technol. **24**, 2523–2546 (2019). https://doi.org/10.1007/s10639-019-09886-3
2. Mcgee, P., Reis, A.: Blended course design: a synthesis of best practice. J. Asynchronous Learn. Netw. **16**(4), 1–22 (2012)
3. Carnell, B., Fung, D.: Developing the Higher Education Curriculum. UCL Press, London (2017). https://doi.org/10.14324/111.9781787350878
4. Graham, C.R.: Emerging practice and research in blended. In: Handbook of Distance Education, 3rd edn., pp. 333–350. Routledge, New York (2013)
5. Hien, M.V., Chang, Z., Nguyet, A.D.: The effect of blended learning on student performance at course-level in higher education: a meta-analysis. Stud. Educ. Eval. **53**, 17–28 (2017). https://doi.org/10.1016/j.stueduc.2017.01.002
6. Dziuban, C., Graham, C.R., et al.: Blended learning: the new normal and emerging technologies. Int. J. Educ. Technol. High. Educ. **15**(1), 3 (2018). https://doi.org/10.1186/s41239-017-0087-5
7. Ghaderizefreh, S., Hoover, M.L.: Student satisfaction with online learning in a blended course. Int. J. Dig. Soc. **9**(3), 1393–1398 (2018)
8. Keržič, D., Tomaževič, N., Aristovnik, A., Umek, L.: Exploring critical factors of the perceived usefulness of blended learning for higher education students (2019). https://doi.org/10.1371/journal.pone.0223767. Accessed 21 Mar 2020
9. Zhao, G., Yuan, S.: Key factors of effecting blended learning satisfaction: a study on Peking university students. In: Tsang, P., Cheung, S.K.S., Lee, V.S.K., Huang, R. (eds.) Hybrid Learning, ICHL 2010, Lecture Notes in Computer Science, vol. 6248. Springer, Heidelberg (2010)
10. Fensel, D., Motta, E., Van Harmelen, F., et al.: The unified problem-solving method development language UPML. Knowl. Inf. Syst. **5**(1), 83–131 (2003). https://doi.org/10.1007/s10115-002-0074-5
11. Samoekan, S.: Factors affecting creative problem solving in the blended learning environment: a review of the literature. Proceed-Soc. Behav. Sci. **174**, 2130–2136 (2015). https://doi.org/10.1016/j.sbspro.2015.02.012
12. Yew, E.H.J., Goh, K.: Problem-based learning: an overview of its process and impact on learning. Health Prof. Educ. **2**(2), 75–79 (2016)

13. Vázquez-Cano, E., López-Meneses, E., Fernández-Márquez, E.: Concept mapping for developing competencies in European higher education. Int. J. Humanit. Soc. Sci. **3**(17), 1–12 (2013)
14. Schwartz, B.M., Gurung, R.A.R.: Evidence-Based Teaching for Higher Education. American Psychological Association, USA (2012)
15. Biggs, J., Tang, C.: Teaching for Quality Learning at University, 4th edn. MacGraw-Hill, Worcester (2008)
16. Amendola, D., Miceli, C.: Online peer assessment to improve students' learning outcomes and soft skills. Ital. J. Educ. Technol. **26**(3), 71–84 (2018). https://doi.org/10.17471/2499-4324/1009

Authoring Interactive-Video Exercises with ELEVATE: The NLS Procedure Case Study

Daniele Dellagiacoma[1(✉)], Paolo Busetta[1], Artem Gabbasov[2], Anna Perini[2], Angelo Susi[2], Eugenio Gabardi[3], Francesco Palmisano[3], Caterina Masè[3], and Cristina Moletta[3]

[1] Delta Informatica SpA, Trento, Italy
{daniele.dellagiacoma,paolo.busetta}@deltainformatica.eu
[2] Fondazione Bruno Kessler (FBK), Trento, Italy
{agabbasov,perini,susi}@fbk.eu
[3] Azienda Provinciale per i Servizi Sanitari (APSS), Trento, Italy
{eugenio.gabardi,francesco.palmisano,caterina.mase, cristina.moletta}@apss.tn.it

Abstract. Interactive videos are becoming common in distance learning, especially for training on procedures to be executed in critical situations, as a complement to simulation sessions in real settings. To this end, research in educational technology devotes attention to the definition of methodologies and authoring tools for teachers and instructional designers. The ELEVATE project is developing a tool suite that offers a novel environment for authoring interactive videos to be deployed and used on popular Learning Management Systems (LMS), e.g. Moodle. One of the project case studies concerns continuous training of obstetric and neonatal teams on the Newborn Life Support (NLS) procedures. This case study helps us identify key requirements for a methodology to support instructional designers while creating and managing interactive video exercises. The latter are to be delivered as part of the blended course material made available to professionals on an LMS. This paper presents the NLS procedures case study to illustrate the ELEVATE tool-supported authoring approach.

Keywords: Authoring methodology · Authoring tool · Case study · NLS procedure · Interactive videos · Distance learning · Healthcare students · Midwives training

1 Introduction

Interactive videos are used more and more in distance learning, especially for training on procedural knowledge [9,11]. Their effectiveness is largely investigated, taking into account peculiarities of knowledge domains, characteristics of target students, as well as teacher perspectives, as reported in literature surveys [5,6,13].

Z. Kubincová et al. (Eds.): MIS4TEL 2020, AISC 1236, pp. 164–174, 2021.
https://doi.org/10.1007/978-3-030-52287-2_17

In the context of the ELEVATE (E-LEarning with Virtual interAcTive Experience)[1] project, we focus on interactive videos for training professionals on the execution of emergency management procedures. In our work, an interactive video for training consists of video-recorded scenes and decision points. At each decision point, actions of the trainee who is playing the video determine which scenes are played next. Such interactive videos are used in distance learning as a complement to simulation sessions that are conducted in real setting.

The main objective of the project is to develop a flexible software system, called *ELEVATE Tool Suite*, which supports a collaborative authoring process for the creation and management of interactive videos, and where video-clips can be produced either in real settings or using virtual reality. The tool suite has been described in detail in [3], while in this paper we focus on the tool at work on a real case. That is, we describe key steps of the ELEVATE authoring process that we performed on a project case study carried out in the context of *APSS* (Azienda Provinciale per i Servizi Sanitari), the organization providing public healthcare services and managing the major local hospitals, who acts as a project partner. Professional midwives employed by *APSS* have to periodically train on the Newborn Life Support (NLS) procedure concerned by the case study (e.g. [12]).

Within this ELEVATE case study, we have experimented tools and methodologies for the creation of exercises on decision-making in critical situations. Further experimentation with *APSS* midwives is in progress at the time of writing.

The rest of this paper is organized as follows. In Sect. 2 we shortly describe the ELEVATE authoring tool-suite, as developed so far. In Sect. 3 we present the case study regarding continuous training of professionals on NLS procedures. In Sect. 4 we present the ELEVATE approach for developing interactive video exercise, using the NLS procedures case as illustrative case study. We summarise related work in Sect. 5. Section 6 points out future work and provides concluding remarks.

2 The ELEVATE Tool Suite

The ELEVATE project concerns a tool suite for creating and delivering interactive videos. ELEVATE follows a previous project that explored the use of virtual reality (VR) and "intelligent" behaviours of artificial characters for training purposes [1,4]. While VR exercises are technically appealing and engaging for trainees, costs and skills required to produce and run them are too high for most purposes and most healthcare organizations. ELEVATE investigates novel ways to produce and interact with training videos that keep budget and technical skills within reach of an average organization, while supporting massive distribution and engaging their target audience. In short, an ELEVATE video contains a

[1] The ELEVATE project - https://elevate.deltainformatica.eu/, is a research and innovation project, proposed and coordinated by Delta Informatica, and funded by Provincia Autonoma di Trento, L.P. 6/1999.

Fig. 1. Main components of the *ELEVATE Tool Suite*

number of decision points, shown to the trainee as actions that can be applied during the course of the video; the trainee's choices (or lack thereof) determine which scenes are to be played. An instructional designer focuses on the decision-making to be learned or verified by means of the video, while the recording of scenes is left to people familiar with video and multi-media authoring tools. These may use VR to produce clips. Data collected during video playing allows both the evaluation of the performance of a student and the refinement of the videos.

The main components of the *ELEVATE Tool Suite* [2,3] are summarized in Fig. 1, together with an overview of the supported authoring and exercise management process. After the training requirements are defined by the responsible of a course and shared with an instructional designer, the EDT (Exercise Design Tool) allows the creation of the structure of an exercise, which resembles a decision tree, but supports cycles and context-dependent choices; an example is shown in Fig. 2. The instructional designer uses EDT to delegate the production of clips to a video director. Videos stored in a media library can be used in many exercises, thus allowing the creation of variants or entirely new exercises from the same material, e.g. targeting different audiences or satisfying different training objectives. The video director produces clips with the environment deemed the most suitable, including live recording or another component of the suite, EVE (ELEVATE Video Editor). EVE allows the creation of a virtual environment by means of a game engine, its animation with artificial intelligence-enhanced virtual characters and a simple script, and the recording of the resulting scenes [2]. Trainees play video exercises with an ELEVATE video player that can be embedded in any Learning Management System (e.g., Fig. 1 shows Moodle) supporting SCORM (Sharable Content Object Reference Model). The data collected during exercises, which include decisions as well as key performance indicators, such as the time taken to select choices, can be queried by a trainer or a course supervisor by means of EMT (Exercise Management Tool). In addition to various course management tasks, EMT allows the analysis both of individual sessions and of aggregated data over a number of runs and users.

3 The NLS Procedure Use Case

The *APSS* department in charge of the management of continuous education and training of professionals follows standard guidelines for education in the healthcare domain, especially concerning training on critical emergency management procedures, such as those described in [12]. The adopted educational approach recognizes a central role to practical experience that professionals have to develop in dedicated simulation sessions, which are performed in realistic settings. Distance learning is also considered, but mainly for teaching descriptive knowledge regarding the domain of interest. Exploring how to exploit distance learning also for teaching procedural knowledge, as an effective complement to simulations in realistic setting, is the underlying motivation of the ELEVATE project.

Specifically, the case study considered in ELEVATE concerns the procedure for reanimating newborns, which is derived from the international guidelines proposed by ILCOR[2], ERC[3] and IRC[4] and described in the NLS algorithm of 2015 [12]. Healthcare professionals, such as midwives, have to periodically train on this procedure. They can access textual learning material online and must attend training sessions, upon schedule.

ELEVATE aims at introducing a new training possibility that can extend the already existing educational program. That is, the online training exercises provided by ELEVATE can complement the training performed in real setting, by allowing trainees to play online interactive videos. During this experience, a trainee is proposed a chain of scenarios in which she is requested to take decisions on actions to be taken, according to the reanimation procedure to be learned. These exercises have to be created by instructional designers, together with domain experts, who play a key role in the collaborative authoring process that is supported by the *ELEVATE Tool Suite*. Once deployed on the Moodle platform as part of a blended course, the exercises are used by the trainees. Anonymised data that can be extracted from session logs can be analysed for improving the exercises.

4 Authoring Interactive Video Exercises

The collaborative authoring process, which is supported by the *ELEVATE Tool Suite* is composed of a set of steps, including the definition of training requirements, the design and validation of appropriate video-based exercises, their deployment on an LMS, and the analysis of data extracted from session logs with the aim of understanding if the exercises need to be improved. The process is meant to be flexible, supporting iterative development, and traceability, a feature that becomes necessary to be able to manage change requests in an

[2] ILCOR (International Liaison Committee on Resuscitation) - https://www.ilcor.org/.

[3] ERC (European Resuscitation Council) - https://www.erc.edu/.

[4] IRC (Italian Resuscitation Council) - https://www.ircouncil.it/.

Table 1. Requirements for the training exercise on the NLS procedure.

Goal	Rule to be trained on	Target	Author	Source
G1 - Train professional to place the head of a newborn correctly	The head of a newborn in apnea should be put in the neutral position	APSS midwives	APSS instructional designer	NLS algorithm [12], IRC
G2 - Train professional to choose between inflation breaths and cardiac massage	A newborn in apnea with a heart rate below 60 bpm should be treated with 5 inflation breaths	APSS midwives	APSS instructional designer	NLS algorithm [12], IRC

easy and efficient way. In this section, we focus on the first two key steps and illustrate them with the help of the NLS procedure case study.

4.1 Step 1: Definition of Training Requirements

The objective of this first step is to understand if video-based exercises are needed, e.g. in a given organization training program. Usually domain experts and instructional designers meet and discuss guided by questions such as: (i) what's the specific knowledge domain to be considered and what are the *sources* providing validated knowledge (e.g. manuals, guidelines, training specifications) to be taught? (ii) who are the key *stakeholders* for the educational material to be developed? E.g. the organization that will adopt the educational material for the professional training of its employees, the professional profile of the intended trainees, if the exercises will be used as part of blended courses; (iii) once the procedural knowledge and the target trainee profiles have been identified, what are the learning objectives?

The extracted information can be organized in a table such as Table 1, where *Target* indicates the target trainees, *Author* the name of people involved in the production of the exercise, *Source* the relevant related documents for the exercise to be designed, as protocols and manuals of the training domain under consideration. This information will be included as descriptive information (meta-data) when creating the new exercise project with *ELEVATE Tool Suite*. Other information elicited as part of key requirements for the exercise are the learning goals and the associated rules, that is the behaviour a trainee should be trained on in order to achieve the associated learning objective.

With reference to the NLS case study, the algorithm proposed by IRC (adapted from ERC [12]) has been selected as a starting point. The algorithm is included in the theoretical material, as well as NLS guidelines that the trainees need to learn before the practice in classroom. The algorithm allows to consider three main cases: apnea, cardiac massage and drugs. In Table 1, the first case is considered as an example, where learning objectives are described by two goals, which aim at covering the first part of the newborn life support algorithm, i. e. teaching the trainees the correct position of the head and deciding correctly between inflation breaths and cardiac massage.

The training requirements definition has been carried out by APSS instructional designer and domain experts (e.g., IRC instructors). Moreover, it has been decided that the target of the NLS case study within the ELEVATE project are the midwives, who require a continuous training. This first step, i.e. the definition of the training requirements, has taken around 2 h to be accomplished.

Table 2. Scenarios for the exercise.

ID		S1		S2	...
Description		The newborn's head should be put in some position just after the birth		The newborn should be treated with some method	...
Characte-ristics	**Head position**	flexed		whichever (doesn't matter)	...
	Heart rate	< 60 bpm		< 60 bpm	...
	...	...		...	...
Correct behavior	**ID**	S1CA1		S2CA1	...
	Action	Put the head to the neutral position		Perform 5 inflation breaths	...
	Consequences	It's easier for the baby to breathe		The baby starts to breathe better	...
Incorrect behavior	**ID**	S1IA1	S1IA2	S2IA1	...
	Action	Put the head to the extended position	Leave the head in its initial position	Perform the cardiac massage	...
	Consequences	It's harder for the baby to breathe	It's harder for the baby to breathe	No significant improvement, wasted time	...
Traceability links		G1		G2	...

4.2 Step 2: Interactive-Video Exercise Implementation

Exercise design activities are performed by the instructional designer, upon the analysis of the training requirements, which are followed by video-clips shooting with the help of the director, and association of the videos to the exercise.

Step 2.1. Exercise Objectives and Scenarios Definition. The instructional designer analyses the identified goals (learning objectives) in order to design the scenarios that correspond to them. The scenarios are described in a structured narrative, such as the one presented in Table 2. Each scenario has a unique ID and some specified characteristics that can vary during the process and from context to context. The set of characteristics depends on the domain and on the context of the exercise. In addition, for every scenario the corresponding goals should be added for traceability reasons.

Step 2.2. Exercise Story Definition. The aforementioned information helps define the basic structure of the interactive training video content. We apply suggestion on how to derive an interactive video exercise for procedural knowledge provided in [8], namely we first consider the story that corresponds to the nominal procedure as a linear video. Top of Fig. 3 shows the linear process where there are no decision points. In this case, it represents the standard NLS algorithm. Decision points are then identified on the linear story, with associated branches that can be then designed. The lower part of Fig. 3 shows the interactive videos flow where decision points and branches have been added. For instance, after the baby has been dried, the story splits in three branches that represent different

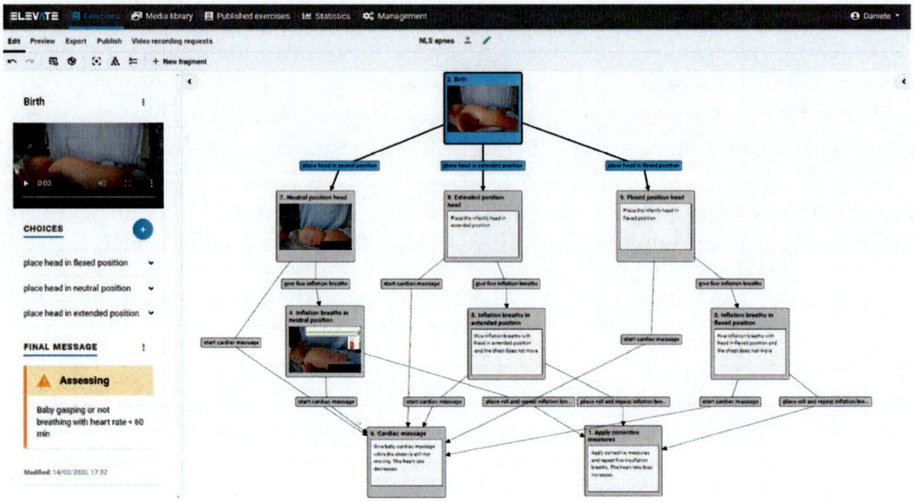

Fig. 2. Exercise design tool snapshot.

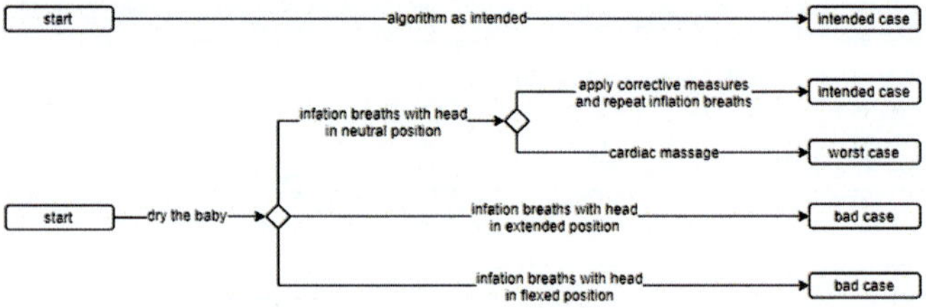

Fig. 3. NLS apnea flowchart.

head positions. Only one branch, the neutral position of the head, is correct whereas the other two lead to a bad ending.

The instructional designer implements the structure of the exercise using the EDT, as depicted in the snapshot of the graphical user interface reported in Fig. 2. When the structure of the exercise has been implemented, each node has to be linked to a video clip. Decision points do not require any video content but are implemented through a proper interactive interface. Appropriate video clips can already be found in the media library of the ELEVATE Tool Suite. In this case, the exercise designer browses the existing videos and selects the most appropriate one for the exercise. Otherwise, if new video clips must be recorded, the director is involved in the process. Here, storyboarding should be used to explain the director what is needed.

Step 2.3. Video Recording. Storyboarding has been used for video in interactive environments for a long time, e.g. [10]. A storyboard can provide a first specification of fragments of videos to be realized. It is usually recommended to do it for two main purposes, namely to ensure the definition of all the scenes that will be recorded, getting a full coverage of the exercise structure, and to save money and time during the implementation phase by avoiding filming unnecessary shots.

With the help of the scenario specification (Table 2), we can easily identify all the scenes that will compose the final training material: each scene corresponds to a fragment in Table 2 and vice versa. We can derive some key information about the fragments, such as the locations and the needed actors of each scene. It allows to address immediately the issues related to privacy and publishing the material.

Table 3. Example of the storyboard.

ID	ST1	...	ST3	ST4	...
Description	A baby in apnea is presented; The baby's head is in the flexed position	...	The midwife performs 5 inflation breaths	The midwife performs cardiac massage	...
Location	Delivery room	...	Delivery room	Delivery room table	...
Actors	Midwife, baby	...	Midwife, baby	Midwife, baby	...
Objects	—	...	Ventilation device	—	...
Sketch		...			...

The director uses the storyboard created by the instructional designer in this step, and takes care of recording the missing video clips for the exercise. The video clips can be recorded either manually in real-life setting or generated by using a video creation tool such as EVE. The director requires the specification which describes the various scenes to be recorded. An example of storyboard for NLS apnea is shown in Table 3. Here, every video clip to be recorded is described through a short description, the location, actors and objects involved. Moreover, the instructional designer can help the director providing sketches for the different scenes to be recorded.

In the NLS case study, all the videos have been recorded in a real-life setting. All the scenes have been recorded more than once in order to select the best shots later. Recording all the videos for the three exercises (i.e., apnea, cardiac massage and drugs) took around 6 h. Replicating the NLS videos, concerning only the apnea part, in a VR environment has taken more than 40 h of work. Most of the VR assets can be reused for the other exercises. Table 4 presents the summary of relevant information collected during the realization of the three exercises for the NLS procedure case study with the *ELEVATE Tool Suite*. Note that some video clips were reused inside and even across the exercises.

Table 4. Summary of the NLS case study.

Exercise	Apnea	Cardiac massage	Drugs	Total
Goals	2	4	3	9
Scenarios	2	4	4	10
Nodes (video fragments)	9	18	10	37
Decision points/choices	7/15	9/18	5/10	21/43
Unique video clips used	9	15	7	31

5 Related Work

Research literature on video-based learning focuses on many different topics, ranging from the analysis of distance learning contexts in which its application can be mostly beneficial [7] to learning analytics with video-based learning, e.g. [5]. Particularly relevant for the work discussed in this paper are studies on interactive video production tools (e.g. the literature survey reported in [13]), which distinguish between video-based learning environments, in which annotation of video is used as principal mechanism, and authoring tools. The *ELEVATE Tool Suite* falls into this second type of environment, and aims at supporting the capabilities that were not so largely addressed by the time considered in the survey (2013), i.e. the creation and management of interactive training exercises to be used in distance learning.

Focusing on how to develop high quality video-based learning material, and on key aspects to be taken into account, in [5] it is recommended to consider and combine two separate perspectives: the technological perspective, which addresses coherency, cohesion and user-friendliness of the software platform interface and of the interaction it enables; and the target audience perspective, i.e. cognitive aspects, personality traits, as well as demographic and cultural characteristics of the students. A more recent study conducted by Wijnker et al. [11], discusses two important practical implications for teachers willing to make or select educational videos, which have been derived from an empirical study involving both teachers and students, using educational videos. The first concerns the goal that teachers intend to achieve by using a video, which should be made explicit. The second, the fact that film theory should be exploited when producing educational video. In ELEVATE we attempt to consider both recommendations.

6 Conclusion and Future Work

This paper reports on the work in progress within the ELEVATE project. ELEVATE concerns a tool suite, described in more detail in other papers (submitted or in publication), for the creation of interactive exercises as collaborations of instructional designers and video authors, who are offered the option of producing video in virtual reality; trainees do their exercises by means of a purposefully made video player, which can be embedded in any standard LMS and that collects data on their performance for further analysis.

This paper has focused on a specific use case, Newborn Life Support training for professional midwives, within which we have experimented a development methodology for interactive videos. This work is done in collaboration with a major healthcare partner whose staff includes an instructional designer, a video director for real-life shooting, and many students (professionals) involved in an experimentation during blended courses. The current tool suite and methodology have had a very good reception so far, both in terms of usability and perceived training value with respect to online courses produced with state-of-the-art authoring and data analysis tools.

While experimentation with trainees is still in progress, the methodology is undergoing a refinement at the light of the lessons learned and will be further experimented in future use cases. ELEVATE is potentially applicable to many types of nurse training requiring experiential learning. Simulation in laboratory or real-life settings is often difficult or impractical, e.g. because of the number of cases to be tried out; an exhaustive set of interactive exercises would complement hands-on experience on a representative number of cases.

References

1. Busetta, P., Dragoni, M.: Composing cognitive agents from behavioural models in presto. In: Proceedings of the 16th Workshop "From Objects to Agents", Naples, Italy, 17–19 June (2015). http://ceur-ws.org/Vol-1382/paper13.pdf
2. Busetta, P., Astegher, M., Dellagiacoma, D., Gabrielli, S., Longato, M., Pedrotti, M.: Artificial actor behaviours for interactive videos: handling AI requirements in ELEVATE. In: RE4AI Workshop (accepted, to be published as part of the online Proceedings of the REFSQ2020 Conference), March 2020
3. Dellagiacoma, D., Busetta, P., Gabbasov, A., Perini, A., Susi, A.: Authoring interactive videos for e-learning: the ELEVATE tool suite. In: 10th International Conference in MIS4TEL. Springer (2020)
4. Dragoni, M., Ghidini, C., Busetta, P., Fruet, M., Pedrotti, M.: Using ontologies for modeling virtual reality scenarios. In: Proceedings of the 2th European Semantic Web Conference, ESWC 2015. LNCS, vol. 9088, pp. 575–590. Springer (2015)
5. Giannakos, M.N., Chorianopoulos, K., Ronchetti, M., Szegedi, P., Teasley, S.D.: Video-based learning and open online courses. iJET **9**(1), 4–7 (2014). https://www.online-journals.org/index.php/i-jet/article/view/3354
6. Kay, R.H.: Exploring the use of video podcasts in education: a comprehensive review of the literature. Comput. Hum. Behav. **28**(3), 820–831 (2012)
7. Kolås, L.: Application of interactive videos in education. In: 2015 International Conference on Information Technology Based Higher Education and Training (ITHET), pp. 1–6. IEEE (2015)
8. Schneider, K., Bertolli, L.M.: Video variants for CrowdRE: how to create linear videos, vision videos, and interactive videos. In: 2019 IEEE 27th International Requirements Engineering Conference Workshops (REW), pp. 186–192. IEEE (2019)
9. Schwan, S., Riempp, R.: The cognitive benefits of interactive videos: learning to tie nautical knots. Learn. Instr. **14**(3), 293–305 (2004)

10. Truong, K.N., Hayes, G.R., Abowd, G.D.: Storyboarding: an empirical determination of best practices and effective guidelines. In: Proceedings of the 6th Conference on Designing Interactive Systems, pp. 12–21 (2006)
11. Wijnker, W., Bakker, A., van Gog, T., Drijvers, P.: Educational videos from a film theory perspective: relating teacher aims to video characteristics. Br. J. Educ. Technol. **50**(6), 3175–3197 (2019)
12. Wyllie, J., Bruinenberg, J., Roehr, C.C., Rüdiger, M., Trevisanuto, D., Urlesberger, B.: European resuscitation council guidelines for resuscitation 2015: Section 7. resuscitation and support of transition of babies at birth (2015)
13. Yousef, A.M.F., Chatti, M.A., Schroeder, U.: The state of video-based learning: a review and future perspectives. Int. J. Adv. Life Sci. **6**(3/4), 122–135 (2014)

Workshop on Social and Personal Computing for Web-Supported Learning Communities (SPeL)

Workshop on Social and Personal Computing for Web-Supported Learning Communities, SPeL

Web-based learning is moving from centralized, institution-based systems to a decentralized and informal creation and sharing of knowledge. Social software, e.g., blogs, wikis, social bookmarking systems, media sharing services is increasingly being used for e-learning purposes, helping to create novel learning experiences and knowledge. In the world of pervasive Internet, learners are also evolving: the so-called "digital natives" want to be in constant communication with their peers, they expect an individualized instruction and a personalized learning environment, which automatically adapt to their individual needs. The challenge in this context is to provide intelligent and adaptive support for collaborative learning, taking into consideration the individual differences between learners.

This workshop deals with current research on the interplay between collaboration and personalization issues for supporting intelligent learning environments. Its aim is to provide a forum for discussing new trends and initiatives in this area, including research about the planning, development, application, and evaluation of intelligent learning environments, where people can learn together in a personalized way through social interaction with other learners.

The workshop is targeted at academic researchers, developers, educationists and practitioners interested in innovative uses of social media and adaptation techniques for the advancement of intelligent learning environments. The proposed field is interdisciplinary and very dynamic, taking into account the recent advent of Web 2.0 and ubiquitous personalization, and it is hoped to attract a large audience.

The topics of interest include:

- Social learning environments
- Theory and modeling of social computing in education
- Web 2.0 tools for collaborative learning
- Personal learning environments
- Lifelong learning networks
- Virtual spaces for learning communities
- Social networks analysis and mining
- Computer-supported collaborative learning
- Personalized and adaptive learning
- Adaptation methods and techniques for groups of learners
- Intelligent learner and group modeling
- Collaborative filtering and recommendations for learners
- Game-based social learning
- Personalized mobile learning applications

- Cloud-based social learning
- Intelligent agent technology for social learning
- Metadata, folksonomies and tagging
- Semantic web and ontologies for personalized learning
- Cognitive, motivational and affective aspects for personalization
- Practice and experience sharing

This year's edition of the workshop includes a total of 6 papers, covering topics such as: computer-supported collaborative learning, recommender systems in education, game-based social learning, practice and experience sharing.

Organization

Workshop Organizers and PC Chairs

Elvira Popescu	University of Craiova, Romania
Sabine Graf	Athabasca University, Canada

Program Committee

Marie-Hélène Abel	Université de Technologie de Compiègne, France
Yacine Atif	Skövde University, Sweden
Kaushal Kumar Bhagat	Indian Institute of Technology, Kharagpur, India
Tharrenos Bratitsis	University of Western Macedonia, Greece
Maria-Iuliana Dascalu	Politehnica University of Bucharest, Romania
Mihai Dascalu	Politehnica University of Bucharest, Romania
Giuliana Dettori	Institute for Educational Technology, ITD-CNR, Italy
Gabriela Grosseck	West University of Timisoara, Romania
Hazra Imran	Athabasca University, Canada
Mirjana Ivanovic	University of Novi Sad, Serbia
Ioannis Kazanidis	Technological Educational Institute of Kavala, Greece
Milos Kravcik	German Research Center for Artificial Intelligence, Germany
Zuzana Kubincová	Comenius University Bratislava, Slovakia
Amruth Kumar	Ramapo College of New Jersey, USA
Frederick Li	University of Durham, UK
Anna Mavroudi	Norwegian University of Science and Technology, Nor-way

Wolfgang Mueller	University of Education Weingarten, Germany
Alexandros Paramythis	Contexity AG, Switzerland
Galena Pisoni	University of Trento, Italy
Ricardo Queirós	Polytechnic Institute of Porto, Portugal
Demetrios Sampson	Curtin University, Australia
Olga Santos	Spanish National University for Distance Education, Spain
Marco Temperini	Sapienza University of Rome, Italy
Stefan Trausan-Matu	Politehnica University of Bucharest, Romania
Rémi Venant	University of Toulouse, France
Riina Vuorikari	Institute for Prospective Technological Studies, IPTS, European Commission

Extending and Evaluating a Collaborative Note-Taking Application: A Pilot Study

Elvira Popescu(✉), Sorin Ilie, and Constantin Stefan

Computers and Information Technology Department, University of Craiova, Craiova, Romania
{popescu_elvira,ilie_sorin}@software.ucv.ro

Abstract. Taking notes during lectures helps students filter information, record key points, clarify ideas and be more actively engaged in learning. This is particularly important in face-to-face lecture-based scenarios supported by slides, when students need to complete instructor handouts with their personal notes in order to facilitate understanding and foster involvement. In order to offer an alternative for handwritten notes or generic note-taking apps, we developed a mobile application with pedagogically grounded functionalities, called EduNotes, which allows students to collaborate with their peers during the note-taking process. In the current paper, we present an extension of the platform with a web version, which supports the simultaneous presentation of the lecture slides alongside the notes, together with a more advanced note management functionality. Both EduNotes versions were successfully used in an experimental study involving 38 students, which is reported in the paper.

Keywords: Note-taking · Collaborative learning · Lecture-based instruction

1 Introduction

Note-taking is an important process in education, by means of which students record key ideas and concepts [11]. Beside the role of "external memories", notes lead to an "internal storage" by contributing to memorization [3, 7]. Note-taking helps students personalize knowledge construction, building connections between the newly received information and their existing knowledge [12]. Furthermore, it fosters active engagement during lectures [14] and encourages reflection and filtering of the lecture content [3].

The traditional lecture-based scenario, in which teachers deliver lectures supported by slides, is still a very popular instructional approach. In this context, even if students are provided with the presentation slides, they still need to take their own notes to facilitate understanding and higher-order learning [1, 16]. Indeed, studies show that provision of partial or guided instructor notes that students can complete with their personal notes is the best approach, offering a scaffold for student note-taking, but still encouraging attendance, attention and engagement [4]. One solution is to provide handouts that students can take notes on, but this is paper-wasting and not very practical; furthermore, students cannot easily share their notes or collaborate with their peers. A

Z. Kubincová et al. (Eds.): MIS4TEL 2020, AISC 1236, pp. 179–186, 2021.
https://doi.org/10.1007/978-3-030-52287-2_18

better solution would be to provide an approach for digital note-taking and note sharing between students. In this context, we designed and implemented a mobile application for note-taking during lectures, called EduNotes, which was presented in [13]. Its advantage over general-purpose, commercial mobile apps for note-taking [10] is represented by the functionalities dedicated for lecture use in classroom settings. First of all, the tool is integrated with the lecture process by enabling students to take notes associated to each lecture slide; furthermore, it supports social interaction and collaboration between learners, by providing functionalities for note sharing, commenting, tagging or rating. In addition, question notes can outline the areas where the lecture is too vague and facilitate the process of clarification, while the option to answer peers' questions has the potential to increase learner engagement [13].

EduNotes mobile application was well received by the students [13], so we decided to extend it with a web version, which allows the simultaneous presentation of the lecture slides alongside the notes. In addition to classroom use, this web application is also suitable for after-class use, allowing the students to manage the notes taken and download all lecture notes associated to the slides.

To the best of our knowledge, no other similar educational platform provides both a mobile and a web version. Indeed, the landscape of existing note-taking systems specifically designed for education is quite limited [13]. A notable example is Tsaap-Notes [16], a web-based application for collaborative note-taking which offers the following main functionalities: posting/deleting/replying to a note, marking a note as favorite, adding hashtags to notes, accessing all existing notes; teachers can also add "notes as questions", in order to assess the current level of understanding of their students. Another example is GroupNotes mobile application [15], which gives students the possibility to jointly take digital notes during a lecture; color codes are used to differentiate between each learner's contribution and students take various roles within a group: note-taker, reviewer, commentator and questioner. In addition, some earlier initiatives for educational note-taking, designed for PDAs, have been proposed in [8, 17].

On the other hand, the literature includes also a related but distinct class of social annotation tools [6, 9], which allow adding notes, comments and highlights to an electronic resource that can be subsequently shared; examples include [2, 5, 18]. By contrast, EduNotes focuses on the process of taking notes during face-to-face lectures, in classroom settings, and not on document annotation. The platform provides a wide range of pedagogically grounded functionalities and has several advantages over similar systems: a simple means for associating notes to lecture slides, flexible sharing options, advanced filters, live notifications, note rating feature, file attachment option as well as different note types (including lecture summary, questions for peers and associated best answers) [13].

In this paper, we aim to introduce the web version of EduNotes, as well as evaluate the usability and usefulness of both versions of the platform, by means of a pilot study. The first objective is covered in Sect. 2, which describes EduNotes web application in terms of functionalities and architecture; the second objective is covered in Sect. 3, which reports on the experimental study conducted with 38 computer science students. The paper ends with conclusions and future research directions.

2 EduNotes Web Application

EduNotes web version was designed as a complementary tool for the Android-based mobile version, which was presented in [13]. The mobile application is well suited for taking notes in the classroom, when teachers give lectures supported by slides and students use their smartphones for adding and sharing notes with peers. The web application can also be used for distance learning, as well as at home, for individual study, as it allows the simultaneous delivery of the lecture slides alongside the notes.

Both versions provide several functionalities for the students, such as creating, visualizing and sharing notes. Thus, the learner can write a note associated to a slide and add specific tags or file attachments; the note can be private, public, or shared with a specific peer or group. The student can also create a summary note, which is associated to the whole lecture and condenses its key points, facilitating recap and comprehension [3, 14].

Collaboration between learners is also supported by EduNotes: they have the option of starting from an existing peer note, mark it as favorite, add comments and ratings. In addition, students can ask and receive help by posting notes of type *question* and providing answers; this helps to clarify parts of the lecture, while also increasing learner engagement. Finally, students receive live note updates and notifications and can visualize all personal and public notes, search, filter and sort them based on various criteria.

In addition, the web version of EduNotes is centered around the lecture slides, which are provided alongside the notes, as shown in Fig. 1. This is more intuitive for the students, who can thus visualize the slide content while writing or reading the notes. Furthermore, there is a synchronization option, which allows the student to keep its current slide always in sync with the one presented by the lecturer. Moreover, the learner can download all lecture notes associated to the slides. This makes the web application also suited for after-class use, providing students with the annotated lecture slides, based on the notes taken in class.

From a technical point of view, extending EduNotes with a web-based client was facilitated by the flexible service-oriented design. The back-end consists of REST services implemented in PHP, which run on an Apache web server[1] and interact with a MariaDB[2] database server; more details regarding the system architecture can be found in [13]. The web front-end is based on JavaScript and consists of 3 modules: 1) the GUI module containing the HTML layouts; 2) the Service Invoker module that manages all interactions with the REST API; this module also handles all response objects received from the REST API and it is implemented using AngularJS[3]; 3) the Updates Manager module that handles both incoming and outgoing real time update messages.

1 https://www.apachefriends.org.

2 https://mariadb.org.

3 https://angularjs.org/.

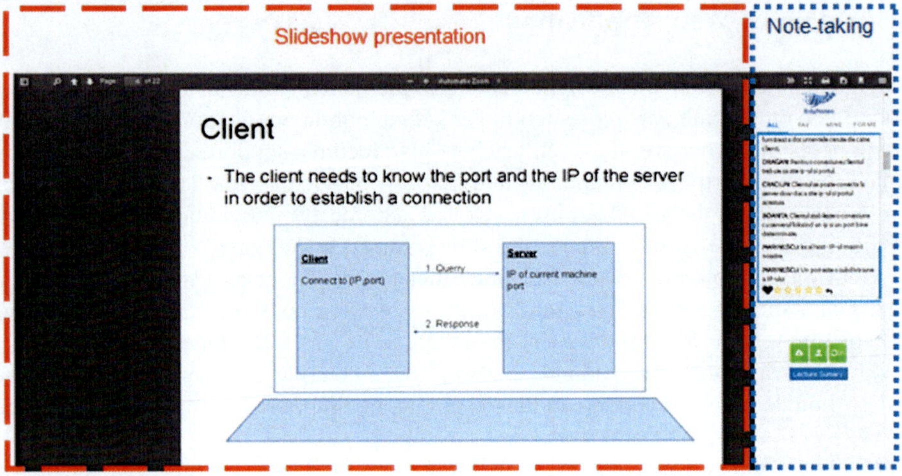

Fig. 1. EduNotes web version: note-taking bar (dotted blue area) next to the associated slide (dashed red area)

3 Experimental Study

We conducted a small study with 38 students from the University of Craiova, Romania, in order to test our EduNotes platform. The experiment took place over the course of three weeks, during a summer practice assignment for second year computer science students. The teacher delivered three presentations on Java programming for a client-server application; these included several code examples, which were explained verbally by the instructor, while students could use EduNotes in order to write down the explanations when needed. More specifically, learners were asked to use the web version for the first presentation, the mobile version for the second presentation and the version of their choice for the third presentation. 31 of the students tested both the web and mobile versions, 6 students used the web application exclusively (as they did not have an Android-based mobile phone) and one student used the mobile application exclusively (as he did not bring his laptop to the classroom). For the third presentation, when students could choose the version they preferred, most of them (37 out of 38) selected the web version; according to the learners, this choice was largely motivated by the fact that they already had laptops readily available for the practical programming activities; otherwise, if only mobile phones had been available, the mobile app would have provided a suitable alternative. At the end of the three weeks, the students were asked to fill in several questionnaires regarding their experience using EduNotes, as detailed next.

First of all, we were interested in learners' perspective on collaborative note-taking in general. Thus, the majority of the students liked the idea of taking lecture notes digitally (over 86%), of sharing lecture notes with peers (over 83%) and of viewing peers' lecture notes (over 89%).

Subsequently, we wanted to gauge learners' satisfaction with EduNotes tool in particular, based on the opinion survey proposed in [13]; 37 students gave their opinions regarding the web version and 32 students regarding the mobile version. In particular, learners were asked to answer the following questions on a five-point Likert scale:

Q1. It was easy for me to use EduNotes
Q2. I read the notes shared by my peers in EduNotes
Q3. The note-taking process with EduNotes was quick
Q4. The note-taking process with EduNotes was not distracting
Q5. The note-taking process with EduNotes made me pay more attention to the lecture
Q6. Overall, I was satisfied with EduNotes app
Q7. I would like to keep using EduNotes in the future
Q8. I would like to use EduNotes in other courses.

The results are summarized in Fig. 2 for EduNotes web application and in Fig. 3 for EduNotes mobile application.

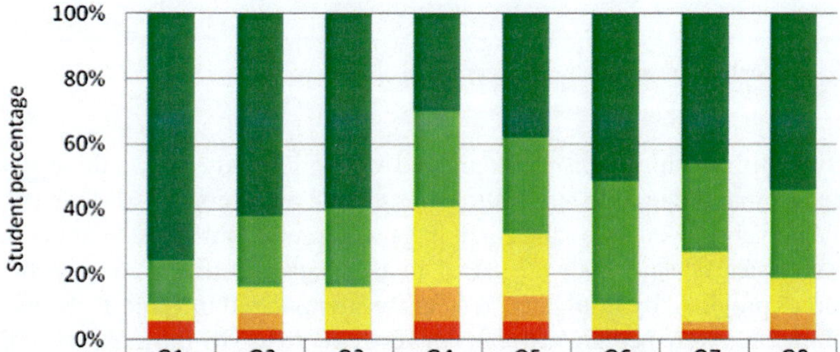

	Q1	Q2	Q3	Q4	Q5	Q6	Q7	Q8
Strongly agree	75.68%	62.16%	59.46%	29.73%	37.84%	51.35%	45.95%	54.05%
Agree	13.51%	21.62%	24.32%	29.73%	29.73%	37.84%	27.03%	27.03%
Neutral	5.41%	8.11%	13.51%	24.32%	18.92%	8.11%	21.62%	10.81%
Disagree	0.00%	5.41%	0.00%	10.81%	8.11%	0.00%	2.70%	5.41%
Strongly disagree	5.41%	2.70%	2.70%	5.41%	5.41%	2.70%	2.70%	2.70%

Fig. 2. Percentages of students' answers on the opinion survey regarding EduNotes web version

As can be seen in the figures, students' opinions were very similar regarding the two versions of EduNotes. According to Q1, most of the students found the applications easy to use; the majority of learners chose to read the notes shared by their peers (Q2). The note-taking process was deemed quick (Q3), not very distracting (Q4) and helpful for paying attention to the lecture (Q5). All in all, most students were satisfied with both versions of the platform (Q6), and willing to continue using them (Q7), also for other courses (Q8).

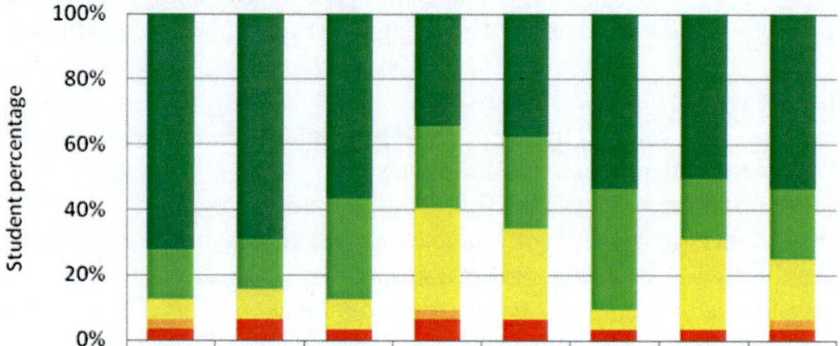

	Q1	Q2	Q3	Q4	Q5	Q6	Q7	Q8
Strongly agree	71.88%	68.75%	56.25%	34.38%	37.50%	53.13%	50.00%	53.13%
Agree	15.63%	15.63%	31.25%	25.00%	28.13%	37.50%	18.75%	21.88%
Neutral	6.25%	9.38%	9.38%	31.25%	28.13%	6.25%	28.13%	18.75%
Disagree	3.13%	0.00%	0.00%	3.13%	0.00%	0.00%	0.00%	3.13%
Strongly disagree	3.13%	6.25%	3.13%	6.25%	6.25%	3.13%	3.13%	3.13%

Fig. 3. Percentages of students' answers on the opinion survey regarding EduNotes mobile version

Overall, the results improved compared to the pilot study in [13]; this can be explained by the fact that the mobile application was then in beta version, so students' experience was negatively impacted by the small bugs which were still present at the time.

Subsequently, students were asked about the ease of use of each of the main functionalities provided by EduNotes; results are summarized in Table 1. As can be seen, all functionalities were perceived as easy or very easy to use by the majority of the students, for both versions of the platform. Writing summary notes and adding notes to favorites were found easier in the web application, while searching for notes and setting note privacy were found easier in the mobile application. Adding tags to notes was the least used functionality, not being tried by a quarter of the students.

As far as the usefulness of the functionalities is concerned, positive results were also recorded, with the majority of the students perceiving them as helpful or very helpful. Taking notes and viewing notes shared by peers were considered the most useful features, while adding tags to notes was reported as the least useful.

Finally, students were asked to provide suggestions for improving EduNotes, as well as additional functionalities that they believe would be useful. The features proposed by the learners include: an option for downloading lecture slides on the mobile version, a larger note-taking sidebar in the web version, text highlighting and drawing tools associated to notes.

Table 1. Student perceived ease of use for EduNotes main functionalities; for each functionality, the first line includes percentages of answers for the web version and the second line for the mobile version (highest values in boldface)

Functionality	Did not use it	Very difficult	Difficult	Neutral	Easy	Very easy
Take note	2.70%	0.00%	0.00%	16.22%	32.43%	**48.65%**
	6.25%	0.00%	3.13%	9.38%	21.88%	**59.38%**
View notes shared by others	0.00%	5.41%	0.00%	5.41%	24.32%	**64.86%**
	0.00%	0.00%	0.00%	9.38%	21.88%	**68.75%**
Search note	10.81%	0.00%	0.00%	21.62%	27.03%	**40.54%**
	3.13%	0.00%	3.13%	12.50%	**40.63%**	**40.63%**
Ask question	10.81%	5.41%	0.00%	13.51%	27.03%	**43.24%**
	12.50%	0.00%	6.25%	12.50%	25.00%	**43.75%**
Write summary	2.70%	2.70%	0.00%	24.32%	29.73%	**40.54%**
	6.25%	0.00%	9.38%	21.88%	**34.38%**	28.13%
Rate notes	13.51%	5.41%	0.00%	13.51%	10.81%	**56.76%**
	15.63%	0.00%	3.13%	18.75%	15.63%	**46.88%**
Add note to favorites	16.22%	5.41%	0.00%	10.81%	18.92%	**48.65%**
	21.88%	0.00%	6.25%	18.75%	15.63%	**37.50%**
Add tag	24.32%	8.11%	0.00%	13.51%	18.92%	**35.14%**
	25.00%	0.00%	6.25%	18.75%	18.75%	**31.25%**
Set note privacy	21.62%	10.81%	2.70%	10.81%	16.22%	**37.84%**
	15.63%	3.13%	6.25%	12.50%	**31.25%**	**31.25%**
Add reply to note	13.51%	5.41%	2.70%	13.51%	21.62%	**43.24%**
	15.63%	0.00%	3.13%	9.38%	**37.50%**	34.38%
View note replies	8.11%	5.41%	2.70%	10.81%	32.43%	**40.54%**
	3.13%	0.00%	6.25%	12.50%	28.13%	**50.00%**

4 Conclusion

We presented a web application designed to support collaborative note-taking in lecture-based settings. The tool was developed as an extension of the mobile app introduced in [13]. Both versions of the platform were well received by the students, as evidenced in a pilot study involving 38 participants. From the instructor point of view, we noticed that students need consistent encouragements to be more actively involved with the lecture and take more notes; furthermore, a positive class climate is needed to foster students' confidence in sharing their notes with peers. Clear guidelines for digitally taking notes in a collaborative way and initial scaffolding are also important for ensuring the success of the process.

As future work, EduNotes could be extended with a module dedicated to the instructor, for monitoring learners' notes and activity; furthermore, the features suggested by the students could be integrated in the platform. In addition, the system could be used on a larger scale, in various course contexts, and more in-depth analyses could be performed to investigate the impact of collaborative note-taking on the learning process.

References

1. Al-Zaidi, M.S., Joy, M., Jane, S.: Exploring the use of micro note-taking with social interaction features for education. In: Proceedings EDULEARN13, pp. 6098–6106 (2013)
2. Atrash, A., Abel, M.H., Moulin, C.: Notes and annotations as information resources in a social networking platform. Comput. Hum. Behav. **51**(B), 1261–1267 (2015)
3. Boch, F., Piolat, A.: Note taking and learning: a summary of research. WAC J. **16**, 101–113 (2005)
4. Boye, A.: Note-taking in the 21st century: tips for instructors and students (2012). https://www.depts.ttu.edu/tlpdc/Resources/Teaching_resources/TLPDC_teaching_resources/Documents/NotetakingWhitepaper.pdf
5. Gao, F.: A case study of using a social annotation tool to support collaboratively learning. Internet High. Educ. **17**, 76–83 (2013)
6. Ghadirian, H., Salehi, K., Ayub, A.F.M.: Social annotation tools in higher education: a preliminary systematic review. Int. J. Learn. Technol. **13**(2), 130–162 (2018)
7. Kiewra, K.A.: Note taking and review: the research and its implications. J. Instr. Sci. **16**, 233–249 (1987)
8. Landay, J.A., Davis, R.C.: Making sharing pervasive: ubiquitous computing for shared note taking. IBM Syst. J. **38**(4), 531–550 (1999)
9. Novak, E., Razzouk, R., Johnson, T.E.: The educational use of social annotation tools in higher education: a literature review. Internet High. Educ. **15**, 39–49 (2012)
10. Nuckles, B.: 7 best note taking apps (2016). http://www.businessnewsdaily.com/6065-best-note-taking-apps.html
11. Paek, S., Fulton, L.A.: Elementary students using a tablet-based note-taking application in the science classroom. J. Digit. Learn. Teach. Educ. **32**(4), 140–149 (2016)
12. Peper, R., Mayer, R.: Generative effects of note-taking during science lectures. J. Educ. Psychol. **78**, 34–38 (1986)
13. Popescu, E., Stefan, C., Ilie, S., Ivanovic, M.: EduNotes - a mobile learning application for collaborative note-taking in lecture settings. In: Proceedings ICWL 2016, Lecture Notes in Computer Science, vol. 10013, pp. 131–140. Springer (2016)
14. Ruby, P., Ruby, R.: Note taking skills: everybody needs them. J. Bus. Econ. **5**(4), 443–448 (2014)
15. Shen, H., Reilly, M.: Personalized multi-user view and content synchronization and retrieval in real-time mobile social software applications. J. Comput. Syst. Sci. **78**(4), 1185–1203 (2012)
16. Silvestre, F., Vidal, P., Broisin, J.: Tsaap-notes - an open micro-blogging tool for collaborative notetaking during face-to-face lectures. In: Proceedings ICALT 2014, pp. 39–43 (2014)
17. Singh, G., Denoue, L., Das, A.: Collaborative note taking. In: Proceedings WMTE 2004, pp. 163–167 (2014)
18. Su, A.Y.S., Yang, S.J..H., Hwang, W.Y., Zhang, J.: A web 2.0-based collaborative annotation system for enhancing knowledge sharing in collaborative learning environments. Comput. Educ. **55**(2), 752–766 (2010)

Enhancing Learning Opportunities for CS: Experiences from Two Learning Systems

Mikko Apiola[1(✉)], Mikko-Jussi Laakso[1], and Mirjana Ivanovic[2]

[1] Department of Future Technologies, University of Turku, Turku, Finland
mikko.apiola@utu.fi
[2] Department of Mathematics and Informatics, University of Novi Sad, Novi Sad, Serbia
mira@dmi.uns.ac.rs

Abstract. Online educational technologies (ET) collect a lot of data. Analysis of that data is called learning analytics (LA) or educational data mining (EDM). In this paper we present and combine research results from development and research work with two ET systems, one in University of Novi Sad, Serbia and another in University of Turku, Finland. We combine the most important findings from our research: improvements to learning outcomes, beneficial features such as automatic assessment, continuous feedback and personalised learning paths. We also discuss ongoing research on eye-tracking, sensors, and educational metrics. In addition, challenges in pedagogical reforms are considered.

Keywords: Computer science education · Learning analytics · Educational technology

1 Introduction

The rapid and widespread increase of educational technologies, e-learning, and establishment of databases of student information have created massive amounts of educational data [23,27]. Various types of modern educational environments, such as flipped classrooms, blended learning environments, virtual and mobile learning, and game-based learning collect massive amounts of learning data, too [27]. Learning contents typically range from non-interactive tasks to a variety of quizzes, interactive and collaborative lectures, and personalised learning opportunities [2,13]. Online educational technologies may track learners' characteristics, such as prior knowledge [35], and other learning habits, and may offer more personalised learning [19] or guidance on beneficial learning practices [2]. Modern approaches include learning records, learning preference cards, mobile- and eye-tracking devices, and flexible classroom designs, and may use eg. simulated social presence (e.g. [7]) in modern and blended learning environments.

Z. Kubincová et al. (Eds.): MIS4TEL 2020, AISC 1236, pp. 187–196, 2021.
https://doi.org/10.1007/978-3-030-52287-2_19

Educational data mining (EDM) and learning analytics (LA) are research tracks, which use a mixture of statistical, computational and qualitative methods in order to analyse educational data [1,30]. Their aim is to improve educational processes by collecting, processing, reporting, and continuously working on digital data [1]. LA may result in more personalised, adaptive, and interactive learning, which may enhance learning outcomes, effectiveness of teaching and learning, as well as reforming of educational practices [1,27]. Indeed, one big challenge for educational institutions is how to deal with the exponentially growing amount of educational data wisely, by transforming the data into insights about educational processes, and related wise educational decisions [27].

In addition to the opportunities of LA, there are challenges, too. For example, transparency in clarifying exactly how information is handled and processed, and how it is not handled and processed, is a challenge [27]. Other challenges include generating feedback that promotes learning, visualising recommendations, transparency about data storage, security and privacy, data infrastructures, scalability, and ethics [27]. Also, challenges include analysis, which is restricted to the contexts of one course in one institution, lacking generalisability of findings [2,11]. In this paper, we discuss experiences from two educational technology systems, used to teach introductory computing courses, one in Serbia and another in Finland. We have the following ideas. First, we want to widen approaches to learning analytics by suggesting more cross-systems analytics. Second, we want to identify positive pedagogical use cases for teachers and students. Third, we want to consider potential future avenues for LA research in CSE.

2 Background: Digital Learning in CS

The increasing importance of LA is a well recognised issue globally, as demonstrated by the EDUCAUSE-report[1], which presents opportunities for modelling learning collaborations within large-scale data collections. A recent report[2] highlights the importance of analytics technologies and adaptive learning technologies. A common definition of LA is "the measurement, collection, analysis and reporting of data about learners and their contexts, for purposes of understanding and optimising learning and the environments in which it occurs" [22]. The following five items summarise the "promise" of learning analytics.

- Assisting in creating new educational content as the combination of technical, information and social networks.
- Innovating and transforming widely used pedagogical approaches and educational models.
- Constant testing and evaluation of curricula and recognition of proper changes and improvements.
- Improving effectiveness of organizational resource provision and crucial educational decision-making.

[1] https://www.nmc.org/organization/educause-learning-initiative/.
[2] https://library.educause.edu/resources/2018/8/2018-nmc-horizon-report.

– Help teachers and other key stakeholders to control quality of teaching, status and reputation of the educational institution.

It is important to consider, from a practical perspective, how these five promises reflect in real life in teaching with digital learning systems, how LA has helped, and what are the challenges and avenues for the future. In the following subsections, we will briefly introduce our educational technology systems. Our concerns include, how to address the demands for new pedagogical approaches, such as increasing "the four C's"; collaboration, creativity, communication, and critical thinking among students of CS [15,28]. Other important aspects include identification of concrete ways, in which LA data can be used beneficially, and discussion about the challenges and possible downsides and threats of LA.

2.1 Programming Tutoring System (ProTuS)

Programming Tutoring System (ProTuS) has been developing, transforming and improving for more than a decade at University of Novi Sad [18], in several versions. Imagined as a general-purpose educational framework it encompasses adequate mechanisms for storing and managing typical learning concepts as tutorials, examples, tests and different exercise types, communication, reports generation and so on. The only completely realized and tested version is about introductory Java programming course [19] for learning Java basics i.e. for learners with no object-oriented programming experience.

From the pedagogical point of view the learning topics are organized in such a form that a number of appropriate tests are attached to each lesson in order to test acquired knowledge. Based on the tests' results, the system further determines the level of learners' progress, and updates his or her learner model and generates further personalization and adaptation of learning paths through the learning material. Accordingly, for advanced learners some lessons could be skipped but for learners with lower level of knowledge some additional explanations and solved tasks can be offered.

These different approaches have been tested and results of experiments show the system's usability, however the best results have been achieved by combination of these approaches and creating a hybrid recommendation method [18]. This method fuses two techniques: collaborative tagging and mining sequential patterns in order to recommend to a specific learner optimal navigational paths through the learning material. ProTuS has been in use and under testing for more than a decade, primarily for first year students at Faculty of Sciences and High Business School at University of Novi Sad. Several times the system has also successfully been used and tested for secondary school pupils within specially organized seminars for talented pupils [17].

2.2 VirtuaL Learning Environment (ViLLE)

ViLLE is a learning tool, with extensive digital tools for learning, for creating learning materials, and means for collecting data for learning analytics research (see [13, pp. 17–34]). ViLLE is used in many Finnish schools, especially in South-West Finland [21]. ViLLE is also used in universities, polytechnics institutes, and in various learning environments of computer science [2,21].

3 Thematic Overview of Research

3.1 Learning Analytics Research in ProTuS

Rapid development of new technologies, such as digital learner records, learner cards, sensors, mobile devices, flexible classroom design, and others are changing the ways of learning and teaching. On the other hand, use of smart content and modern learning environments, services and resources need innovative approaches to handle such data to achieving attractive utilities: operative self-learning, useful peer groups, and available class time for creativity. Based on recommender systems improvement [36], cultural considerations in analytics [34], reputation mechanisms, and participating learning [9], ProTuS has evolved as follows. In last several years ProTuS platform has been intensively enhanced and tested [12,16,17] with promising results. Recently, the original group of developers has been divided in two groups that have been continuing to consider different ways and possibilities to enhance and develop system in the future. These innovation groups are the University of Novi Sad (UNS) group, and Norwegian University of Science and Technology (NTNU) group.

Both groups are oriented towards making the system smarter. Smart learning as a form of technology-enhanced learning actively provides to learners the necessary learning guidance and scaffolding and also help-seeking behaviour at the right time and in the proper way. Smart learning environments should utilise important learning elements like: personalisation and adaptation of the learning process, beneficial effects of learning analytics results, and more comprehensive smart knowledge management.

One important characteristic of ProTuS system is that it allows the adjustment of learning materials according to identification of individual learning styles. For Java lessons, different forms of learning materials have been prepared according to different learning styles, such as Visual/Verbal, Active/Reflective, Sequential/Global, to allow to learners the best way to acquire knowledge for appropriate topics and to enjoy learning. The learning style for the learner is identified at the beginning of use of the system by the Felder-Silverman questionnaire. But later during learning activities and being more experienced with system and learning material, the learner can change his or her originally identified learning style to a more preferred one.

Recently the UNS group has been considering inclusion of new features into ProTuS. The considerations went in the direction of increasing active and collaborative learners' participation and reaching better learning achievements and results. In a research [16] the possible influences of big data techniques in enhancement and improvement of interactive education environments was researched. On the other hand, as adding new modules with new functionalities is easy in ProTuS, we considered how enhancing ProTuS with Eye Trackers may bring learning advantages. Generally speaking, eye tracking technologies and adequate metrics (applying LA methods) can support different phases of the recommendation process: automatic identification of learning style; determination of efficiency of user interface and possible automatic adjustment of it, tracking a learner's course progress, identification of important areas of presented topic, automatic tagging of the terms the learner focuses on, determination of a learners' mental state like concentration, stress [14], and tiredness [12].

Recent innovations in ProTuS are presented in [37]. The novel advanced feature in this system is oriented towards inclusion of visualization and learning dashboard for students. Interactive visualizations of learners' heterogeneous data sources aggregate: logs of learners' interactions; grades achieved from assignments and lab's task; logs and grades from third-party systems (e.g. Mastery-Grids [10]). Current version of the system offers to the learner reports about his/her activities and shows them visualized comparison of his/her individual progress with the progress of other learners [37].

To conclude, based on current trends in interactive e-learning environments that support smart learning content and open learner models, the main research direction for further improvement of systems are as follows:

- Using LAD (LA dashboards) applications that support both parties: learners and teachers.
- Visual learning analytics and interactive visualization tools that supports building interactive visualizations [20]. They can offer valuable insights into the learning process and participation of students in a course.
- Use of additional technologies and devices, like eye trackers, virtual reality headsets [17] and so on.

3.2 Learning Analytics Research in ViLLE

Learning analytics in ViLLE can be categorised to (1) experimental designs, (2) learning metrics, and (3) educational metrics [2]. These three categories are presented in Table 1.

Experimental Research for Comparing Learning Performance. This class of research forms one pillar of R&D in ViLLE [2]. Examples of evaluation approaches include investigating effectiveness of visualisation in programming tasks at high school level [13, pp 35–39], investigating educational impacts in university programming courses [13, 41–47], and courses on object oriented programming [13, 48–49].

Learning Metrics. This class contains research that pedagogically investigates data from exercise times, number of submissions, log data, attendance data [2]. Research designs include investigating relationships between educational outcomes and task time usage, issues in collaboration and groupwork behaviour in learning activities [5, 25, 26], investigating associations in previous skills and learning outcomes, and misconceptions in coding tasks [35].

Table 1. Three classes of Learning Analytics in ViLLE [2]

(1) Experimental designs
Experimental designs are used to investigate impact of different features or pedagogies of ViLLE to eg. learning outcomes.
(2) Learning metrics (live learning data and achievements)
Variables, such as time usage, number of assignment submissions, log data, attendance information, groupwork data, and patterns of code writing are analysed for statistical associations.
(3) Educational metrics
Educational metrics are used to investigate eg. motivation, problem solving behaviours, attitudes, mindsets, or stress.

Educational Metrics. This class of research includes experiments with educational metrics [2]. Current experiments span the testing of Problem Solving Inventory (PSI), which has been investigated in courses on introductory programming and courses in algorithms. Results have revealed, for example, that PSI scores are valuable, to a certain extent, for predicting students' final exam performance. Other instruments under experimentation are The Nature of Attitudes Towards Learning Mathematics Questionnaire (NALMQ) [38], mindset [3,4], and mood sensors [14]. At the moment, educational metrics is at piloting stage, thus, teachers do not yet have the opportunity to automatically collect such data. However, teachers have means of preparing their own questionnaires.

4 Discussion

In this article we have given an overview of research and development of two well-developed learning systems, which are used in computing education in two institutions of higher education in Europe. In this section we summarise findings from our research.

4.1 Concrete Benefits of Digital Learning

In both systems and institutions, increases in learning outcomes in introductory computer science courses have been shown. Other helpful pedagogical uses include the possibility to identify students, who are in risk of dropping out. This has made it possible to develop new models of student counselling, with positive impact to course attendance. Other positive sides include pedagogical ease for teachers, who can more easily track learning progressions, challenges, and needs of support of students. Also, identification of students' learning habits and preferences has helped in designing individualised learning paths.

4.2 Most Promising Research Approaches

We are bringing educational metrics research into CSE and LA by investigating instruments that have been developed by educational psychologists. In the future, learning systems may provide learning content and exercises that are better adapted to knowledge and skills. On the other hand, adaptive learning can provide many challenges for those who master a specific topic well. Third, work on eye-tracking, and moodmetric sensors provides many research opportunities. Also, experiments with modern pedagogies and many educational settings, including those in the global south (e.g. [33]) pose many opportunities for future R&D activities.

4.3 Ethical and Pedagogical Considerations

Pressing ethical considerations in LA include how to use the data pedagogically wisely in respect of privacy and students' rights, to avoid a feeling of being controlled and not to lose the joy of learning. Some of these issues are highlighted by Picciano [24], who considers several highly relevant questions. First, how should learners be informed about the monitoring of their activities, and how should the data be used pedagogically wisely? Picciano [24] points out that transparency is important so students understand and have control on what data is collected and how it is used.

4.4 Next Generation of Learning Analytics

It is clear that the next generation of learning analytics must overcome the burden of legacy systems or conventional design. It must be kept in mind that technological solutions and related learning analytics may limit the ways in which a more appropriate solution might disrupt learning, when, for example, behavioristically oriented teaching machines lead to learning that is only based on rote memorising via automated drills. With this in mind, it is essential to look into the future generation of learning analytics, which deals with innovation pedagogy [6], choice-based assessment [29], analytics for knowledge creation [8], modern approaches such as social learning analytics [32], and epistemic network analysis [31]. It is also extremely important to include a serious ethical discussion and related action into the development process of future learning analytics.

5 Conclusions

A large number of research in CSE focuses on introductory programming courses (CS1 or CS2). Even though this is an important line of research, one challenge is that a lot of that research is often conducted within the context of one institution and one course. This is a central critique towards learning analytics in digital learning within CSE research, too [11]. We have shown several positive pedagogical use cases of learning analytics, and discussed challenges and ongoing research efforts. We are thinking about the possibilities of building inter-systems research approaches, where features and approaches could be studied in multiple systems [2]. Our fantasy is to build a global network of learning analytics for cross-inspiration, innovation, and joint research.

References

1. Aldowah, H., Al-Samarraie, H., Fauzy, W.M.: Educational data mining and learning analytics for 21st century higher education: a review and synthesis. Telematics Inform. **37**, 13–49 (2019)
2. Apiola, M., Karunaratne, T., Kaila, E., Laakso, M.: Experiences from digital learning analytics in finland and sweden: a collaborative approach. In: 2019 42nd International Convention on Information and Communication Technology, Electronics and Microelectronics (MIPRO), pp. 627–632, May 2019
3. Apiola, M., Laakso, M.J.: The impact of self-theories to academic achievement and soft skills in undergraduate CS studies: first findings. In: Innovation and Technology in Computer Science Education (ITiCSE 2019), pp. 16–22. ACM. Aberdeen (2019)
4. Apiola, M., Laakso, M.J.: The role of implicit theories in computer science and engineering studies: first findings and review. In: Proceedings of IEEE Frontiers in Education 2019 Conference. IEEE (2019)
5. Apiola, M., Lokkila, E., Laakso, M.J.: Digital learning approaches in an intermediate-level computer science course. Int. J. Inf. Learn. Technol. **36**(5), 467–484 (2019)
6. Apiola, M., Tedre, M.: Deepening learning through learning-by-inventing. J. Inf. Technol. Educ. **12**, 185–202 (2013)
7. Borup, J., West, R.E., Graham, C.R.: Improving online social presence through asynchronous video. Internet High. Educ. **15**(3), 195–203 (2012)
8. Chen, B., Zhang, J.: Analytics for knowledge creation: towards epistemic agency and design-mode thinking. J. Learn. Anal. **3**(2), 139–163 (2016)
9. Clow, D., Makriyannis, E.: iSpot analysed: participatory learning and reputation. In: Proceedings of the 1st International Conference on Learning Analytics and Knowledge, LAK 2011, pp. 34–43. ACM, New York (2011)
10. Guerra, J., Hosseini, R., Somyurek, S., Brusilovsky, P.: An intelligent interface for learning content: combining an open learner model and social comparison to support self-regulated learning and engagement. In: Proceedings of the 21st International Conference on Intelligent User Interfaces, IUI 2016, pp. 152–163. ACM, New York (2016)
11. Ihantola, P., Vihavainen, A., Ahadi, A., Butler, M., Börstler, J., Edwards, S.H., Isohanni, E., Korhonen, A., Petersen, A., Rivers, K., Rubio, M.A., Sheard, J., Skupas, B., Spacco, J., Szabo, C., Toll, D.: Educational data mining and learning analytics in programming: literature review and case studies. In: Proceedings of the 2015 ITiCSE on Working Group Reports, ITICSE-WGR 2015, pp. 41–63. ACM. New York (2015)
12. Ivanović, M., Klašnja-Milićević, A., Ivković, J., Porta, M.: Integration of eye tracking technologies and methods in an e-learning system. In: Proceedings of the 8th Balkan Conference in Informatics, BCI 2017, pp. 29:1–29:4. ACM, New York (2017)
13. Kaila, E.: Utilizing educational technology in computer science and programming courses. Ph.D. thesis, Turku Centre for Computer Science (University of Turku, Department of Future Technologies), February 2018

14. Kajasilta, H., Apiola, M.V., Lokkila, E., Veerasamy, A., Laakso, M.J.: Measuring students' stress with mood sensors: first findings. In: Herzog, M.A., Kubincová, Z., Han, P., Temperini, M. (eds.) Advances in Web-Based Learning-ICWL 2019, pp. 92–99. Springer, Cham (2019)
15. Kivunja, C.: Teaching students to learn and to work well with 21st century skills: unpacking the career and life skills domain of the new learning paradigm. Int. J. High. Educ. **4**(1), 1–11 (2015)
16. Klašnja-Milićević, A., Ivanović, M., Budimac, Z.: Data science in education: big data and learning analytics. Comput. Appl. Eng. Educ. **25**(6), 1066–1078 (2017)
17. Klašnja-Milićević, A., Ivanović, M., Vesin, B., Budimac, Z.: Enhancing e-learning systems with personalized recommendation based on collaborative tagging techniques. Appl. Intell. **48**(6), 1519–1535 (2018)
18. Klašnja-Milićević, A., Vesin, B., Ivanovi, M., Budimac, Z., Jain, L.C.: E-Learning Systems: Intelligent Techniques for Personalization, 1st edn. Springer, Cham (2016)
19. Klašnja-Milićević, A., Vesin, B., Ivanović, M., Budimac, Z.: E-learning personalization based on hybrid recommendation strategy and learning style identification. Comput. Educ. **56**(3), 885–899 (2011)
20. Kuosa, K., Distante, D., Tervakari, A., Cerulo, L., Fernández, A., Koro, J., Kailanto, M.: Interactive visualization tools to improve learning and teaching in online learning environments. Int. J. Distance Educ. Technol. **14**(1), 1–21 (2016)
21. Laakso, M.J., Kaila, E., Rajala, T.: ViLLE - collaborative education tool: designing and utilizing an exercise-based learning environment. Educ. Inf. Technol. **23**(4), 1655–1676 (2018)
22. Long, P.D., Siemens, G., Conole, G., Gašević, D.: Proceedings of the 1st international conference on learning analytics and knowledge (lak11). ACM, New York (2011)
23. Peña-Ayala, A.: Learning analytics: a glance of evolution, status, and trends according to a proposed taxonomy. Wiley Interdisciplinary Reviews: Data Mining and Knowledge Discovery **8**(3) (2018)
24. Picciano, A.G.: The evolution of big data and learning analytics in american higher education. J. Asynchronous Learn. Netw. **16**(3), 9–20 (2012)
25. Rajala, T., Kaila, E., Holvitie, J., Haavisto, R., Laakso, M.J., Salakoski, T.: Comparing the collaborative and independent viewing of program visualizations. In: 2011 Frontiers in Education Conference (FIE), pp. F3G–1–F3G–7, October 2011
26. Rajala, T., Salakoski, T., Kaila, E., Laakso, M.J.: How does collaboration affect algorithm learning? A case study using TRAKLA2 algorithm visualization tool. In: 2010 2nd International Conference on Education Technology and Computer, vol. 3, pp. V3–504–V3–508, June 2010
27. Romero, C., Ventura, S.: Educational data mining and learning analytics: an updated survey. Data Min. Knowl. Discov. Wiley Interdisc. Rev. **10**, e1355 (2020)
28. Sahami, M., Roach, S., Cuadros-Vargas, E., Hawthorne, E.K., Kumar, A., LeBlanc, R., Reed, D., Seker, R.: ACM/IEEE-CS computer science curricula 2013: implementing the final report. In: Proceedings of the 45th ACM Technical Symposium on Computer Science Education, SIGCSE 2014, pp. 175–176. ACM, New York (2014)
29. Schwartz, D.L., Arena, D.: Measuring What Matters Most: Choice-Based Assessments for the Digital Age. MIT PRESS (MacArthur Foundation) (2013)
30. Shaffer, D.W.: Quantitative Ethnography. Cathcart Press, Madison (2017)
31. Shaffer, D.W., Hatfield, D., Svarovsky, G.N., Nash, P., Nulty, A., Bagley, E., Frank, K., Rupp, A.A., Mislevy, R.: Epistemic network analysis: a prototype for 21st-century assessment of learning. Int. J. Learn. Media **1**(2), 33–53 (2009)

32. Shum, S.B., Ferguson, R.: Social learning analytics. J. Educ. Technol. Soc. **15**(3), 3–26 (2012)
33. Tedre, M., Apiola, M., Oroma, J.: Developing IT education in Tanzania: empowering students. In: Proceedings of the FIE 2011 Frontiers in Education Conference, Rapid City, SD, USA, 12th–15th October 2011
34. Vatrapu, R.: Cultural considerations in learning analytics. In: Proceedings of the 1st International Conference on Learning Analytics and Knowledge, LAK 2011, pp. 127–133. ACM, New York (2011)
35. Veerasamy, A.K., D'Souza, D., Lindén, R., Laakso, M.J.: The impact of prior programming knowledge on lecture attendance and final exam. J. Educ. Comput. Res. **56**(2), 226–253 (2018)
36. Verbert, K., Drachsler, H., Manouselis, N., Wolpers, M., Vuorikari, R., Duval, E.: Dataset-driven research for improving recommender systems for learning. In: Proceedings of the 1st International Conference on Learning Analytics and Knowledge, LAK 2011, pp. 44–53. ACM, New York (2011)
37. Vesin, B., Mangaroska, K., Giannakos, M.: Learning in smart environments: user-centered design and analytics of an adaptive learning system. Smart Learn. Environ. **5**(1), 24 (2018)
38. Wong, K.Y., Chen, Q.: Nature of attitudes toward learning mathematics questionnaire. In Dindyal, C. (ed.) Mathematics Education: Expanding Horizons (Proceedings of the 35th Annual Conference on the Mathematics Education Research Group of Australasia) (2012)

A Dynamic Recommender System for Online Judges Based on Autoencoder Neural Networks

Paolo Fantozzi[1,2] and Luigi Laura[1,3](✉)

[1] Italian Association for Informatics and Automatic Calculus (AICA), Milano, Italy
fantozzi@diag.uniroma1.it
[2] Sapienza University of Rome, Rome, Italy
[3] International Telematic University Uninettuno, Rome, Italy
luigi.laura@uninettunouniversity.net

Abstract. In recent years, we have witnessed the raising popularity of programming contests such as International Olympiads in Informatics (IOI) and ACM International Collegiate Programming Contest (ICPC). In order to train for these contests, there are several Online Judges available, in which users can test their skills against a usually large set of programming tasks.

In the literature, so far few papers have addressed the problem of recommending tasks in online judges. Most notably, as opposed with traditional Recommender Systems, since the learners improve their skills as they solve more problems, there is an intrinsic dynamic dimension that has to be considered: when recommending movies or books, it is likely that the preferences of the users are more or less stable, whilst in recommending tasks this does not hold true. In this paper we present a dynamic Recommender System (RS) for Online Judges based on an Autoencoder (Artificial) Neural Network (ANN).

Keywords: Autoencoder Neural Networks · Recommender Systems · Programming contests

1 Introduction

During a Programming Contest (PC), participants are faced a set of tasks that require writing computer programs. Recent literature have emphasized the importance and the effectiveness of programming contests in the process of learning computer programming [3,4,6,10,14,21].

In order to train and improve their skills, learners use Online Judges (OJs), also known as Programming Online Judges, i.e., web based e-learning tools where a user can submit solutions to a programming task. Usually there is a large list of available tasks, and the user can select the one he will try to solve; after reading its statement, that includes the required formatting of input and output

Z. Kubincová et al. (Eds.): MIS4TEL 2020, AISC 1236, pp. 197–205, 2021.
https://doi.org/10.1007/978-3-030-52287-2_20

or the use of a programming interface, the user writes a code to solve the task. The code is submitted to the OJ, that verifies both the correctness, usually by testing it against a certain number of test cases, and the efficiency, by checking that the running time and/or the memory usage is under some limit. In Fig. 1 is shown an example of a programming task.

However, choosing the right task is becoming a complex problem: University of Valladolid Online Judge has more than 200 k users and 2 k tasks, whilst SPOJ accounts approximately 600 k users and 6 k (public) tasks.

Thus, it is important to help users selecting their next task by using a Recommender System (RS). However, as observed in [5], there are some peculiarities of Online Judges that prevent the use of a general Recommender System:

- users slowly improve their abilities, one task after the other, so the general concept of user *preferences* does not apply: recommending a movie or a novel differs significantly from recommending a task; a user will probably still like a novel after one year, whilst he might find a task too easy after the same amount of time.
- Users with *similar* skills, i.e. users to whom we might want to suggest the same set of tasks, might behave very differently in OJs, thus preventing us from considering them *similar*. For example, one might solve all the tasks involving a given skill, while the other might just solve one task, related to that skill, and then move on to tasks involving different skills.

In this paper, to the best of our knowledge, we propose the first dynamic recommender system for Online Judges, based on an Autoencoder Neural Network (ANN). We trained and tested the ANN using data from the Online Judge used in the Italian Olympiads in Informatics (Olimpiadi Italiane di Informatica - OII) [12], targeted at secondary school students training. The preliminary experimental results confirm the effectiveness of our approach.

This paper is organized as follows: the next section provides the necessary background related to programming contests, online judges, and recommender systems, whilst our approach is detailed in Sect. 3, including the results of a preliminary experimental evaluation. Concluding remarks are addressed in Sect. 4.

2 Related Works and Background

In this section we discuss related work and the necessary background concerning programming contests, online judges, and recommender systems.

Programming Contests and Online Judges. As mentioned before, a programming contest is a competition in which contestants are faced with a set of programming tasks, also called problems, to be solved in a limited amount of time and/or with a limited amount of memory usage.

A single task can be broken into different subtasks of increasing complexity: basic techniques might be enough to solve, within the given time and/or space limits, some of the subtasks whilst the most difficult ones might require very specific algorithmic techniques and data structures.

OIS2019 – Round 3

Online, December 19th, 2018 **paths • EN**

Railway Schedule (`paths`)

The railway network in the *Pordenone* county consists of N train stations connected by $N - 1$ tracks (X_i, Y_i) so that from every station is possible to reach any other station: in other words, the tracks form a *tree*.

This choice makes the transportation system extremely inefficient: trains going in opposite directions cannot cross each other on a single track, so they need to perform lengthy and complex manoeuvres to pass each other. The new administration founded its campaign trail on changing this situation once and for all... and now it's time to keep promises!

Edoardo, the local leading expert in logistics, already has a mind-blowing idea for fixing the situation: making each track one-way, so that no crossings will ever occur! Of course, the tricky part is choosing the orientations so that the service remains acceptable for the majority of the population. After inspecting the traffic patterns, Edoardo discovered that most people travel between one of M pairs (A_i, B_i) of stations. Thus, an orientation of the tracks will be considered *acceptable* by the population only if for each such pair, either a path from A_i to B_i or a path from B_i to A_i should exist. However, many acceptable orientations exist and Edoardo cannot choose among them, otherwise his system would be deemed as unfair: the only solution is to use *all* of them in a periodic schedule of daily track orientations. Help Edoardo design such a schedule by counting how many acceptable orientations exist! Since this number may be large, report it modulo 1 000 000 007.

Fig. 1. An example of a problem from a programming contest; this task is taken from the final contest of the 2019 edition of the Italian Team Olympiads in Informatics (OIS) [2].

Popular programming contests include: the International Olympiads in Informatics (IOI), the ACM International Collegiate Programming Contest (ICPC), Google Code Jam and Facebook Hacker Cup.

Online Judges are web based platforms that provide a large number of programming tasks to be solved. There are several popular OJ platforms, including the already mentioned University of Valladolid Online Judge, Sphere Online Judge (SPOJ), CodeChef, and Peking University Online Judge.

Yera and Toledo [23] present a brief survey on OJs, whilst more information on tools and techniques for automatic evaluation of solutions submitted to OJs can be found in [1,7].

Recommender Systems in OJs. Despite the large amount of literature devoted to RS, the peculiarities of recommendation in OJs, where the relation user-item is way more complex than the typical RS cases, prevent from using standard techniques and forces the development of ad-hoc methods. This aspect is detailed in the paper of Audrito et al. [5], where the authors propose a first approach on building a RS by tackling the problem of ranking tasks in Online Judges.

Indeed, so far few research focused in the recommendation of tasks in OJs: we mention the traditional collaborative filtering method with a new similarity measure adapted to the case [18], and an approach based on fuzzy logic [23]. Caro and Jimenez considered user-based and similarity-based approaches In [8]. Di Mascio et al. proposed a framework that can allow recommendations and that can foster motivation in students by means of a lightweight, badge-based, gamified approach [13].

Recommender Systems and Artificial Neural Networks. The use of deep learning techniques for recommender systems is divided in two categories that we call *classical*, that uses the standard architectures of neural networks, and *hybrid*, that consists in using more than one type of architecture at the same time.

Since a recommendation task is similar to a dimensionality reduction task, many of the state-of-the-art techniques use some kind of Autoencoder to map the input in a smaller space, that will be the representation of the correlations in the recommender system. In particolar, Sedhain et al. [16] introduce the using of a vanilla Autoencoder to build a recommender system. They use a partial masked input (the same techniques we use in this work) and try to reconstruct it in output, the elements added in the output will be the recommended elements. Strub and Mary [17] extends the work of Sedain et al. [16]: they use a denoising Autoencoder instead of the vanilla Autoencoder to build a more robust system.

Chen and de Rijke [9] follow a similar approach, but they use a Variational Autoencoder to perform top-N recommendation. They encode both the user ratings and some side information in the compact space in the Autoencoder. Zhang et al. [24] generalize the Contractive Autoencoder paradigm into matrix factorization framework. Li et al. [15] combine a probabilistic matrix factorization with Marginalized Denoising Stacked Autoencoders to perform collaborative filtering. This work can be considered as a general framework to use these kinds of techniques; in this context, several works, including [19,20,22], can be viewed as special cases of this framework.

3 Recommending Tasks Using Autoencoder Neural Network

In this section we briefly discuss our approach from a theoretical point of view, and then present the result of a preliminary experimental evaluation. Summing up, our goal is to provide recommendations to the users regarding the next task to deal with among all the tasks in a system; we assume that:

- if a user obtains a score for a task it means that it is the max score possible for that user in that task;
- a user should solve the problems sorted by their grade of difficulty for the user; thus, a user should never try to solve a problem that is much harder than the last one he solved;

- it is possible to deduce the score for a new problem based on the scores the user obtained in other problems;
- given a snapshot of the scores for the users it is not important the order followed by the user to solve the tasks to foresee the score for another task;

Based on the above assumptions, we decided to build a model that takes as input the current scores of a user and provides probable scores for the same user for other tasks. Then we can choose between the forecasted scores and suggest the task with the highest score between them.

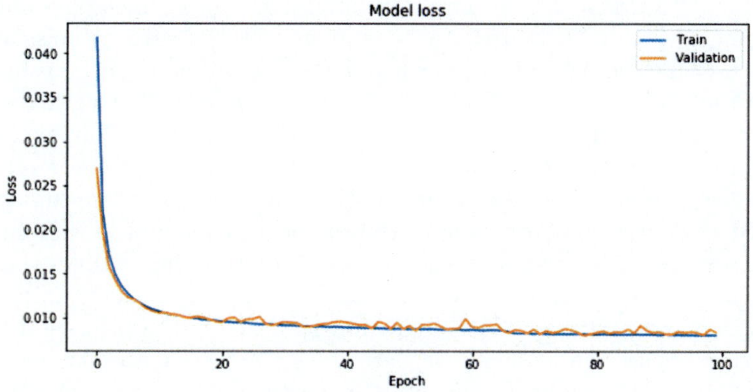

Fig. 2. Model loss.

If we consider the score as a judgment of the user for the item (i.e, the programming task), then we can just exploit the already known techniques for recommending items. We chose to use an autoencoder to build the model. Since that we want to use just the scores of the users, without any information from other sources, we use a masking mechanism on the input: we mask a fraction of the scores in input as a non-solved task, and then we perform backpropagation from the complete scores. In this way, in the bottleneck layer, there should be a compact representation of the similarities of the tasks.

3.1 Experimental Evaluation

In order to test our approach, we considered the dataset of the submissions to the OII Training platform [12] in a defined time range. The submissions were in the form:

`< user_id, task_id, datetime, score >`

where each submission corresponds to a possible solution to a task from a user that performs a certain score, based on many test cases. We filtered out all the scores equal to zero because we can't know if they were just users testing the behaviour of the platform. Then we considered only the best score for each task, for each user, to ignore all the attempts to solve the problem before the user found the solution.

We built a user x task matrix where each cell contains the best score of the user for the task. We have 3148 users and 409 different tasks; thus our baseline dataset is a 3148×409 matrix with 43051 non-empty cells. The matrix has an average of 105 submissions for each task and 13 for each user.

From this baseline dataset, we built a user x task matrix where each cell contains the best score of the user for the task. The result is a 42155×409 matrix with 2035447 non-empty cells (we had to drop few submissions from the baseline dataset). As before, the dataset has an average of 105 submissions for each task and 13 for each user. The max number of users which have submitted to the same task is 1070 and the max number of tasks with submissions from the same user is 336. We consider the zero cells as a problem with no submission from the user. We built the X and Y matrices imposing that each row in the Y matrix is the row next to the same row in the X matrix. This means that, for each user, we put the first $n-1$ rows in X and the last $n-1$ rows in Y. After the preliminary experiments, it was clear that data was too small, thus we performed an operation of data augmentation. So we have multiplied many times the same rows of the matrix with the result of a matrix with 8 times the rows of the original. To avoid the presence of the same data both in the training set and in the test set, we performed data augmentation only on the training data, after the splitting phase. We load all the data on a Google Colab instance with an available GPU. Then we have splitted the data on train and test set with a ratio of 0.8/0.2. We used Tensorflow to build a Sequential model with 11 layers: the input layer, two dense 128 neurons layer, two dense 64 neurons layers, a dense 32 neurons layer, two dense 64 neurons layer, two dense 128 neurons layers, and an output layer with dimension equal to the input layer. All the activations for the layers are ReLU with a constraint of a max value of 1.0 (the max value of the score). We used an Adam optimizer with a learning rate of 0.001 and a mean squared error loss function. We trained the model for 100 epochs with a batch size of 128, and we have imposed a validation split of 0.2. The resulting learning curve is the shown in Fig. 2, whilst the accuracy curve measured is depicted in Fig. 3.

Since that the standard accuracy doesn't represent well the error of the model (i.e., we have a sparse matrix) we computed a sum of the squared errors on each samples in the test set. The resulting values follow the distribution shown in Fig. 4; in Table 1 we report some stats of the SSE distribution.

Overall, the results seem promising, and we plan to evaluate the effectiveness of this approach in the next Italian Training Camp for IOI, where we will ask learners to evaluate the tasks suggestions of three distinct RSes:

1. the dynamic approach described above, based on Autoencoder Neural Networks;
2. a *static* approach, also based on Autoencoder Neural Networks;

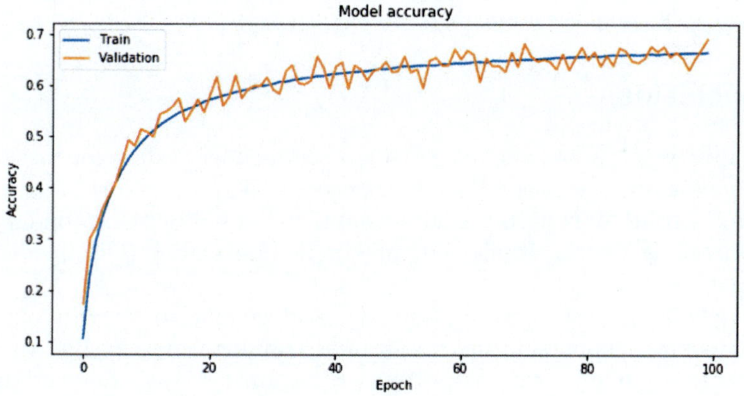

Fig. 3. Model accuracy.

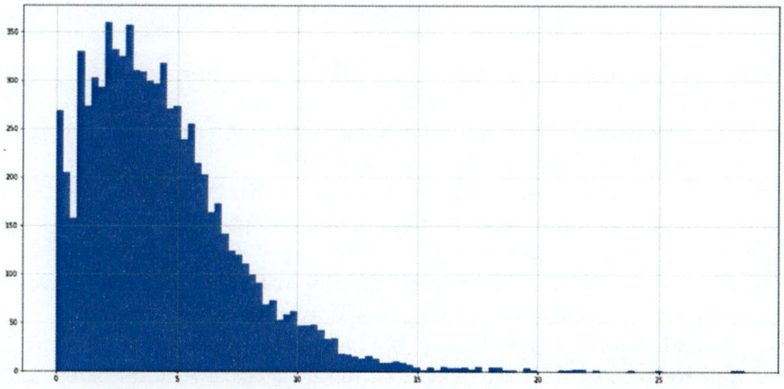

Fig. 4. Distribution of sum of squared errors (SSE).

Table 1. Statistics of the distribution of SSE (Fig. 4)

Mean	4.40
Std deviation	3.10
Min	0.00
25% (1st quartile)	2.12
50% (2nd quartile)	3.86
75% (3rd quartile)	5.98
Max	28.56

3. a more traditional Recommender System, i.e. the one provided by the python scikit library *Simple Python RecommendatIon System Engine* (SurPRISE[1]).

4 Conclusions

In this paper we proposed the design of a recommender system for tasks suggestions in Online Judges, based on a Autoencoder Neural Network.

We trained the ANN with the data from the OJ used by the secondary school students training for the Italian Olympiads in Informatics (Olimpiadi Italiane di Informatica - OII) [11,12].

Our preliminary results seem promising, and we plan to carry on our investigations by asking students, in the next Italian Training Camp for IOI, to evaluate the suggestions made by this ANN based RS against a static one and against a more classical recommender system built using the Simple Python RecommendatIon System Engine (SurPRISE)). After this experimental evaluation, based on the results, we plan to implement a RS inside the OII Training platform.

References

1. Ala-Mutka, K.M.: A survey of automated assessment approaches for programming assignments. Comput. Sci. Educ. **15**(2), 83–102 (2005)
2. Amaroli, N., Audrito, G., Laura, L.: Fostering informatics education through teams olympiad. Olympiads Inform. **12**, 133–146 (2018)
3. Astrachan, O.: Non-competitive programming contest problems as the basis for just-in-time teaching. In: 34th Annual Frontiers in Education, FIE 2004, pp. T3H/20–T3H/24, vol. 1, October 2004
4. Audrito, G., Demo, G.B., Giovannetti, E.: The role of contests in changing informatics education: a local view. Olympiads Inform. **6**, 3–20 (2012)
5. Audrito, G., Mascio, T.D., Fantozzi, P., Laura, L., Martini, G., Nanni, U., Temperini, M.: Recommending tasks in online judges. Methodologies and Intelligent Systems for Technology Enhanced Learning. AISC, vol. 1007, pp. 129–136. Springer, Cham (2019)
6. Blumenstein, M., Green, S., Fogelman, S., Nguyen, A., Muthukkumarasamy, V.: Performance analysis of game: a generic automated marking environment. Comput. Educ. **50**, 1203–1216 (2008)
7. Caiza, J., Del Alamo, J.: Programming assignments automatic grading: Review of tools and implementations. In: INTED2013 Proceedings, 7th International Technology, Education and Development Conference, IATED, 4–5 March, 2013, pp. 5691–5700 (2013)
8. Caro-Martinez, M., Jimenez-Diaz, G.: Similar users or similar items? comparing similarity-based approaches for recommender systems in online judges. In: Aha, D.W., Lieber, J. (eds.) Case-Based Reasoning Research and Development, vol. 10339, pp. 92–107. Springer, Cham (2017)
9. Chen, Y., de Rijke, M.: A collective variational autoencoder for top-n recommendation with side information. In: Proceedings of the 3rd Workshop on Deep Learning for Recommender Systems, pp. 3–9 (2018)

[1] http://surpriselib.com.

10. Dagienė, V.: Sustaining informatics education by contests. In: International Conference on Informatics in Secondary Schools-Evolution and Perspectives, pp. 1–12. Springer (2010)
11. Di Luigi, W., Fantozzi, P., Laura, L., Martini, G., Morassutto, E., Ostuni, D., Piccardo, G., Versari, L.: Learning analytics in competitive programming training systems. In: 2018 22nd International Conference Information Visualisation (IV), pp. 321–325, July 2018
12. Di Luigi, W., Farina, G., Laura, L., Nanni, U., Temperini, M., Versari, L.: oii-web: an interactive online programming contest training system. Olympiads Inform. **10**, 195–205 (2016)
13. Di Mascio, T., Laura, L., Temperini, M.: A framework for personalized competitive programming training. In 2018 17th International Conference on Information Technology Based Higher Education and Training (ITHET), pp. 1–8, April 2018
14. Garcia-Mateos, G., Fernandez-Aleman, J.L.: Make learning fun with programming contests. In: Transactions on Edutainment II, pp. 246–257. Springer (2009)
15. Li, S., Kawale, J., Fu, Y.: Deep collaborative filtering via marginalized denoising auto-encoder. In: Proceedings of the 24th ACM International on Conference on Information and Knowledge Management, pp. 811–820 (2015)
16. Sedhain, S., Menon, A.K., Sanner, S., Xie, L.: Autorec: autoencoders meet collaborative filtering. In: Proceedings of the 24th International Conference on World Wide Web, pp. 111–112 (2015)
17. Strub, F., Mary, J.: Collaborative filtering with stacked denoising autoencoders and sparse inputs (2015)
18. Toledo, R.Y., Mota, Y.C.: An e-learning collaborative filtering approach to suggest problems to solve in programming online judges. Int. J. Distance Educ. Technol. **12**(2), 51–65 (2014)
19. Van den Oord, A., Dieleman, S., Schrauwen, B.: Deep content-based music recommendation. In: Advances in Neural Information Processing Systems, pp. 2643–2651 (2013)
20. Wang, H., Wang, N., Yeung, D.-Y.: Collaborative deep learning for recommender systems. In: Proceedings of the 21th ACM SIGKDD International Conference on Knowledge Discovery and Data Mining, pp. 1235–1244 (2015)
21. Wang, T., Su, P., Ma, X., Wang, Y., Wang, K.: Ability-training-oriented automated assessment in introductory programming course. Comput. Educ. **56**, 220–226 (2011)
22. Wang, X., Wang, Y.: Improving content-based and hybrid music recommendation using deep learning. In: Proceedings of the 22nd ACM International Conference on Multimedia, pp. 627–636 (2014)
23. Yera Toledo, R., Caballero Mota, Y., Martínez, L.: A recommender system for programming online judges using fuzzy information modeling. Informatics **5**(2), 17 (2018)
24. Zhang, S., Yao, L., Xu, X.: Autosvd++: an efficient hybrid collaborative filtering model via contractive auto-encoders. In: Proceedings of the 40th International ACM SIGIR Conference on Research and Development in Information Retrieval, SIGIR 2017, pp. 957–960. Association for Computing Machinery, New York (2007)

Lessons Learned from Implementing Blended Learning for Classes of Different Size

Galena Pisoni(✉)

Department of Information Engineering and Computer Science, University of Trento, via Sommarive 9, 38122 Trento, Italy
galena.pisoni@unitn.it

Abstract. Research investigating why blended learning, despite many of its advantages, is usually difficult to scale up for classes of different size has been needed for a long time. Research data were obtained from 3 courses with different numbers of students: one course with 4 students, one course with 26 students, and one course with 94 students and the data were compared to identify factors that influence scaling up of blended education. All of the groups used Moodle as LMS. This study investigates differences in operational, instructional and technological factors between these three courses which adopted blended learning, in an effort to understand the challenges and obstacles inherent in the successful implementation of blended learning for classes of different size. Findings indicate that no significant differences exist in the effort required to set the course for each of the three groups, neither in managing the registration of students, however, significant differences existed in handling subsequent issues arising from selected delivery format options. Discussions about improving on-line and blended delivery methods are elaborated upon based on the research findings. We also discuss implications for deployment of blended learning for Universities.

Keywords: Blended learning · Online education · Class size · Blended model · I&E education

1 Introduction

Blended or hybrid instruction uses a combination of face-to-face and online learning activities that are integrated in a planned and pedagogically valuable way, where some of the face-to-face time is replaced by online activities [12,27]. The main reason for using blended learning is to improve student activation and participation in class, which is usually more difficult to obtain in medium to large courses.

Adapting to blended instruction is a difficult task. The classroom is the place where goals and tasks are laid out and rules are specified, and student

Z. Kubincová et al. (Eds.): MIS4TEL 2020, AISC 1236, pp. 206–215, 2021.
https://doi.org/10.1007/978-3-030-52287-2_21

involvement is heavily influenced by the course structure [24]. In this paper we describe the challenges encountered and factors influencing the adoption of blended learning for different class sizes.

Small classes are generally perceived as desirable and researchers have favoured smaller classes over larger ones for better teacher effectiveness, teacher-student interaction, and student achievement [13,16]. Students in smaller classrooms naturally gain more intense individual attention from teachers which, in turn, improves their chances of learning [4,9].

On the other hand, in big classes teacher have less time to dedicate to single students and it is more difficult logistically to provide classroom activities that require the students to engage in the learning process. In larger classes, it is more common for students to come to classes less prepared and they are less willing to participate in the lesson. Students seem happy to passively absorb knowledge and when questions are asked, few students volunteer answers, and when they do, it is usually the same students [19].

Previous research describes lessons learned from implementing blended learning at scale and the authors found that for blended learning to be successfully implemented at institutions there must be an alignment of institutional, faculty, and student goals [1,7,18].

Other works studied projects in order to understand the bottom-up change process used in blended learning. They report that the faculty involved in the course re-design process must share a common vision to enable blending and that the institutions should not be overly strict in imposing how to implement blended learning [5]. Others observe that more "autonomous" blended learning styles, where students watch the content and take quizzes or assignments after watching videos or peer evaluate themselves are better suited to bigger classrooms, where approaches based on a flipped classroom may be better for medium to small classes [20].

Another study investigated six institutions at various stages of blended learning implementation and the researchers conducted interviews with senior leaders of different universities. They claim that at the initial stage of the blended learning implementation it is important to have an "advocate" for blended learning and that when blended learning begins to mature, more robust support systems must be in place to support blended learning at scale [6].

A relevant answer to the question of how to scale the blended learning implementation requires careful evaluation of the different factors influencing the use of blended learning, namely *operation factors* (how easy or difficult it is to set up the online course and how this time is influenced by the different class sizes), *instructional factors* (what kind of pedagogical methods are more suited for successful implementation of blended learning at scale), and *technological factors* what kind of issues appear from an LMS point of view). This paper tries to provide such an analysis of the problem at stake.

In this paper we present three different blended learning settings and investigate operational, instructional and technological factors that lead to a (successful) blended learning implementation. We compare the settings and reflect on what factors contribute to effective blended learning and how easy (or difficult) it was to deliver the modules for classes of different sizes. The three settings that we describe in this paper come from three different universities in the network of EIT Digital. The small size class contained 4 students, the middle size class contained 26 students, and the large class contained, in total, 94 students. We used the same course on Innovation and Entrepreneurship Basics delivered in all of the places at the same time for all of the blended learning implementations.

This paper is structured as follows: Sect. 2 presents the educational con- text, the course structure, and the session to which the students were exposed in the three settings of this study. Section 3 presents the overall reflections on instructional, organizational and technological factors that influenced the implementation of blended learning for the different classroom sizes, in Sect. 4 we discuss the implications of our study for the future design of blended classes for classes of different size, and in Sect. 5 we outline our next steps.

2 Course "Innovation and Entrepreneurship Basics"

All three of the Universities are part of the European Institute of Innovation and Technology (EIT) and more specifically of the EIT Digital network of Universities. The EIT Digital Master School is a joint initiative by the leading technical universities and business schools in Europe with the aim of training IT graduates at Masters level, with strong innovation and entrepreneurial competences. Students have the opportunity to collaborate with industrial partners and bring innovation to market as part of their Innovation and Entrepreneurship studies.

Each partner university in the EIT Digital network implements an Innovation and Entrepreneurship (I&E) minor, for which the Universities must implement three harmonized I&E courses: I&E Basics, Business Development Lab, I&E study and one elective course. All the courses for their implementation rely on blended learning [20,21]. The blended settings we describe in this paper are used on the course of I&E Basics.

All the universities followed the same guidelines on how to implement the courses and have the same intended learning outcomes (ILOs). The main objective of the course is to enable the students to:

- Get an in-depth understanding of the general process and roles involved in developing an idea and starting up a new technology-based company,
- Develop the ability to systematically explore customers and markets,
- Understand alternative technological solutions already or nearly in the market (tech watch), so as to identify the potential value technology in the value chain of existing companies.

Different ways of implementing entrepreneurship education have been described in the literature [2,3,11,23]. Entrepreneurship education provides a mix of experiential learning, skill building and, most importantly, mindset shift which present an additional challenge in how to adequately teach this online [8,10,15,17]. In our scenario all the course organizers are supported in their blending efforts and are provided full support in setting up their personal blended courses. The personalized courses are usually copies of template courses that contain a pre-packaged session on the themes of the course as per the guidelines. The templates were compiled following different educational practices in the literature on how to design relevant entrepreneurship learning experiences [14,22,25,26].

The list of sessions includes:

- Introduction to Digital transformation
- Opportunity Generation
- Business Model Development
- Human Resources
- Commercialization Strategies
- Leadership
- Design thinking

Each session is accompanied by a quiz, an assignment (that can be opened also in peer review modality), and a proposal for activities that teachers can do with the students in class.

2.1 Small Size Class

This course organizer decided, for its implementation of the course, to use only one of the sessions from the blended blue-print, that is the session on the theme of "Introduction to Digital Transformation" and asked the students to do a peer review assignment for it. The class had only 4 students.

2.2 Middle Size Class

In the implementation of the course at this university, the teachers decided that they would open all the available sessions with mandatory requirements for the students to deliver a quiz for the sessions on "Business Model Development", and a mandatory requirement to deliver a peer review assignment for the assignment on "Introduction to Digital Transformation". The class had 26 students.

2.3 Large Size Class

The teaching staff at this university decided to use for their blended learning implementation three of the sessions as provided in the template of the blended course, the sessions on "Leadership", "Design Thinking" and "Introduction to Digital Transformation". For all of the sessions the students were asked to deliver peer review assignments. The class totalled 94 students.

2.4 Method

The author and involved teachers collected observations and took notes during the execution of the courses on the three aspects: operational, instructional and technical challenges and the obstacles we observed in our scenario. We converted these into guidelines and suggestions for future designers on how to create valuable technology enhanced learning (TEL) experiences for classes of different sizes.

3 Factors Influencing the Implementation of Blended Learning at Scale

3.1 Operational Factors Influencing Blended Learning

Care was taken by central support staff of EIT Digital in the creation of the courses as well as in the addition of students to them. A template course was created, copies were made and specific adaptations were made for each university. Central support staff also took care of student additions to the online courses created.

What was observed is that the creation of the template significantly reduced the time needed to set up the course for each university, so there were no significant difference observed in the time needed to set a course for blended learning between the three different universities with 4, 26 and 94 students respectively. The version of Moodle we used for the pilots allows for bulk enrolment of students to courses, so no significant differences in time to add students to the course was observed.

3.2 Instructional Factors Influencing Blended Learning

Two of the Universities relied on more "autonomous" delivery of the online modules, with non-mandatory follow up of the contents in class (the small and large size class respectively), while the middle size class university employed a flipped classroom.

Instructors that implemented the flipped classroom had more tasks when teaching a blended lesson compared to the teachers that opened the modules to the students in autonomous modality. In addition to delivering the contents and facilitating discussion, they needed to monitor the students in their online part and some of them expressed that it was not easy to maintain all the different roles in the new blended setting.

From our experience the blended learning experiences relying on peer review assignments were significantly more problematic when compared to quizzes for several reasons.

First, some students were insensitive when marking other students and some, for instance, marked the assignments of their peers quite low although the quality of the assignment was good. Due to this many students complained to the teachers on the received mark or even on the quality of the review received, and this created an excessive amount of communication that had to be handled, which was especially difficult in the case of the large class where the course organizers had three peer review assignments and 94 students. We expected 'MOOC-like behaviour' in which students would need no further assistance, given that the level of instructions provided was purposely high and given that the students were Master's level students. However, this wasn't the case and one learning we deduct from the pilot is that in settings like ours, having the teacher as a person who the students can address, and who acts as an independent "judge" on the spot, created an environment in which the students felt even more encouraged to ask for a better mark or further explanations. Another lesson learned from this was that in order to prevent differences in sensitivity in students, and to put all the students on the same level of understanding of the task, additional training needs to be adequately planned to teach the students how to properly perform the review task.

The second observed problem with peer review assignments was that, by design, they required that one moved to the review phase only when all the students have finished their assignments This presents a problem in practice, especially in big classes, as there were students that were late, requiring additional communication with the students to ensure that they could deliver the assignments on time. Again, this was most noted in the case of the large class, where the fact that three peer review assignments were used made the task more difficult.

One of the teachers involved commented that they had a lot more work to do since they had to check with the central support staff for delivery every week. They also needed to prepare each week for the questions raised by the students in this respect.

In general, in all of the settings the students were engaged in learning. One student commented that at the beginning engagement was low due to the lack of clear instruction from the teacher, however, as the assignment became clearer, the engagement level improved significantly.

3.3 Technical Factors Influencing Blended Learning

The support needs varied with the instructors' technical abilities and the complexity of the setting decided at the beginning. The planned settings had a considerable effect on the instructional effort and overall success of the blended setting for the I&E Basics course across the three universities as described below.

We used the Workshop tool in Moodle for the peer review assignment, and experienced problems with the setting of the tool (small and middle size class respectively), and the platform support team needed to react to correct the experienced situation. The problems revolved around how many "aspects" students needed to review, that is the tool that allows the teachers to set out how many "aspects" the students needed to provide feedback on to their peers. By default the tool expects at least two aspects, while the teachers were convinced that the tool would work with only a single aspect (text field) presented to the students. Additionally, the aspect option caused a further problem as each aspect must receive a mark, and the marks of the two aspects must be merged to provide one final mark on the assignment to the students. The involved teachers are not tech savvy, which additionally limited any possibility of solving issues ad hoc, and the teachers opened the peer review assignments and only then understood that there were problems in the settings of the Workshop tool. Another lesson learned was that there's a need to thoroughly check the settings before the assignments are opened to students. This has been included in the new practices of the group, that is to ask the central staff to check the assignments before opening them to the students.

The settings of the quizzes caused further discussion as well. The quizzes in Moodle are, by default, set to show the right answer for each quiz question. The teachers wanted to implement a system in which the students take the quiz 5 times and the platform stores only the highest score among the trials, and the fact that the correct answers were presented to the students made the job easy and already at the second or third trial the students were obtaining the highest scores. Teachers commented that quizzes should show only the total mark at the end of the quiz and not the individual mark (and correct solutions) for each question. This was not possible with the version of Moodle we have.

The general impression was that the technological support was average and could be further improved. Teachers both, acknowledged and criticized the support the technology was providing in their work. One teacher commented for instance: "Couldn't have done it without a support team, especially for the peer review assignments". The common problems reported were mainly of a technological nature and problems with understanding the settings of the assignments and quizzes at all the locations.

4 Discussion

Our aim with this paper was to understand the factors for scaling and delivering blended learning to classes of different sizes. In our study, students were exposed to blended learning with standardized online content on I&E themes and we tried to understand how easy or difficult it is to use blended learning for courses of different class sizes.

Although the instructional model differed between the different settings, the online contents presented were the same, and the assignments whether followed in each of the modalities (in flipped classroom or in more "autonomous" MOOC-like approach) were the same, which made the differences between the settings less evident from a course design point of view and made it easier for us to compare the different settings. What we observed is that there were no significant differences in the time needed to set up courses, or to handle the addition of students. We found, however, that the handling issues resulting from blended learning (and more in particular from setting up peer review assignment and quizzes) can be significantly more demanding in courses with large numbers of students, compared to courses with small classes. What we observed additionally, is that although peer review assignments require more support time, they bring improved learning and teachers felt that students learned more in the sessions where they per- formed peer review assignments compared to the sessions for which they took a quiz as an assessment method.

From our experience, it also seems that instructors with strong technological skills may have more success in implementing blended learning at different scales. It seems that student-facilitated approaches with more students are more likely to require further assistance. Consequently, teachers said that the implementation as we did it, where they were supported in the blended learning implementation was key to success, and that they needed to have someone to handle technical issues while they carry on with instruction.

We did not measure the impact of our approach on the learning outcomes of the students in the three classes. However the impression of the central EIT Digital staff was that the students' learning depended on the teacher involvement and the teacher's interest in the subject matter of the online content. For instance, the teachers of the middle and large classes were equally passionate about the topics presented, and, consequently, the quality of the submissions seemed to be, on average, higher than that of the small class. Further experiments should seek to understand if indeed there is a correlation between teacher involvement and student success in such settings as ours.

Despite the challenges, the authors of this article express optimism in implementing blended learning in large classes. The approach we took, in preparing templates from which teachers created copies and customized them for their delivery made the platform tasks easier for teachers allowing them to focus on teaching which they highly appreciated.

Additionally, the experiments forced the faculty of the three universities to coordinate and share practices more with the central EIT Digital staff, enriching the original version of I&E Basics course, and thus contributing to the possible improvement of the course as well.

In the experiment we also noticed that, the fact that all the universities delivered the same contents, would be a perfect scenario to develop and practice shared assignments between the classes from the different settings (especially around topics important for innovation and entrepreneurship education such as communication and leadership) and this is what we aim to try in our next steps.

5 Conclusions and Next Steps

Our paper describes the operational, instructional and technological factors that are at stake when one implements blended learning at scale. Our findings indicate that no significant differences existed in the effort of setting the course between the three groups, neither in managing the registration of students, however, significant differences existed in handling subsequent issues arising from the selected delivery format options, as well as from different instructional methods used.

One limitation of the current research is that it used observational data, making it difficult to isolate the personal bias of the teachers or the author.

Future studies should aim to further understand the effect of class size by considering additional factors. First the exact times needed to perform the different tasks on the platform, as well as detailed logging of the difficulties observed by the teachers, while the course is taking place should be set and used. Second, course-design factors should be minimized, and in order to understand only the effect of the class size, the different types of tasks should be minimized, and a symmetrical design study should be employed in which the different settings would be exposed to the same type of assignments. Third and last, we will try to understand the effect on the mark and in general on the learning dimension, and if and how the different class size influenced the learning, in addition to investigation into student satisfaction. A broader assessment of this kind would be especially interesting in other online course settings.

Acknowledgment. The authors would like to thank the local teaching staff for their participation and all the students who took part in the blended learning pilots. The pilots were supported financially by EIT Digital under the Innovation and Entrepreneurship Improvements Education projects.

References

1. Boling, E.C., Hough, M., Krinsky, H., Saleem, H., Stevens, M.: Cutting the distance in distance education: perspectives on what promotes positive, online learning experiences. Internet Higher Educ. **15**(2), 118–126 (2012)
2. Buchta, K., Jakubiak, M., Skiert, M., Wilczewski, A.: Universities' social responsibility vs. graduates' human capital. Eur. J. Serv. Manag. **26**(2), 15–22 (2018)
3. Buchta, K., Jakubiak, M., Skiert, M., Wilczewski, A.: The analysis of students' expectations as a marketing challenge of a modern university. Przedsiskobiorczo i Zarzdzanie **20**(6, cz. 2 Zmiany w myśleniu marketingowym), 41–51 (2019)
4. Burruss, N.M., Billings, D.M., Brownrigg, V., Skiba, D.J., Connors, H.R.: Class size as related to the use of technology, educational practices, and outcomes in web-based nursing courses. J. Prof. Nurs. **25**(1), 33–41 (2009)
5. Carbonell, K.B., Dailey-Hebert, A., Gijselaers, W.: Unleashing the creative potential of faculty to create blended learning. Internet High. Educ. **18**, 29–37 (2013)
6. Dziuban, C., Hartman, J., Juge, F., Moskal, P., Sorg, S.: Blended learning enters the mainstream. The handbook of blended learning: Global perspectives, local designs **195**, 206 (2006)
7. Graham, C.R., Woodfield, W., Harrison, J.B.: A framework for institutional adoption and implementation of blended learning in higher education. Internet High. Educ. **18**, 4–14 (2013)
8. Halberstadt, J., Schank, C., Euler, M., Harms, R.: Learning sustainability entrepreneurship by doing: providing a lecturer-oriented service learning framework. Sustainability **11**(5), 1217 (2019)
9. Hattie, J.: The paradox of reducing class size and improving learning outcomes. Int. J. Educ. Res. **43**(6), 387–425 (2005)
10. Jakubiak, M., Buchta, K.: Determinants of entrepreneurial attitudes in relation to students of economics and non-economics. Studia i Materiały (2/2016 (21), cz. 1), 17–30 (2016)

11. Jie, S., Harms, R.: Cross-cultural competences and international entre preneurial intention: a study on entrepreneurship education. Educ. Res. Int. **2017**, 17 (2017)
12. Kenney, J., Newcombe, E.: Adopting a blended learning approach: challenges encountered and lessons learned in an action research study. J. Asynchronous Learn. Networks **15**(1), 45–57 (2011)
13. Kokkelenberg, E.C., Dillon, M., Christy, S.M.: The effects of class size on student grades at a public university. Econ. Educ. Rev. **27**(2), 221–233 (2008)
14. León, G., Leceta, J.M., Tejero, A.: Impact of the EIT in the creation of an open educational ecosystem: UPM experience. Int. J. Innov. Sci. **10**(2), 178–206 (2018)
15. León, G., Tejero, A., Dévora, N., Pau, I.: University as a platform: an evolutionary process towards an open educational ecosystem in Europe (2020)
16. Lin, C.H., Kwon, J.B., Zhang, Y.: Online self-paced high-school class size and student achievement. Educ. Tech. Res. Dev. **67**(2), 317–336 (2019)
17. Maresch, D., Harms, R., Kailer, N., Wimmer-Wurm, B.: The impact of entrepreneurship education on the entrepreneurial intention of students in science and engineering versus business studies university programs. Technol. Forecast. Soc. Chang. **104**, 172–179 (2016)
18. Moskal, P., Dziuban, C., Hartman, J.: Blended learning: a dangerous idea? Internet High. Educ. **18**, 15–23 (2013)
19. Owston, R.: Blended learning policy and implementation: Introduction to the special issue (2013)
20. Pisoni, G.: Strategies for pan-european implementation of blended learning for innovation and entrepreneurship (i&e) education. Educ.Sci. **9**(2), 124 (2019)
21. Pisoni, G., Renouard, F.: Integrating online education in innovation and entrepreneurship (i&e) doctoral training program. In: 2019 17th International Conference on Emerging eLearning Technologies and Applications (ICETA), pp. 633–638. IEEE (2019)
22. Resei, C., Friedl, C., Żur, A.: Moocs and entrepreneurship education-contributions, opportunities and gaps. Int. Entrepreneurship Rev. **4**(3), 151–166 (2018)
23. Tejero, A., Pau, I., León, G.: Analysis of the dynamism in university-driven innovation ecosystems through the assessment of entrepreneurship role. IEEE Access **7**, 89869–89885 (2019)
24. Weaver, R.R., Qi, J.: Classroom organization and participation: College students' perceptions. J. High. Educ. **76**(5), 570–601 (2005)
25. Żur, A.: Changing entrepreneurship learning ecosystems: Massive open online courses. opportunities and limitations. Przedsikebiorczo-Edukacja **14**, 473–482 (2018)
26. Żur, A.: Two heads are better than one-entrepreneurial continuous learning through massive open online courses. Educ. Sci. **10**(3), 62 (2020)
27. Zydney, J.M., McKimmy, P., Lindberg, R., Schmidt, M.: Here or there instruction: Lessons learned in implementing innovative approaches to blended synchronous learning. TechTrends **63**(2), 123–132 (2019)

A Pilot Study to Inform the Design of a Supportive Environment for Challenge-Based Collaboration

Galena Pisoni[1](✉) and Hannie Gijlers[2]

[1] Department of Information Engineering and Computer Science, University of Trento, via Sommarive 9, 38122 Trento, Italy
galena.pisoni@unitn.it

[2] University of Twente, Drienerlolaan 5, 7522 Enschede, NB, The Netherlands
a.h.gijlers@utwente.nl

Abstract. In the course "An introduction to FinTech: from mobile payments to blockchains", students are introduced to the broad area of FinTech, from both technical and business side. The course consists of two technical lectures that explain the students the basic technology for authentication and blockchain and two entrepreneurship lectures that discuss the fintech business ecosystem. In the rest of the course the students work on a project provided by companies in which they put into practice the obtained knowledge and provide a solution to a real life challenge. The course is face-to-face, with two weekly sessions of 2 h each. The students work in teams and each week they need to show progress/work towards the challenge resolution. The students are not constrained in any way and are asked to self organize for the team work. In this study we present the results we obtained from questionnaires that we delivered before the course on course expectations (that we used also to form the teams), during the course on teams' dynamics, and after the course to measure students' perception on collaboration strategies. We end the paper with a reflection on the design space for an application that can support this type of collaborative work online.

Keywords: Collaborative work · Team dynamics · Challenge-based education

1 Introduction

In order to meet the requirements of the 21st century labor market educational strategies need to adapt and focus more on strategies that promote self-regulated learning, problem solving and collaboration (teamwork). In the current paper we focus on teamwork in the context of solving authentic problems. Authentic problems presented by stakeholders form industry invite groups of students to demonstrate their knowledge and skills through the development of an innovative

Z. Kubincová et al. (Eds.): MIS4TEL 2020, AISC 1236, pp. 216–225, 2021.
https://doi.org/10.1007/978-3-030-52287-2_22

product addressing the presented problem. In this context students not only apply their knowledge and learn to deal with stakeholders within companies but they also develop their ability to operate in (multi) disciplinary teams. Challenge-based education is an instructional approach that addresses the challenges of preparing university students for the 21st century labor market.

Challenge-based education is an alternative to traditional education in training engineering graduates to become independent learners, critical thinkers, problem solvers, lifelong learners as well as team players [10]. This educational model is relatively new, dates from around 2011, and it is built on problem-based learning, and it represents the next step forward. Challenge-based learning does not only require a problem as the center of the learning process, it also requires the challenge proposer to be involved as a stakeholder and to intensively cooperate with the students and guide them while they are working on the project. This approach facilitates students in developing practical competences through the resolution of a real business case [10,15].

In challenge-based learning scenarios lecturers participate as mentors. They are in charge of the knowledge construction process and monitor the development of skills and competences. In such a context, the lecturers monitor the level of knowledge the learners gain during the learning process and at the same time make the students and challenge proposers collaborate with each other during the learning and work on the challenge. In challenged based education the methods of teaching and training as well as the specifics of the contents and tasks are constantly adapted based on the goals achieved by the students.

Teaching entrepreneurship is challenging due to its practical and experiential nature of the topic [4,6,11]. Entrepreneurship education is growing in popularity as it helps the transition from education to the labour market, and different initiatives aim at spreading entrepreneurship in the European higher education system [11,20]. However, key educational and didactic issues on how to implement entrepreneurship education adequately still remain open [5,9].

Specifically tailored challenge-based courses in different entrepreneurship areas are starting to appear, with different possible course designs and course structure of such examples [21]. Still, besides all the advantages of the challenge-based learning, there are actually little in our universities that implement it, and this paper showcases one example of how one such course.

Learning to work in teams is an important skill in today's (or modern) organizations and companies, because employees are often required to complete tasks and projects in flexible and multidisciplinary teams [1,2]. Although students can acquire important skills in teamwork and teamwork is positively associated with performance [14], grouping students in teams does not guarantee productive interaction [19].

Moreover, students' often have mixed feelings about teamwork [16]. Students expressing negative opinions about teamwork often refer to issues related to free-riding/unequal participation and issues related to receiving a grade as an entire team. Free-riding occurs when students do not contribute their fair share to the team. In situations where the process and students' individual efforts toward the overall team task are not taken into account in the assessment this is detrimental for students' attitude towards teamwork [7].

Furthermore, teams might experience difficulties reaching consensus, or dealing with diverse opinions and individual schedules. Research indicates that students' attitude towards teamwork, their beliefs in the effectiveness influence not only their willingness to cooperate with others [13] but also the outcomes of the team process [8]. However, if during a team assignment students have positive experiences like task-focused communication, and equal participation of the team members this positively influences their attitude towards future team assignments [17].

These findings stress that it is important to gain understanding of how students' attitudes and experiences develop over the course of a team activity, to ensure effective and satisfying team learning experiences. These insights can be used by educational designers and software developers to design adequate supportive measures, that allow for timely intervention when students experience problems.

Within the context of the current study we explore students expectations, experience with teamwork, as well as the reported difficulties after the completion of the task in the context of challenged-based team assignment. The results will be used to inform the design of an interactive learning environment or app that facilitates and supports effective team work in the challenge based education.

Our aim is to study team dynamics in students coursework, follow team work with standardized questionnaires, and from the experience draw conclusions about how to support teachers in following teams with different dynamics with a tool specifically designed for this. We contribute to the field by first exploring where difficulties in group work appear and then provide suggestions on how to track them in teams.

This paper is structured as follows: in Sect. 2 we present the method, the course setting and the participants, in Sect. 3 we present the results from the questionnaires as well as the perceptions regarding teamwork during the course work, as well as the observed challenges and benefits from the team work after the course as well as their impact on student learning and quality of the deliverables, in Sect. 4 we discuss design space for an application that can support this type of collaborative work online, and in Sect. 5 we conclude the paper and outline our next steps.

2 Methods

The approach we took to study team dynamics was the following: we first delivered an introductory questionnaire to understand better the starting expectations and motivations of the students towards the course and group work, later, during the course, we administered the midterm questionnaire, the aim was to track students perceptions of the activities, and at the end of the course we provided the final questionnaire to invite students to reflect on their participation in the group; for all of the three phases we relied on standardized questionnaires from the literature.

2.1 Participants

Third year Bachelors students, and 1st and 2nd year Master school students from Innovation program, following the Financial Technology course, participated in the study. Students were assigned to teams of 4 or 5, based on their responses on a questionnaire administered before the course, teams were formed consisting of students with a varying attitude towards teamwork. Due to the number of students that signed up and completed the course, the data from 25 participants were taken into account in our analysis. More specifically, the data originated from 6 teams, including 2 teams of 5 students, 3 teams of 4 students, and one team of 3 students.

2.2 Cases

Case 1. Pay with a Smile. Pay with a Smile (PwaS) aims to implement an evolutionary next step in the digital payments: develop a system that makes it possible to pay with biometrics (mainly face recognition, but later other methods are considered as well). PwaS doesn't require users to have a bank card or phone present. The payment is initiated by the merchant's device, face recognition is done by the back end system. PwaS aims for small retail stores (for frequent recurrent purchases) and tries to attract the widest audience possible. The students were asked to develop solutions around: i) the business model for the company (e.g. offered directly to merchants or use banks/acquirer companies as middlemen), ii) to come up with new use cases for use of biometrics and biometric identification in different scenarios, based in smooth interactions, and good user experience for final users, and iii) to suggest how services can preserve privacy and be GDPR-compliant. The teams were asked to choose between one of the three options of projects.

Case 2. mID 2.0. mID 2.0 is a new product (app) that securely stores identity credentials on smartphones. Depending on the security features of your smartphone hybrid storage is an option (e.g. partly cloud storage). With mID 2.0 users can use their smartphones as identity token/document. It replaces smartcards, tokens, documents... Service providers can securely verify the mID and provide access to services, e.g. a bank account, car rental, entry to a building. mID can be uses in face2face situations, human2machine or online. The students were asked to deliver: i) a reasonable business rationale for a bank to include mID 2.0 into a banking app, ii) to perform an analysis in terms of strengths, weaknesses, opportunities, treats in the European domain for a product like mID 2.0 and iii) to come up with a reasonable pricing model for a product like mID 2.0. Also in this case the students were asked to choose between one of the three options for projects.

2.3 Questionnaires

Three questionnaires were administered before participation in the team assignment expectations related to the team task were assessed, during the task students' perceptions about the collaborative learning processes and attitudes during the task were assessed, and after completion of the assignment students reflected on the challenges they experienced during the task. All questionnaires were administered using surveyhero (www.surveyhero.com) platform.

Introductory Questionnaire. Before the actual team work started students completed an adapted and contextualized version of the Students' Appraisal of group assignments questionnaire (SAGA), developed by [18]. Students completed the following six scales: 1) cognition, 2) motivation, 3) affect, 4) interpersonal, 5) team management and 6) group assessment. Students specified their level of agreement with each statement on a 4-point Likert scale (strongly disagree to strongly agree). An overview of the items in the questionnaire can be found in [18]. The alpha for the questionnaire as a whole reached .80 and Cronbach's alpha s of the subscales ranged between .44 and .79.

Midterm Questionnaire. At the midterm we used a translated and adapted version of PCLA questionnaire, an overview of the items can be found in [12] to get a global understanding of the cooperation as it is perceived by the students during the collaboration. The questionnaire consisted of a total of 15 items (cronbach's alpha = .82), 10 items addressing the group processes Cronbach's alpha = .73) and 5 items addressing the attitude towards the attitude/perceived utility value (Cronbach's alpha = .72). Sample items included: "All group members participated actively" (group processes); "By working in a group on the assignment, I have learned more than I would have learned individually" (attitude). Cronbach's alpha for the questionnaire as a whole reached .82.

End of Term Questionnaire. At the end of the course questionnaire was administered to gain understanding of the challenges individual team members were experiencing during the collaboration. Based on the work of [3] the questionnaire addressed the following four different aspects of the collaborative process: 1) Planning; 2) task execution; 3) progress monitoring; 4) and team functioning with respect to interaction and communication. For each of the four aspects challenges were formulated. An example of such a challenge for planning would be: We have different goals/standards for our work. A total of 22 challenges were presented to the students who had to report on a 5-point likert scale (from 1, not a problem to 5, a very big problem) to what extent these challenges were a problem in their team. Cronbach's alpha for the questionnaire as a whole reached .95. Items were recorded to make sure that high scores reflected a positive self-report.

3 Results

Mean scores on the SAGA questionnaires were compared with a Kruskall-Wallis test, results indicated no significant differences between teams on the test as a whole ($p = .858$), and the subscales cognitive ($p = .311$); motivation ($p = .749$); affect ($p = .749$); interpersonal ($p = .738$); team management ($p = .850$) and group assessment ($p = .337$). Indicating that on average teams did not differ in their expectations about the task (Fig. 1).

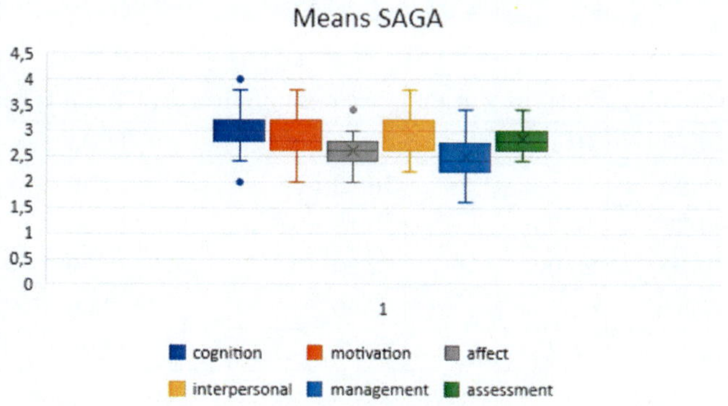

Fig. 1. The results from SAGA questionnaire.

The scale for the SAGA ranged between 1 and 4, and overall we can see that the mean scores are fairly positive, above the midpoint of the scale. The results of a one way anova for within subjects effects shows that there was a significant effect of type of scale $F(5, 105) = 10.88$, $p = .00$. Suggesting that students responded differently to the subscales. The items that were lower rated seem to relate to the reported affect.

Follow up pairwise comparisons revealed that between most scales significant differences were found, but that the cognition scale did not differ significantly from the interpersonal scale, the motivational scale did not differ from the affective, interpersonal and assessment scale, the affective scale did not differ significantly from the team management scale and the interpersonal scale also did not differ significantly from the assessment scale.

During the execution of the task students completed the PCLA questionnaire. The results of a nonparametric independent sample Kruskal-Wallis test revealed no significant differences between teams on both the scale addressing the group processes ($p = .232$) and the attitude towards teamwork ($p = .209$)

The end of the term questionnaire focussed on evaluation regarding the planning, task execution, progress monitoring and team functioning. The results of a Kruskall- Wallis test indicated significant differences between teams with respect to their scores on planning. Follow up analysis using Bonferroni corrections for multiple testing revealed that team 5 differed significantly from team 4 (adjusted $p = .016$), the differences between the other teams were not significant. The differences between teams with respect task execution ($p = .051$) and progress monitoring ($p = .057$) are bordering significance. Again the highest means are observed in team 4 and the lowest in team 5. Suggesting that team 4 evaluated the team task positively compared to team 5.

The results of the end of term questionnaire showed that team 5 reported difficulties related to different ideas about the standards for their work, as well as how they should work together. They also indicate problems with unmotivated team members, unequal participation, and report an unsupportive group climate (Table 1).

Table 1. Mean ranks for the teams on each of the Subscales of the end of term questionnaire.

Group	Planning	Task execution	Progress monitoring	Team functioning
1	10.75	10.12	12.75	11.25
2	15.90	14	16.40	12.60
3	9.25	10.75	10.12	15.50
4	20.20	20.00	19.20	18.60
5	2.67	3.50	3.50	3.00
6	14.12	15.25	11.25	13.25

Correlations between tests administered at the beginning, mid-term and end of term questionnaires were calculated. A positive and significant correlation between the reported group process at the midterm and the reported team functioning at the end term questionnaire was found ($r = 409$, $p = .047$).

Our impressions from the course confirmed the results obtained from the questionnaires, and it was indeed group 4 that was most cohesive, seemed to have best collaboration among the members and provided an in depth solution to the business challenge on several possible dimensions, whereas group 5 required most of the teachers attention, quite often requiring the teacher to intervene and to ask little proactive members to contribute to team work, or to send reminders to the whole team so that the team delivers the weekly deliverable together with the rest of the teams. The other groups performed sufficiently well and there was no need for any further teacher intervention in the group dynamics.

From our findings we can understand that the mid term questionnaire was most important for predicting the final outcomes of the team. As such we strongly suggest frequent administration of short questionnaires, most probably on a weekly basis. For future editions of the course that would be the aim, i.e to make it mandatory for each team to deliver the weekly submission and individually each team member to compile the questionnaire.

We next observed although many tools exist to help students to work together and assign tasks to members, a tools that allows teachers to monitor team dynamics and provide alerts when problematic dynamics occur or in general to allow for a teacher to have a comprehensive view on team dynamics over time doesn't exist, and in the next section based on our findings we draw guidelines for such tools.

4 Discussion of Guidelines

Designing task and assessment procedures that stress students' individual accountability and create a positive interdependence between students' can positively influence participation. However, for instructors it might be challenging to accurately assess the group processes. Software, supporting group work should address these challenges and provide features that not only support online collaboration and communication, but also increase students' awareness about their own as well others part in the group task.

We discuss further design principles for software to support group work for challenge-based courses, based on our initial results.

Contrasting the most and least successful group suggests that in the successful group students were able to reach agreement on the division of work and the quality criteria standards that they wanted to obtain, in the least successful group problems arose related to these issues. Our results are inline with the results of other studies investigating collaborative learning.

Team formation support based on cognitive and interpersonal, affect and management, and assessment and interpersonal similarity (or distance) between students

In order to tackle these problems the proposed tool should support students from different backgrounds that might have different views on collaboration in meeting each other online. Based on the approach that we used in our pilot provided is with information that suggests that it is important to take into account the different motivation traits of the students, teams can be formed based on cognitive and interpersonal affect and management, and interpersonal similarity. This can be evaluated based on introductory administration of the SAGA questionnaire with a mandatory follow-up and in-person interaction.

Group process and attitude towards work traced all the time

In our case we had a short term course over a period of one month, which made it difficult to administer the PCLA questionnaire more than once. However, we already saw first differences between the successful and less successful groups emerging at the midterm point, suggesting that in intensive courses, it is possible to identify emerging difficulties after two week of collaboration. In longer courses it might be relevant to ask the students to fill in the questionnaire periodically, to trace process and attitude towards work in students and capture variations in the students and their attitudes towards the team work. This feature would help teachers substantially in obtaining insights about team dynamics and each member's attitude towards the team work. Moreover, periodical assessment informs the teachers also about the effect of intervening.

Students need to be supported in planning of activities and the tool should allow for easy task execution and progress monitoring

Students told us that for them it was difficult to plan the distributed activities and that it was difficult to follow the "status" of different tasks that they had assigned to each other. This was especially important for tasks that required previous tasks to be finished in order to start working on them. One of our suggestions is to allow the students to create and trace tasks as well as their status, to have the possibility to assign task to team member, and to have an overview of the whole process of work towards the challenge, i.e in which phase each task is. Such a view will make it easier also for the teacher and the challenge provider to trace the status of the work towards the challenge too. Students reported use of different tools that allow for collaborative work in order to make their coordination on the task easier: some of them reported to have used Google Docs and Google Sheets for sharing of ideas and data, Hangouts and Skype for audio and video calls, and Trello for task management and task coordination.

A complete solution should allow students tools for video audio communications, shared writing space and sharing of data documents. Next to all these functionalities, it is very important for good teams functioning and dynamics to be able to plan, prioritize and assign tasks, this will additionally help identifying the problems with free-riding and quality of work. The display should also represent which sub-tasks are input for other tasks, like a Gantt chart. In addition, the tool should introduce to the students the basics of good group work, and support them with suggestions on how to achieve this (for instance, how to deal with free-riders, etc.).

5 Conclusions

In this paper we presented the results from an exploratory study on a challenge-based course regarding teamwork of the students, and provided recommendations and suggestions for a tool that can support team work for tracing project work during a course. It is assumed that this type of tool will enhance teamwork skills that are important to fully prepare graduates for the workplace.

Although the number of participants is small and results have to be interpreted with caution, the data provided us insights for the development of a tool. The experiment forced the students to coordinate and communicate often and highlighted aspects relevant pertinent to expectations, motivation, team dynamics and attitude important that need to be in place to have a successful implementation of a challenge-based course.

Future pilots are planned and foreseen with different challenge-based courses.

Acknowledgment. The authors would like to acknowledge and thank the companies (E-group and Innovalor) for their participation and all the students who took part in the course.

References

1. Buchta, K., Jakubiak, M., Skiert, M., Wilczewski, A.: The analysis of students' expectations as a marketing challenge of a modern university. Przedsiskobiorczo i Zarzdzanie **20**(6, cz. 2 Zmiany w myśleniu marketingowym), 41–51 (2019)
2. Cappelli, P., Keller, J.R.: Classifying work in the new economy. Acad. Manag. Rev. **38**(4), 575–596 (2013)
3. Hadwin, A.F., Bakhtiar, A., Miller, M.: Challenges in online collaboration: effects of scripting shared task perceptions. Int. J. Comput. Support. Collab. Learn. **13**(3), 301–329 (2018)
4. Halberstadt, J., Schank, C., Euler, M., Harms, R.: Learning sustainability entrepreneurship by doing: providing a lecturer-oriented service learning framework. Sustainability **11**(5), 1217 (2019)

5. Jakubiak, M., Buchta, K.: Determinants of entrepreneurial attitudes in relation to students of economics and non-economics. Studia i Materiały (2/2016 (21), cz. 1), 17–30 (2016)
6. Jie, S., Harms, R.: Cross-cultural competences and international entre preneurial intention: A study on entrepreneurship education. Educ. Res. Int. **2017** (2017)
7. Le, H., Janssen, J., Wubbels, T.: Collaborative learning practices: teacher and student perceived obstacles to effective student collaboration. Cambridge J. Educ. **48**(1), 103–122 (2018)
8. Lembke, S., Wilson, M.G.: Putting the "team" into teamwork: alternative theoretical contributions for contemporary management practice. Hum. Relat. **51**(7), 927–944 (1998)
9. León, G., Tejero, A., Dévora, N., Pau, I.: University as a platform: an evolutionary process towards an open educational ecosystem in Europe (2020)
10. Malmqvist, J., Rådberg, K.K., Lundqvist, U.: Comparative analysis of challenge-based learning experiences. In: Proceedings of the 11th International CDIO Conference, Chengdu University of Information Technology, Chengdu, Sichuan, PR China (2015)
11. Maresch, D., Harms, R., Kailer, N., Wimmer-Wurm, B.: The impact of entrepreneurship education on the entrepreneurial intention of students in science and engineering versus business studies university programs. Technol. Forecast. Soc. Chang. **104**, 172–179 (2016)
12. Mouw, J., Saab, N., Pat-El, R., van den Broek, P.: Student-and task-related predictors of primary-school students' perceptions of cooperative learning activities. Pedagogische Studiën **96**(2), 98–122 (2019)
13. Rudawska, A., et al.: Students' team project experiences and their attitudes towards teamwork. J. Manag. Busi. Adm. Central Europe **25**(1), 78–97 (2017)
14. Stump, G.S., Hilpert, J.C., Husman, J., Chung, W.t., Kim, W.: Collaborative learning in engineering students: gender and achievement. J. Eng. Educ. **100**(3), 475–497 (2011)
15. Tejero, A., Pau, I., León, G.: Analysis of the dynamism in university-driven innovation ecosystems through the assessment of entrepreneurship role. IEEE Access **7**, 89869–89885 (2019)
16. Thompson, D., Anitsal, I., Barrett, H.: Attitudes toward teamwork in higher education: a comparative study of religiously affiliated universities and secular-based universities. In: Proceedings of the Academy of Educational Leadership, vol. 13, Citeseer (2008)
17. Ulloa, B.C.R., Adams, S.G.: Attitude toward teamwork and effective teaming. Team Performance Manag. **10**(7–8), 145–151 (2004)
18. Volet, S.: Significance of cultural and motivation variables on students' attitudes towards group work. In: Student Motivation, pp. 309–333. Springer (2001)
19. Volet, S., Vauras, M., Salo, A.E., Khosa, D.: Individual contributions in student-led collaborative learning: Insights from two analytical approaches to explain the quality of group outcome. Learn. Individ. Differ. **53**, 79–92 (2017)
20. Żur, A.: Changing entrepreneurship learning ecosystems: Massive open online courses. opportunities and limitations. Przedsikebiorczo-Edukacja **14**, 473–482 (2018)
21. Żur, A.: Two heads are better than one-entrepreneurial continuous learning through massive open online courses. Educ. Sci. **10**(3), 62 (2020)

Intelligent Pedagogic Agents (IPAs) in GEA2, an Educational Game to Teach STEM Topics

Lauren S. Ferro, Francesco Sapio, Massimo Mecella, Marco Temperini(✉), and Annalisa Terracina

Department of Computer, Control, and Management Engineering, Sapienza University of Rome, Rome, Italy
{lsferro,sapio,mecella,marte,terracina}@diag.uniroma1.it

Abstract. In this work we present an Intelligent Pedagogic Agent (IPA) we developed to act within the *Gea2: A New Earth* educational game. The IPA was designed and developed based on a pilot study, where information was gathered from students and teachers of four classes of two high school institutions. The IPA has the ability to be conversational in Natural Language, and is also capable of autonomous intervention through the use of unsolicited hints during learner's gameplay. The autonomous intervention is based on both the detection of player's emotions, and the state of player's game.

Keywords: Intelligent Pedagogic Agent (IPA) · Serious game · Educational game · Game Based Learning · Interactive virtual tutor

1 Introduction

The modern world is digital and the current generation of students fall under the dome of *digital natives*, where for them, technology as it is today, has always existed. One context that has seen the value of technology is education. Gamification, and Game Based Learning [5,11] have surged ahead in the last two decades as a means to face the change, and to foster motivation and engagement in learners. Many educational games have arisen; in particular we refer here to *Gea2: a New Earth* [27], that was recently developed as a complex and composite system, deemed to support teaching and learning in *Science, Technology, Engineering, and Mathematics* (STEM) in High School settings.

Beside more traditional components of the game, such as the *Inventory*, or the *Chat*, and less traditional ones such as *Note Sharing* and *Animated Experiments* by 3D mini games, *Gea2* tries to provide the user with adaptive interaction, mainly by means of the *Unsolicited Hint System*.

Tuning and adaptation of the learning experience to the personal traits and current state of the learner is a winning strategy in education in general, and

Z. Kubincová et al. (Eds.): MIS4TEL 2020, AISC 1236, pp. 226–236, 2021.
https://doi.org/10.1007/978-3-030-52287-2_23

in Technology Enhanced Learning (TEL) in particular. The idea of adapting the learning tasks in educational games has emerged as well, although still less developed and firmly established. Following Conati in [4], there are two main challenges:

1. *Assessing students' knowledge and learning from the interaction with the game* (whereas the connection between learner's game action and understanding of the underlying domain might be not clear).
2. *Providing individualised interventions without interfering with the experience* (as, often, to get success, the system ought to provide help in ways that will not resemble, or remind the learner of, traditional educational activities).

In this paper, we describe how did we try to tackle these challenges, with the development of the Intelligent Pedagogic Agent (IPA) we integrated in *Gea2: A New Earth* to provide the learner with unsolicited hints based on the learner's state of interaction in the game. The detailed description of the game *Gea2: A New Earth* is outside the scope of this paper (some info here [27]). After a brief background, this work will explain how our IPA is able to answer in Natural Language and provide unsolicited hints, but exploring some the implemented Dialogue Management system as well as emotions have bee gathered from the students and implemented into a Bayesian Network model.

2 Background

Research within the field of Artificial Intelligence aims at developing intelligent educational software, such as computer-aided teaching tools known as Intelligent Tutoring Systems (ITS).

We define an ITS as a computer learning environment that helps the student master deep knowledge/skills by implementing powerful intelligent algorithms that adapt to the learner at a fine grained level and that instantiate complex principles of learning (following Graesser et al. in [9]).

In *ITS*s, the processes of tracking knowledge (called user modelling) and adaptively responding to the learner incorporate computational models in Artificial Intelligence and Cognitive Science [9]. Successful *ITS*s have been developed for programming languages [1], physics [6] and information technology [16]. *Auto Tutor* [8] helps college students learn about computer literacy, physics, and critical thinking skills by holding conversations in natural language. Other natural language *ITS*s, that have shown learning gains, include *DeepTutor* [23], *iSTART* [10], and *My Science Tutor* [30].

In this work, we have focused our attention on a sub branch of *ITS*, mainly on system capable of making a conversation in natural language: namely Conversational Agents (CAs).

2.1 Conversational Agent

Conversational *ITS*s have several advantages over other types of *ITS*s. They encourage deep learning as students need to explain their reasoning and reflect on their basic approach to solving a problem. The impact of conversational *ITS*s can be augmented by the use of animated agents [7]. For instance, among the most popular chatbot technologies are *ALICE* [29] and *ChatScript* [24].

Populating applied games with conversational agents is not novel, since agents have been used to create alive virtual environments. Some examples in this context are by [19] and [2], who use characters to guide the user in a virtual archaeological museum set in *Second Life* and in a virtual reconstruction of a real museum. The interactions in these examples are only passive, since the user is only allowed to receive information from the guide. Finally, [18] implements interactive storytelling through an embodied agent, who from time to time during gameplay, asks the user a question to verify his learning status, accepting an answer among four predetermined choices for the learner to select from. Other examples of successful conversational *ITS*s are *Why2* [28] and *GuruTutor* [20].

2.2 Intelligent Pedagogic Agent

*ITS*s can deal with emotional learning as well. Emotions are important for students since they have impact on learning and influence their ability to process information. Thus, it is important for teachers to create a positive and emotionally safe environment for the optimal learning of students. Therefore, learning how to manage feelings and relationships allow to develop a kind of *Emotional Intelligence*, that enables people to be successful. This expands on Gardner's [3] theory of multiple intelligence's, with particular reference to the *intra-personal* and *inter-personal* intelligence's defined there, that deal with understanding oneself and others. One of the most influential papers on emotional learning [14] demonstrates the *Persona Effect*: learning is facilitated by an animated pedagogic agent that is a life-like persona and expresses affect. The rationale for this research has been the media equation hypothesis [22], which suggests that learners respond to pedagogic agents as if they were social actors.

Shen et al. [25] describe the development of an affective e-Learning model, and demonstrate the machine's ability to recognise learner emotions from physiological signals [4], that presents an approach to a possible modelling of user affect designed to assess a variety of emotional states during interactions. Zakharov et al. [31] describes a pedagogic agent capable to act according to the detected learner's cognitive and affective states is presented.

3 Intelligent Pedagogic Agent in *Gea2: A New Earth*

The *IPA* integrated in *Gea2: A New Earth* is basically characterized by two abilities:

- It is an *Intelligent Virtual Tutor* (*IVT*), i.e. a virtual agent capable of a conversation in Natural Language (*NL*). In other words, it can reply in NL to questions written in NL. It is implemented as a *Dialogue Management System* (DMS), as described in Sect. 3.1.

- It provides *unsolicited hints* to the learner, by evaluating the game progression and the player's emotions. This makes of it an IPA, i.e. an *IVT* capable of actuating interventions in the game, to support specific user needs. This is discussed in Sect. 3.2.

3.1 Dialogue Management System

In our system, the communication is implemented through *Natural Language Processing* (NLP), by means of an algorithm based on an ad hoc text retrieval problem solver and on a Naïve Bayes text classifier with an inner product-based threshold criterion. Basically, the sentence coming from the learner is analysed, classified, and an answer (or a reaction) is then selected from a *Knowledge Database* (KDB). As the problem of selecting the right answer from a KDB can be expressed formally like an information retrieval problem [13], we implemented our approach as a variation of a text retrieval algorithm. In our case, the user's query corresponds to the user's question, and the documents of the collection where to retrieve from, are the sentences stored in the KDB. For this retrieval problem we used an hybrid approach based on the collaboration between two methods available in literature: the *Vector Space Model* [26], and the *Language Modeling Theory* [21]. To design our approach we relied on the research work done by Mori et al. [17].

The core of our IPA is an implementation of Naïve Bayes Text Classification [15], which is a probabilistic classification method based on Language Modeling under the hypothesis of words' conditional independence. Moreover, it is important to give the user an answer extracted from the KDB only if there is a certain amount of confidence that the chosen answer is appropriate to the query. Otherwise, the system has to reject the identified answer and it has to return a sentence telling that the IPA cannot answer the user's question.

The first task of Algorithm 1 consists in a sentence lemmatization, which transforms all the terms of the sentence in a form suitable for word analysis and computation. We then evaluate the Interrogative/Adjective Pronoun (*IAP*) of the question, in order to assign a higher classification score to the sentences from the *KDB* with a matching *IAP*. Instead of excluding all the sentences in the KB with a different *IAP*, we just penalise them with a lower classification score; in fact, a situation could occur where an answer might be the best one to respond to the user's question even if each of the expected questions that were associated with it in the KB had a different *IAP*. We then proceed with the Naïve Bayes Text Classification, and filter the result by applying the threshold criterion.

3.2 Emotions and Schooling: Process of IPA Development

In our research we decided to detect emotions in text, so students can feel confident and immerse in a *emotionally safe* environment. In fact, we focused on the creation of a *trusted relationship* between students and their virtual tutors, the *IPAs*: "*social intelligence in pedagogic agents may be important not just to gain*

Algorithm 1. NAÏVE BAYES TEXT CLASSIFIER ALGORITHM

Require: input question = Q; NPC stored questions = NPC_{qs}

Q_{lem} = **lemmitise**(Q)
set $score$ = **evaluate**(Q_{lem}, $NPCqs$)
sort by $score$ $Qlem$
for $i = 0$; $i{+}{+}$; $Qlem\_n$ **do**
    $Pr_i = Pr_i(Q_{lem})$ = **Pr_Bayes_classifier**($sorted\_Q_{lem}$)
    if Pr_{li} ¿ $treshold$ **then**
        return ("Normal",**NPCanswer**)
    else
        return ("No match", **random**("Interlocutory"))
    end if
end for

user acceptance, but also to increase the effectiveness of the agent as a pedagogic intervention" [25].

In order to understand students' emotions, we preferred to start from the bottom, asking students to investigate their feeling in school context. We conducted a pilot study with an experimental group of students to collect a set of emotions and expected behaviour of the *IPA*. We believed that the strength of this approach is the collection of emotions' coming directly from students/stakeholders, reducing the risk to model the *IPA* in a way that could be not in line with their expectations (Fig. 1).

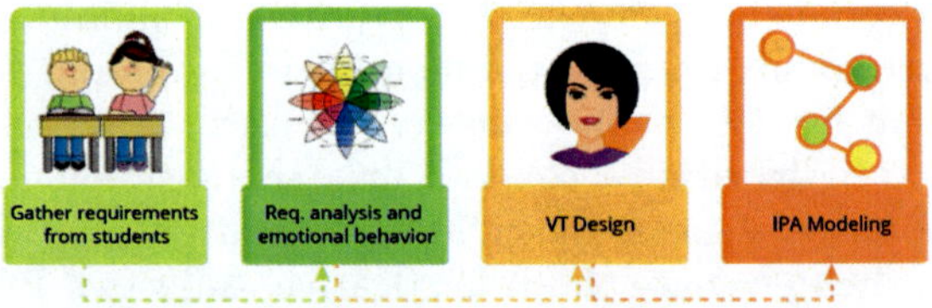

Fig. 1. Emotions & Schooling pilot study

With an experimental group of students, we have worked on emotions, starting to analyse what they feel during real school scenarios, like a test session or an oral presentation. The research on emotion and schooling has been conducted with a high school class of 20 students aged 15 years. The research has been assisted by two teachers: one from literature and one from computer science, for a total of seven meetings (of the duration of 2 h) with the students. This has been only the pilot study that helped shaped the IPA for *Gea2: A New Earth*; overall evaluation with multiple classes in multiple schools for the game is outside the scope of this paper, and left for some other work. Figure 2 reports some of the support materials used for the research study. The basic emotions schema used to introduce students to emotions understanding [12] is in Fig. 2(a). To match

(a) Basic emotion schema (b) *Lie to me* TV series (c) Crazy Talk interface

Fig. 2. Emotions and schooling: supports material

face expressions to emotions we were inspired by the famous TV series *Lie to me* (Fig. 2(b)). Figure 2(c) shows the Crazy Talk tool that we used to animate the IPA interface. We let the students use a simple worksheet during lectures, to annotate their emotions and wishes about the IPA behavior.

3.3 Unsolicited Hints

The *IPA* evaluates some parameters inside the game and decides when to provide help. The rationale behind, is to keep the player involved in the game and raise his attention and motivation. Two distinct aspects of players behaviour are monitored in the game, which in turn result in 5 independent variables dealing with game progression and emotions: 1) *Players' advancement in the game* (3 variables), and 2) *Emotions felt by players* (2 variables).

Players' Advancement in the Game: To analyse the progress in games' sessions we monitor the following game's variables:

- ***Time***: elapsed time since the beginning of the game
- ***Score***: overall score obtained by the player
- ***Required help***: questions asked by player to *IPA*

In our model, each one of the *progress* variables has a state in {*Low*, *Medium*, *High*}. To set properly the interval for each variable, we have taken the maximum value of each variables' state and then divided it into three parts. Some values can be inferred during assessment (like average time session), some others are fixed (like maximum score). Hence, to each state we assigned a probability.

Players' Emotions: We tried to infer players' emotions by analysing:

- ***Emoticons***: use of emoticons in chat messages among team members
- ***Sentiment***: output of sentiment analysis performed on player/*IPA* dialogue

We partitioned each node (emoticons and emotions) into three states: *Negative*, *Neutral*, *Positive*. Sentiment analysis performed on player/*IPA* chat automatically assigns sentiment tags to each sentence. These tags and their percentage are used to classify the expressed sentiment as *Negative*, *Neutral* or *Positive*.

Table 1. Emoticons' value

Emoticons	Value	Image	Emoticons	Value	Image
acceptance	+1		fear	-2	
anger	-4		joy	+4	
anticipation	+ 3		sadness	-3	
disgust	-3		surprise	+2	

Emoticons were assigned a value in $[-4, 4]$ (see Table 1). The emoticon to emotion relation has been identified during the pilot study with the students.

To classify the player's sentiment we accumulated the used emoticons (*emot*) and assigned the *Etag* as follows

$$\begin{array}{lcl} emot \leq -1 & \rightarrow & Etag = negative \\ -1 \leq emot \leq +1 & \rightarrow & Etag = neutral \\ emot \geq +1 & \rightarrow & Etag = negative \end{array}$$

If both present, the sentiment variables are combined as shown in Table 2.

Table 2. Emotions (Emo) vs Emoticons (Emot)

Emo	Emot		
	Neg	*Pos*	*Neu*
Neg	Neg	Neu	Neg
Pos	Neu	Pos	Pos
Neu	Neg	Pos	Neu

Bayesian Network and Unsolicited Hint Algorithm: To combine all the observable variables we used the Bayesian Network (BN) represented in Fig. 3. Light blue nodes are *fathers* and *visible* in the network; green nodes are *not visible* but *predictable*.

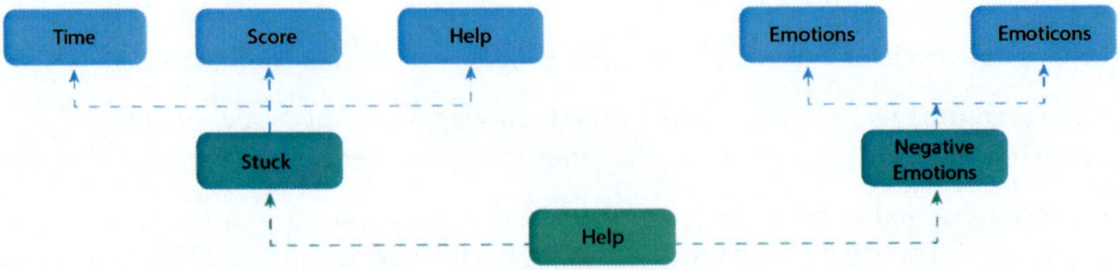

Fig. 3. Unsolicited hint Bayesian Network variables nodes

In the BN, combinations of probability bring to different states: the player can end up in a *stuck* state, where (s)he appears to be not progressing, or in experiencing a *negative* emotional state, or in a state where, the presence of both *stuck* and *negative* states leads to the probability for an unsolicited hint to be given. In our algorithm, we consider the 5 variables independent and we check the value of the variables every Δ_t, so that:

$$Pr(time \wedge score \wedge help) = Pr(time)Pr(score)Pr(help)$$

$$and$$

$$Pr(emo \wedge emot) = Pr(emo)Pr(emot)$$

During the game session, the system checks the 5 variables' values and calculate the probability of the *Help* node. Algorithm 2 runs on the client side and

Algorithm 2. UNSOLICITED HINT CLIENT SIDE

```
Require: time = t, score = s, help = h, teamChat = tc, IPAChat = ic
for t = t + Δt do
    UnsHint(t,s,h,tc,ic) = (hi,hint)
    if hi= true then
        provide hint;
    else
        return
    end if
end for
```

Algorithm 3. UNSOLICITED HINT SERVER SIDE

```
Require: t,s,h,tm,ic;
for i = [0,5]; i++ do
    for all a = inputVar do
        if (a = tm — ic) then
            b = Emoticons(tm) — Sentiment(ic)
            Pr_i(a) = PrCalc(b)
        else
            Pr_i(a) = PrCalc(a)
        end if
    end for
end for
Pr(stuck) = Σ_i^{1-3} Pr_i
Pr(negative) = Σ_i^{4-5} Pr_i
Pr(hint) = PrCalc(stuck)PrCalc(negative)
if random ¿ Pr(hint) then
    Return (true, ProvideHint)
else
    Return (false,"")
end if
```

calls *UnsHint* (by providing the runtime parameters) contained in Algorithm 3, which runs on the server side.

The function *Emoticons* parses the chat among teams' members and evaluate emoticons usage and function *Sentiment* analyses the sentiment expressed by the player in chatting with his/her *IPA*. *PrCalc* calculates the probability of the node given the input value; whereas function *ProvideHint* takes an hint from the *NPC*'s database.

4 Conclusion

The Intelligent Pedagogic Agent (IPA) that we have developed for *Gea2: a New Earth* has been greatly used during the tests of the game when it has been deployed in multiple classes and schools (outside the scope of this paper). It emerged that the IPA was a really useful tool for the teachers to collect information. Subsequently, the students also found that talking with the IPA was very engaging and helpful.

As a result, similar approaches have the potential to empower educational games with a new level of features that can help the teachers to keep track of the learning progress as well as the emotional state of the student, while the game autonomously provides the students with targeted intervention to increase the rate of learning retention.

References

1. Aleven, V., McLaren, B.M., Sewall, J., Koedinger, K.R.: The cognitive tutor authoring tools (CTAT): preliminary evaluation of efficiency gains. In: International Conference on Intelligent Tutoring Systems, pp. 61–70. Springer (2006)
2. Chittaro, L., Ieronutti, L., Ranon, R.: Navigating 3D virtual environments by following embodied agents: a proposal and its informal evaluation on a virtual museum application. PsychNol. J. **2**(1), 24–42 (2004)
3. Clinchy, B.M., Gardner, H.: Frames of Mind: The Theory of Multiple Intelligences. Hachette, London (1984)
4. Conati, C.: Probabilistic assessment of user's emotions in educational games. Appl. Artif. Intell. **16**(7–8), 555–575 (2002)
5. Deterding, S., Dixon, D., Khaled, R., Nacke, L.: From game design elements to gamefulness: defining "gamification" . In: Proceedings of the 15th International Academic MindTrek Conference, pp. 1–6. ACM (2011)
6. Freedman, R.: Atlas: a plan manager for mixed-initiative, multimodal dialogue. In: AAAI 1999 Workshop on Mixed-Initiative Intelligence, pp. 1–8 (1999)
7. Graesser, A., Rus, V., D'Mello, S., Jackson, G.T., Robinson, D., Schraw, G.: Autotutor: learning through natural language dialogue that adapts to the cognitive and affective states of the learner. In: Recent Innovations in Educational Technology that Facilitate Student Learning, pp. 95–125 (2008)
8. Graesser, A.C., Chipman, P., Haynes, B.C., Olney, A.: AutoTutor: an intelligent tutoring system with mixed-initiative dialogue. IEEE Trans. Educ. **48**(4), 612–618 (2005)

9. Graesser, A.C., Conley, M.W., Olney, A.: Intelligent tutoring systems. In: APA (2012)
10. Jackson, G.T., Boonthum, C., McNamara, D.S.: The efficacy of iSTART extended practice: low ability students catch up. In: International Conference on ITS, pp. 349–351 (2010)
11. Kapp, K.: The Gamification of Learning and Instruction: Game-Based Methods and Strategies for Training and Education. Wiley, New York (2012)
12. Lane, R.D., Nadel, L.: Cognitive Neuroscience of Emotion. Oxford University Press, Oxford (1999)
13. Lease, M.: Natural language processing for information retrieval: the time is ripe (again). In: Proceedings of the ACM 1st Ph. D. Workshop in CIKM, pp. 1–8. ACM (2007)
14. Lester, J.C., Converse, S.A., Kahler, S.E., Barlow, S.T., Stone, B.A., Bhogal, R.S.: The persona effect: affective impact of animated pedagogical agents. In: Proceedings of the ACM SIGCHI Conference on Human Factors in Computing System, pp. 359–366. ACM (1997)
15. Manning, C.D., Raghavan, P., Schütze, H.: Introduction to Information Retrieval. Cambridge University Press, New York (2008)
16. Mitrovic, A., Martin, B., Suraweera, P.: Intelligent tutors for all: constraint-based modeling methodology, systems and authoring. University of Canterbury (2007)
17. Mori, D., Berta, R., De Gloria, A., Fiore, V., Magnani, L.: An easy to author dialogue management system for serious games. J. Comput. Cult. Herit. **6**(2) (2013)
18. Neto, J.N., Silva, R., Neto, J.P., Pereira, J.M., Fernandes, J.: Solis' curse: a cultural heritage game using voice interaction with a virtual agent. In: International Conference on Games and Virtual Worlds for Serious Applications, pp. 164–167. IEEE (2011)
19. Oberlander, J., Karakatsiotis, G., Isard, A., Androutsopoulos, I., et al.: Building an adaptive museum gallery in second life. In: Proceedings of the Museums and the Web (2008)
20. Olney, A., D'Mello, S., Person, N., Cade, W., Hays, P., Williams, C., Lehman, B., Graesser, A.: Guru: a computer tutor that models expert human tutors. In: Intelligent Tutoring Systems, pp. 256–261. Springer (2012)
21. Ponte, J.M., Croft, W.B.: A language modeling approach to information retrieval. In: Proceedings of the 21st Annual International ACM SIGIR Conference on Research and Development in Information Retrieval, pp. 275–281. ACM (1998)
22. Reeves, B., Nass, C.: The media equation: how people treat computers, television, & new media like real people & places. Comput. Math. Appl. **5**(33), 128 (1997)
23. Rus, V., D'Mello, S., Hu, X., Graesser, A.: Recent advances in conversational intelligent tutoring systems. AI Mag. **34**(3), 42–54 (2013)
24. Santangelo, A., Augello, A., Gentile, A., Pilato, G., Gaglio, S.: A chat-bot based multimodal virtual guide for cultural heritage tours. In: PSC, pp. 114–120 (2006)
25. Shen, L., Wang, M., Shen, R.: Affective e-learning: using "emotional" data to improve learning in pervasive learning environment. Educ. Technol. Soc. **12**(2), 176–189 (2009)
26. Singhal, A.: Modern information retrieval: a brief overview. IEEE Data Eng. Bull. **24**(4), 35–43 (2001)
27. Terracina, A., Fabiani, F., Ferro, L.S., Litardi, D., Sapio, F., Zendri, G., Mecella, M.: Conquering an exo-planet through the use of a virtual role playing game assisted by an emotionally intelligent pedagogical agent. In: European Conference on GBL, p. 666. Academic Conferences International Limited (2016)

28. VanLehn, K., Graesser, A.C., Jackson, G.T., Jordan, P., Olney, A., Rosé, C.P.: When are tutorial dialogues more effective than reading? Cog. Sci. **31**(1), 3–62 (2007)
29. Wallace, R.S.: The anatomy of alice. In: Parsing the Turing Test, pp. 181–210. Springer (2009)
30. Ward, W., Cole, R., Bolaños, D., Buchenroth-Martin, C., Svirsky, E., Weston, T.: My science tutor: a conversational multimedia virtual tutor. J. Educ. Psychol. **105**(4), 1115 (2013)
31. Zakharov, K., Mitrovic, A., Johnston, L.: Towards emotionally-intelligent pedagogical agents. In: International Conference on Intelligent Tutoring Systems, pp. 19–28. Springer (2008)

Workshop on Technology - Enhanced Learning for Future Citizens (TEL4FC)

Workshop on Technology - Enhanced Learning for Future Citizens, TEL4FC

"Why is my Personal Assistant suggesting me this?" "Who/what is driving my car?" "Where is the software I am using?", "What do the data show?" "Who owns my photos?", "How can I participate and contribute to the government decisions?" are only few of the questions that the newcomer citizens from 2030, and after, will be mostly faced with during their lifetime as engaged citizens.

The way in which Technology-Enhanced Learning research can contribute to the development of a mature and technologically aware citizenship is the main objective of this workshop. In fact, as the development of digital and audiovisual tools has profoundly changed the Knowledge Society, the institutions and the schools are faced with the challenge to carry students to a digital citizenship and to a society more and more based on da-ta, information and digital processes. This scenario requires strong abilities to search, select, utilize information, recognition of fake news, individuation of automated systems, etc., in short, knowing well the "the good, the bad and the ugly" of our current and future technological society.

TEL4FC workshop aims to address the educational needs and technological awareness of the future citizens, i.e. the youth that will experience their citizenship in a rapidly evolving Knowledge Society. Artificial Intelligence, Machine Learning, Data Science, Social Networks, Cloud Computing, are only some of the crucial technological challenges, already permeating our world at any level and in many fields. But also Media Literacy, Ethics, Sociology, and History are substantial fields that shape the comprehension and the behavior of a citizen, when engaged in a technologically connected and globalized world.

Their integration in a Technology-enhanced educational framework could represent an opportunity of increasing the citizens' digital awareness and gradually setting up a transversal mindset, both ecological and sustainable for all learners: a universal intangible cultural heritage: the future of our society depends on an engaged, informed, and critically-thinking population.

The workshop is meant to connect researchers, educators and technologists involved in different and diverse areas, such as education, digital government, pedagogy, social and collaborative systems, cultural heritage, ethics, etc. to promote interdisciplinary re-search around the citizenship and the role of Technology-Enhanced Learning in shaping the future of our society, through the construction of future citizens that are fully aware of potentials and risks.

We fully present in the workshop 6 papers on different topics such as Data Science, Cybersecurity, Computational Thinking, Digital Participation and Cultural Heritage. The authors presented their research work in the perspective of real challenges for future citizens, and their contribution is very interesting for how they variously decline the motivations of the workshop. As they show a strong attitude in imagining and designing new ways of learning in the Knowledge Society, we definitely welcome those contributions in our trans-disciplinary workshop.

The workshop organization was supported and partially funded by the Dipartimento di Informatica of the University of Salerno, Italy, and we thank the Program Committee members, below, for their precious contributions in reviewing the papers.

Organization

Workshop Organizers and PC Chairs

Agnese Addone	Università degli Studi di Salerno, Italy
Vittorio Scarano	Università degli Studi di Salerno, Italy

Program Committee

Shaaron Ainsworth	University of Nottingham, UK
Jerry Andriessen	Wide & Munro, The Netherlands
Michael Baker	Centre National de la Recherche Scientifique, France
Pietro Boccadoro	Politecnico di Bari, Italy
Serena Cangiano	SUPSI, Switzerland
Françoise Détienne	Centre National de la Recherche Scientifique, France
Rute Vera Maria Favero	Universidade Federal do Rio Grande do Sul, Brazil
Åke Grönlund	Örebro University, Sweden
Lorenzo Guasti	INDIRE, Italy
Letizia Jaccheri	Norwegian University of Science and Technology, Norway
Beatrice Ligorio	Università degli Studi di Bari, Italy
Margaret Low	University of Warwick, UK
Flavia Marzano	Link Campus University, Italy
Jens Moenig	SAP SE - Knowledge and Education, Germany
Alfonso Molina	The University of Edinburgh, UK
Enrico Nardelli	University of Roma "Tor Vergata", Italy
Mario Pireddu	Università degli Studi della Tuscia, Viterbo, Italy
Cynthia Solomon	MIT Media Lab, USA
Luca Tateo	University of Oslo, Norway

Awareness of Cybersecurity: Implications for Learning for Future Citizens

Jerry Andriessen(✉) and Mirjam Pardijs

Wise and Munro, The Hague, The Netherlands
jerry@wisemunro.eu

Abstract. In real life contexts, predicting how technology will actually be used is often difficult. This can be a problem in the context of cybercrime: attempts to steal our identities, our money, or simply to frustrate our work by infecting our computers, rely on exceptions, derogations, people not paying attention, or not understanding what is at stake. There are various ways for the criminal mind to operate, but our concern is that future users need to develop more awareness of cybersecurity, meaning not only knowing the risks of using technology, but also how to handle threats that may happen in cyberspace. We will report two qualitative studies that show that cybersecurity awareness is about evolving resilience in a collaborative practice, and therefore cybersecurity awareness is collaborative. This has implications for Technology Enhanced Learning for future citizens.

Keywords: Cybersecurity · Awareness · Learning

1 Learning to (Inter-)Act with Technology

Technology is rarely used in the way that its designers intended. In addition, the use of technology, especially in common areas of life, is evolving rather than being fixed and stable. Users do not interact with the instrument (the technology), but with the task (and the knowledge) at hand (which may differ from the task they are asked to complete). The actual task they consider and the way they consider it determines their use of the instrument [1]. Moreover, over time, users adapt to the constraints of the instrument, at the same time as they attribute the functions of the instrument [2, 3]. And finally, the way that users consider a task is strongly influenced by the context of use, i.e. the educational context, or the organisational context. The context impacts on the learning practices, for example by setting implicit norms about what (e.g. what knowledge, which attitudes, what behaviour) is seen as more or less important [4].

Understanding the use of technology and its evolution in particular practices relates to different learning needs and technological awareness of future citizens. In formal educational and organisational contexts, what users are supposed to know and learn

The studies reported were undertaken in the context of the CS-AWARE project, financed by the European Union (Horizon 2020, grant 740723), coordinated by the University of Oulu.

Z. Kubincová et al. (Eds.): MIS4TEL 2020, AISC 1236, pp. 241–248, 2021.
https://doi.org/10.1007/978-3-030-52287-2_24

by using the technology, is imposed on them (by researchers, policy makers, and by teachers). In real life, learning is more open. Most technology is appropriated outside of formal contexts, and most learning takes place outside of education. Our conjecture is that appropriation of the technology itself, as an instrument for accomplishing various tasks for formal as well as for informal learning, is the main issue for educating future citizens. This means, they need more awareness of the technology they are using and the impact this has on their daily lives. We will illustrate our position with an example from a cybersecurity project.

2 Awareness of Cybersecurity: The CS-AWARE Project

Awareness of technology in this paper is defined in the context of the CS-AWARE project, which aims to develop awareness technology for cybersecurity. Cybersecurity is one of today's most challenging problems for commercial companies, NGOs, governmental institutions as well as individuals. In addition to technology focused boundaries of classical information technology (IT) security, cybersecurity includes organizational and behavioural aspects of IT systems and also needs to comply to legal and regulatory frameworks. For example, the European Union recently passed the Network and Information Security (NIS) directive that obliges member states to get in line with the EU strategy. Innovative solutions that will help (especially) local public administrations to ease the burden of being in line with cybersecurity requirements are needed.

In the CS-AWARE project [5] we propose a cybersecurity situational awareness solution for local public administrations that, based on an analysis of the context (their system network and its services), provides automatic incident detection and visualization, and enables information exchange with relevant national and EU level NIS authorities like CERTs. Advanced features like system self-healing based on the situational awareness technologies, and multi-lingual semantics support to account for language barriers in the EU context, are part of the solution. As we are writing this, the system has been recently implemented in two municipal contexts, and we are starting deployment activities. In the current paper, we report about two initial studies that are relevant for our understanding of technology use in context.

Awareness technology supports learning for (individual) understanding of some domain, so it is different from awareness that is about users being able to see what other users are doing [6], or for fostering collective intelligence [7]. The latter systems may support different types of collaborative learning. Collaboration is not intrinsic to the design of the CS-AWARE system.

In the cybersecurity domain, awareness is a complex concept, it has been linked to attitudes, behaviour and experience, knowledge and understanding, organisation policy, etc. [8, 9]. Current approaches to awareness of cybersecurity favour a holistic approach, covering both personal and workplace use, and addressing the spectrum from end-users through to cybersecurity specialists and the legal, economic, and technological issues involved [10]. The CS-AWARE approach involves the concept of situational awareness of cyberthreats, comprising the knowledge- and behavioural dimensions of awareness [11]. The following list of components of cybersecurity awareness for system administrators (those who typically handle cybersecurity in municipalities) can be taken as requirements that the CS-AWARE system needs to support, but also comprises contextual requirements for the system administrators who require organisational support.

1. Sufficient technical understanding of the technology within the system network of the organisation and the interdependencies between the network components.
2. Understanding the weaknesses in the system network, the services provided and the software that is used.
3. Continuous monitoring of the components of the complex system for situational awareness of the overall picture and detection irregularities.
4. Accurate information about threats and vulnerabilities, and being able to exploit such information for (a) Perception of cues: what are the crucial clues for a possible cyberthreat? (b) Comprehension: being able to diagnose and explain the meaning of cues; (c) Projection: predicting consequences and risks of particular threats; and (d) Decision making: knowing how to limit damage and resolve the issues.
5. The harmed party must be aware that they are harmed. Actual attacks and damage could be unnoticed because of firewalls, lack of attention, trojans and spyware.
6. Understanding the need for internal and external information sharing. This links to transparency as well as of knowledge management within an organisation.

3 Study 1: Story Telling

For a baseline (before the CS-AWARE system was developed) understanding how a municipal organisation (200 000 inhabitants) handles cybersecurity, we organised a story-telling workshop [12, 13]. We think stories are a good way to capture personal experiences. However, in our workshop, we added the collaborative dimension. In our view, a collaborative story (a story written together) can be taken as a joint understanding by the participants and a coherent integration of their personal experiences.

Our procedure was as follows: We first asked 18 individual participants (system administrators and users from various departments) to think about an event of their personal experience related to cybersecurity. We then asked them to form small groups (4 groups of 3–5 people) to collaboratively broaden and deepen [14] one story according to a number of topics. They discussed what this particular experience told them about the organisation, the system, precise actions undertaken, their emotions and the resulting perspective on cybersecurity in their municipality. This collaborative phase lasted about half an hour for each story. The plenary presentation of the first four stories took about 90 min, as the presentations invoked a lot of discussion, which added some more detail to the stories. We collected 6 stories in total.

An example of a typical story is the following: One of the tasks of this employee (E) is to read all mail the mayor receives, which may comprise several hundreds of emails every day. In spite of strict regulations, E opened an infected email. This mail had no subject, which could have caused suspicion. After E opened the mail, files in each directory on his computer he accessed became inaccessible. The IT-Department was notified, they discovered a type of ransomware that had encrypted all the files in the directories E had accessed. The computer was restored in the IT-Department, where all files that were not yet accessed were retrieved, and the system was reinstalled, and the intact files were restored. Infected files could not be retrieved. E was initially terrified, feared the computer was lost, but now sees this as a lesson learned, to more strictly apply the regulations in such cases.

For interpretation of the stories we consider the user with an issue as part of an activity system [15]. An activity is seen as a system of human "doing" whereby a subject works on an object in order to obtain a desired outcome. In order to do this, the subject employs tools, which may be physical (e.g. an axe, a computer) or semiotic (e.g. a plan, a text). An activity system further pictures an individual as part of a system characterised by particular rules, communities and division of labour. An activity is a hierarchical structure, and its nature changes over time. This change can be related to tensions in the system, so tensions can be a good thing.

Figure 1 shows our interpretation of the activity system for users with cybersecurity issues in this municipality, based on the set of stories that they constructed. First, there is a subject, which is the generic user of techno-services who has the objective of making use of a computer (the tool) as a regular part of the job, but is faced with some obstacle so that the regular activity cannot be performed. This is the initial issue, and it gives rise to a shifted objective: the resolution of the technical issue. The subject is part of the community of all other users within a department. These users may frequently interact, but we do not know the extent to which these interactions concern cybersecurity. The small number of stories that we collected show that our users may have some awareness of risks, but there is no general (formal) procedure for sharing cybersecurity issues with all members of the community, or within a particular department. What users are all supposed to share is the rules: this is a set of explicit safety-regulations set out by the IT-Department for secure cyber behaviour. The most important one for this case is: "all users are warned not to open e-mails from unknown senders or without a subject or with attached files in.zip or.exe format". The *division of labour* is very clear: all cyber issues will be handled by the IT-Department. They are seen as competent, reliable, quick, especially good at estimating the risk for the organisation and at taking the necessary measures. Nevertheless, we may assume implicit *norms* operating in two directions: (1) the IT-department gets the blame (from individual users) for lost files or for lost time; and (2) the users are seen by the IT-Department as risks, possible sources of mistakes, who need to be reminded frequently to stick to the regulations to avoid making these mistakes.

Let us now look at the tensions, expressed by red arrows in the figure. At first sight, one might locate the main source of tension in failing technology. However, in the framework used, tension is in a relationship, otherwise it would not exist. In our interpretation, the first type of tension is not just a malfunction of the technology, it is the fact that *users experience opaqueness of security issues*: generally, users do not understand the nature of the problem, and therefore rely on the expertise of the IT-Department for resolution, but also for awareness about causes and the severity of the threat. Please note, that some users may search and look for a solution themselves, but that does not make a difference for our interpretation. For our users, a second type of tension appears to be that *regular user activities may create security issues*, meaning for example that it can be in the job description of some employees that they have to open all emails, or that they have to sometimes work with files from external sources.

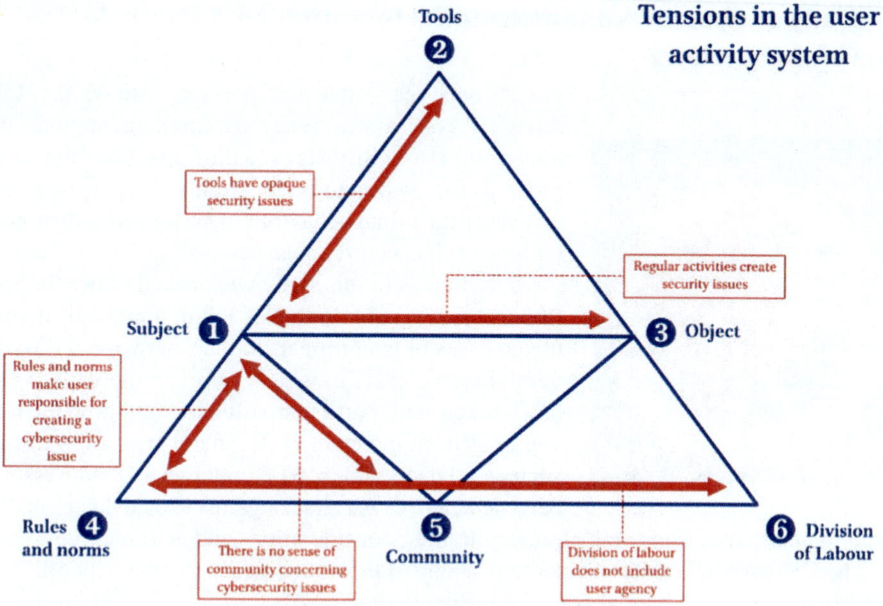

Fig. 1. Tensions in the user activity system

Tension is also found in relations with the underlying components of the organisational activity system: the rules & norms, the division of labour and the community. These fundamental types of tension characterise the practice of dealing with cybersecurity issues within the organisation. The third type of tension is that whilst users may assume responsibility for their mistakes, the division of labour is such that *users experience no agency in dealing with cybersecurity issues*, so their learning is limited. They are merely reminded of adhering to the rules, and as a consequence, a fourth source of tension is that *users experience guilt for having the issue.* Some employees then try and transfer the blame to the IT-department. Fifthly, there is a transparency issue: *the community (of other users in the municipality) remains uninformed about issues and potential dangers.* Actually, there is no notion of community at all. Experiences may be shared informally, but issues are often seen as the mistake of the person who has the issue, and not as an organisational issue.

From a small set of stories (we collected more stories at another municipality) we can derive some general conjectures about the practice of cybersecurity. First, there is not much learning: in the best case, the individual user only learns to avoid the same mistake again. Second, the organisation (other employees) does not learn much either, because information is not shared. There is not sense of community concerning cybersecurity. Thirdly, as a consequence, cybersecurity remains at a low level, in the hands of a few and this situation will not change.

4 Study 2: System Administrators Operating an Awareness Tool, Thinking Aloud

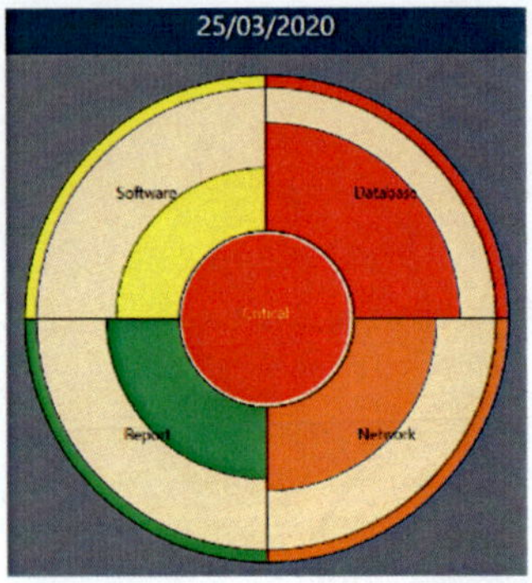

About one year after the previous study, the CS-AWARE system was ready for implementation. We undertook a usability study with users from the same group as for the previous study. This study is used here to discuss the initial behaviour of system administrators with our cybersecurity awareness tool.

Participants in this study were asked to use the tool (the opening screen on the left) with a number of simulated cases of cyberthreats to their network: 1) A vulnerability use case, in which a user in the department has worked with particular software, and now the network has been infected. 2) A general security warning, such as a DDoS attack. 3) A malicious IP address has been detected. 4) An attack against a local database. We asked the participants to think aloud about their considerations and actions.

The CS-AWARE system works with a map of the municipality system network, performs regular scans at critical points of the network, it can recognise and diagnose symptoms of cyberthreats, and pinpoint them to particular locations in the system network, including the dependencies with other network components that may be threatened. The user can consult information about the threat in order to understand the problem. The tool suggests remedies and can (sometimes) implement the resolution of the threat. This is a great improvement over existing firewalls and virus scanners, who often simply block websites or software applications.

Compared to the situation before CS-AWARE, handling of cybersecurity threats is now about understanding the information and undertaking appropriate action, while the procedure for arriving at a diagnosis is taken over by the system. What we saw our participants do was unexpectedly diverse in terms of practices. We found different approaches, related to expertise (the knowledge of system administrators is not uniform), how and what information was consulted (from extensive reading to brief skimming), and differences in ease of accepting the suggested solution by the system.

System administrators working with the CS-AWARE interface for the first time, approached the system as if in their usual practice. They were not focusing on learning, but on dealing with the threat. This is not strange, as the awareness system is supposed to function in actual practice, with real threats. Learning to use the system may involve trying to adapt its use to existing practice. In that practice, decision making still requires interaction with human colleagues. However, diagnosis and understanding of a threat is made much easier by CS-AWARE, so we expect the nature and content of these interactions to change.

A main conclusion from this study is that we expect that using the awareness system will not replace human interventions, but it might change its nature. Learning will not be restricted to increased knowledge of threats, but it will involve the whole process of diagnosing, handling and preventing threats in the entire organisation. However, this transformation will not happen if there is no attention for collaboration.

5 Implications for TEL4FC

The usability study showed that users will not just do automatically what the system tells them to do, even with trusted information about threats. This is because the knowledge of system administrators is highly divergent. Also, their trust in the system appeared very different, some simply follow the suggestion, others need further consultation. What awareness systems provide for learning will therefore depend evolving trust and collaborative practices. There is a danger that learning is not part of the deal, and the system will simply be used to confirm the old situation, albeit with less effort, and hopefully a higher success rate. In addition, the possibility to monitor logs and reports by managers might reveal more of what their system administrators actually do. This sounds familiar to what we already know about technology: it is often used for control rather than for learning and transformation.

The potential gains really are much higher: a transformation of thinking about cybersecurity. Realising this potential will be the challenge for the future. In that case, technology enhanced learning is not only about designing technology to support learning, it is also about designing support for learning to work with technology [16, 17]. We propose that the future of TEL for citizens is not in developing more comprehensive systems that support and evaluate learning, but it is in better understanding the evolving learning needs of citizens who are users of technology in their daily lives and culture. For that. TEL should be about developing learning support for acting and interacting with technology rather than about developing technology support for learning.

What we have seen is that fighting cybersecurity is taken as an individual issue, while it can be fought much more effectively in a more collaborative setting. For example, we can imagine users collaborating on a) detection and understanding of an issue; b) deciding what to do; c) informing other users of what happened. This could be done formally (by following regulations) or informally, because that is the norm or attitude, or both. Also, we can imagine collaboration between the IT-department and individual users, as well as with the larger user community (other users in the organisation) on sharing and reporting incidents, and on explaining and adapting advice and regulations for prevention of cyberincidents. The IT-department could make logs of incidents, consult with other departments about specific incidents and how to deal with them, and to share cybersecurity issues with management and security authorities. What we call collaboration here is more than discussion. The main reason we call for collaboration is because we insist on considering the other in this activity, as equals, as being in the same boat, and on envisaging a joint goal of combatting cybercrime [17]. This is an attitude as much as it is a vision on how to act.

References

1. Tchounikine, P.: Contribution to a theory of CSCL scripts: taking into account the appropriation of scripts by learners. Int. J. Comput.-Support. Collab. Learn. **11**(3), 349–369 (2016). https://doi.org/10.1007/s11412-016-9240-8
2. Rabardel, P.: Instrument mediated activity in situations. In: Blandford, A., Vanderdonckt, J., Gray, P. (eds.) People and Computers XV - Interactions Without Frontiers, pp. 17–30. Springer, Berlin (2001)

3. Rabardel, P.: From artefact to instrument. Interact. Comput. **15**(5), 641–645 (2003)
4. Chevallard, Y.: Readjusting didactics to a changing epistemology. Eur. Educ. Res. J. **6**, 9–27 (2007)
5. Schaberreiter, T., Roning, J., Quirchmayr, G., Kupfersberger, V., Wills, C., Bregonzio, M., Koumpis, A., Sales, J.E., Vasiliu, L., Gammelgaard, K., Papanikolaou, A., Rantos, K., Spyros, A.: A cybersecurity situational awareness and information-sharing solution for local public administrations based on advanced big data analysis: the CS-AWARE project. In: Bernabe, J.B., Skarmeta, A. (eds.) Challenges in Cybersecurity and Privacy – The European Research Landscape, pp. 149–180. River Publishers, Gistrup (2019)
6. Collazos, C.A., Gutiérrez, F.L., Gallardo, J., Ortega, M., Fardoun, H.M., Molina, A.I.: Descriptive theory of awareness for groupware development. J. Ambient Intell. Humaniz. Comput. **10**(12), 4789–4818 (2019). https://doi.org/10.1007/s12652-018-1165-9
7. Piccolo, L.S.G., Liddo, A.D., Burel, G., Fernandez, M., Alani, H.: Collective intelligence for promoting changes in behaviour: a case study on energy conservation. AI Soc. **33**(1), 15–25 (2018). https://doi.org/10.1007/s00146-017-0710-y
8. Safa, N.S., Sookhak, M., Von Solms, R., Furnell, S., Ghani, N.A., Herawan, T.: Information security conscious care behaviour formation in organizations. Comput. Secur. **5**(Supplement C), 65–78 (2015). https://doi.org/10.1016/j.cose.2015.05.012
9. Bitton, R., Finkelshtein, A., Sidi, L., Puzis, R., Rokach, L., Shabtai, A.: Taxonomy of mobile users' security awareness. Comput. Secur. **73**, 266–293 (2018). https://doi.org/10.1016/j.cose.2017.10.015
10. Vasileiou, I., Furnell, S. (eds.): Cybersecurity Education for Awareness and Compliance. IGI Global (2019). https://doi.org/10.4018/978-1-5225-7847-5
11. Schaberreiter, T., Quirchmayr, G., Juuso, A.-M., Ouedraogo, M., Roning, J.: Towards a complex systems approach to legal and economic impact analysis of critical infrastructures. In: 11th International Conference on Availability, Reliability and Security (ARES), pp. 668–676 (2016). https://doi.org/10.1109/ARES.2016.65
12. Kurtz, C.: Working with Stories in Your Community or Organization: Participatory Narrative Inquiry, 3rd edn. Kurtz-Fernhout Publishing, New York (2014)
13. Snowden, D.: From atomism to networks in social systems. Learn. Organ. **12**(6), 552 (2005)
14. Baker, M., Andriessen, J., Lund, K., van Amelsvoort, M., Quignard, M.: Rainbow: a framework for analysing computer-mediated pedagogical debates. Int. J. Comput.-Support. Collab. Learn. **2**(2–3), 315–357 (2007). https://doi.org/10.1007/s11412-007-9022-4
15. Engeström, Y., Miettinen, R., Punamäki, R.-L.: Perspectives on Activity Theory. Cambridge University Press, Cambridge (1999)
16. Piccolo, L.S.G., Pereira, R.: Culture-based artefacts to inform ICT design: Foundations and practice. AI Soc. **34**(3), 437–453 (2019). https://doi.org/10.1007/s00146-017-0743-2
17. Andriessen, J., Baker, M.: On collaboration: personal, educational and societal arenas. Brill/Sense, Leiden (2020)

Roobopoli: A Project to Learn Robotics by a Constructionism-Based Approach

Mauro D'Angelo[1,2] and Maria Angela Pellegrino[1,3(✉)]

[1] Perlatecnica no-profit organization, Maddaloni, Italy
mauro.dangelo@perlatecnica.it
[2] STMicroelectronics, Arzano, NA, Italy
[3] Dipartimento di Informatica, Università di Salerno, Fisciano, Italy
mapellegrino@unisa.it

Abstract. The technological growth follows an exponential trend, and we will hit the knee in the curve in 2029. From that time, challenges will rapidly change, and the job demand will ask for profiles that we can only partially predict. Therefore, there is the need to adapt educational programs to exponential growth as soon as possible to provide children of today (alias future citizens) with the necessary basic skills and the ability to easily combine them to face any challenge the future will pose. We consider computational thinking and robotics as fundamental skills since students should be provided with transversal critical thinking and be technologically-aware to meet job demands. Thus, i) we designed Roobopoli, a project focused on the direct experience of robotics, and ii) we propose an educational procedure to teach robotics by exploiting Roobopoli through a constructionism-based approach. We tested our approach both in formal and informal educational settings by involving students heterogeneous in age, skills, and personal attitudes. The engagement and the interest manifested by participants prove that the constructionism-based approach, in general, and Roobopoli, in particular, are positively accepted by students to learn robotics and to develop computational thinking. In this article, we will present both Roobopoli and the performed experiments by pointing out lessons learned.

Keywords: Automated systems · Robots · IoT · Constructionism · Critical thinking · Computational thinking · Transversal mindset

1 Introduction

The evolutionary progress and technological growth follow an exponential trend [1]. Exponential growth is deceptive: at first, it is nearly flat, and we can linearly approximate it. As long as this approximation is still valid, the skills required for satisfying the job opportunities and the roles in working life are stable and easily predictable. In 2029, we will hit the knee in the curve, and newcomer citizens will face challenges that we can only partially predict. Once

Z. Kubincová et al. (Eds.): MIS4TEL 2020, AISC 1236, pp. 249–257, 2021.
https://doi.org/10.1007/978-3-030-52287-2_25

overcome that point, the growth will vertically rise, and the required roles and offered positions will rapidly change. If we wisely manage these transformations, it could lead to a new age of good work, good jobs, and improved quality of life for all, but if managed poorly, it poses the risk of widening skills gaps and inequality [2].

The problem we want to face is how to provide children of today with skills to meet any challenge the future will pose them. In other words, we are interested in designing an educational procedure that can both develop a transversal mindset to face various challenges and teach technical skills to meet the job trend. It is not a novelty: educational programs have always tried to indulge in the job offer trend. The main issues are that i) job roles can only be partially predicted since they are in symbiosis with the technological growth [2] and ii) we have little time to adapt learning techniques to exponential technological growth.

We propose to provide students with the necessary fundamental skills and to teach them how to quickly and adequately combine acquired skills to address real and complex tasks. Since automated systems and robots are critical skills in the near future [2], in this article, we will focus on teaching robotics. Furthermore, future citizens must get familiar with problem-solving and computational thinking (CT), classified as a fundamental skill, just like reading, writing, and arithmetic [3]. The CT is the thought process involved in abstracting a problem and expressing its solution in a way that a human or a machine can effectively carry out. Therefore, CT not only involves the basic concepts of Computer Science to implement the solution but also problem-solving and system design. Our proposal is to teach robotics and automated systems by a *constructionism-based* approach [4]. The constructionism assumes that children can better gain knowledge when they are actively engaged in creating artifacts and collaboratively solving real tasks. While robotics addresses the technological educational aspect, constructionism leads to the achievement of critical and computational thinking (representing transversal competencies). Our main contributions are:

- an educational procedure to teach robotics and autonomous systems by a constructionism-based approach by addressing both technical and transversal skills. We tested its level of age-independence by involving students heterogeneous in age, and we observed active engagement and positive feedback;
- the design and development of Roobopoli, a project where students can be engaged in a direct experience of automated systems and system design.

The rest of the paper is structured as follows: in Sect. 2, we present related work based on the constructionism framework and its application in teaching robotics; in Sect. 3, we describe Roobopoli and the proposed approach; in Sect. 4, we will report the carried out experiments by focusing on lessons learned; then, we conclude with final considerations and future directions.

2 Related Work

Papert defined the ***constructionism***-based activities [4] by assuming that students learn better when they are actively engaged in creating artifacts.

This approach is based on three dimensions: personal, social, and cultural. *Personal* refers to the individual learning, *social* refers to the collaborative aspect during the creation of shared artifacts, and *cultural* relates to how gender, age, and personal attitudes affect activities and participation. Learners need *objects-to-think-with* to impersonate, robots in our case. Constructionism-based activities have been widely studied in both formal (in traditional classrooms) and informal education (in other contexts) [5]. In our experiments, we mainly consider formal settings, and we perform a preliminary investigation in an informal context.

Teaching robotics by a constructionism-based approach. Robotics has a multidisciplinary nature: it integrates STEM disciplines (i.e., science, technology, engineering, and mathematics), computer science basics, CT, and teamwork skills [6]. Learning through designing, building, and operating robots can promote the development of thinking, problem-solving, self-regulation, and teamwork skills [7]. Therefore, it naturally combines technological awareness and critical thinking.

By following Papert's footsteps [4], there have been several studies on utilizing robots to teach various technical concepts based on the constructionism. These studies observed the effectiveness of the constructionism-based approach in learning robotics if compared with the traditional unidirectional lessons.

The constructionism-based approach has been widely used at the university level [8–10] by developing dedicated courses to provide a robust hands-on experience. Burbaite et al. [9] also involved High-school students in their experiments. The authors [9] interpreted robots as physical entities consisting of hardware (mechanical/electrical parts of the robot) and software (control programs). When the learning object corresponds to the entire mechanism, the main focus is the robot's behaviors and the learning of the correct sequence of control actions to perform tasks such as line following, collision avoidance [9]. This scenario is precisely the one we take into account. Moreover, they consider robots as a *smart thing* (as in IoT) with sensors and actuators to interact with the surrounding intelligent environment and a control program to control their behavior. In their experiments, first, they provide the needed facilities to perform tasks and, then, assign projects to test their ability in solving the given problem by the provided skills. We follow the same pattern.

Arlegui et al. [11] and Scaradozzi et al. [12] worked with the primary school in formal education. Their experiments manifest a significant upgrade in the children's education, in particular in developing general skills useful in the future. They worked on the entire engineer process from design to construction and implementation of a solution for a given challenge. We aim to do the same.

The main differences between our and the previously cited experiments are in the target age and the used equipment. Our target age is 10–18 by considering Middle-school students (from 10 to 13) and High-school students (from 13 to 18), according to the Italian educational plan. Moreover, the previously cited work based the experiments on third-party robots (mainly LEGO-Mindstorms robots), while we propose a project developed by our self. We designed and implemented the Roobopoli city and its vehicles Roobokart. The entire town and its robots are completely programmable based on student preferences. Thus, the oldest students (13–18 years old) have absolute freedom not only in designing the solution they prefer but also customize robots during the assembly phase.

3 Roobopoli: The Project and the Proposed Approach

Roobopoli[1] is a project that aims to create an educational experience in the field of smart cities and autonomous systems. It is designed by our non-profit association Perlatecnica[2] with the technical support of Bluenet[3]. Roobopoli is a tiny smart city where the life of the inhabitants, called *Roobo*, is assisted by modern technologies. The town exhibits the same technology available in real cities, here reproduced in scale for educational, testing and simulation purposes. For instance, traffic lights manage intersections. As simplification, in Roobopoli, traffic lights are bichromatic, i.e, they only assume red and green as color. They are managed by a control unit made with a Nucleo F401RE microcontroller board, that animates the traffic light cycle by a relay. The traffic light is always preceded by a road sign which allows the reading from the sensors of Roobokart. As part of the Roobopoli project, one of the main activities consists in assembling and programming city's vehicles called *Roobokart*, which will have to move independently on the roads of the city. The Roobokart and the city are intelligent, i.e., they are equipped with advanced sensors and boards based on STMicroelectronics micro-controllers. The Roobokart is controlled with an STM32 microcontroller mounted on a Nucleo F401RE board that is connected to a motherboard that includes the sensors used by the Roobokart to carry out its mission(s). Roobokart is equipped with accelerometers and gyroscope, infrared sensors, color sensors, and proximity Time-of-flight.

The entire city and its inhabitants constitute a useful lab for experimenting skills in automated systems and ambient intelligence in real settings, valuable for future profiles. In our experiments, students have to assemble and implement each component of the city and the Roobokart. Each feature of Roobopoli described so far is provided as a task (also referred to as *mission*) that students have to solve by designing and implementing a solution. The project, as described, is the result of all the assigned missions. As designed, Roobopoli is not bounded to any specific mission, but it can be used in any desired task.

Proposed approach based on Roobopoli. Participants are split into groups to satisfy both the personal and the social needs required by the constructionism. In our experiments, we organize groups homogeneous in age. However, it does not give guarantees on the homogeneity in skills. Therefore, the first phase of our approach is the ***training phase*** that assures that everyone becomes familiar

[1] Further details about Roobopoli are available by following http://www.roobopoli.org.

[2] http://www.perlatecnica.it/.

[3] https://bluenetita.com.

with all the technical skills and problem-solving attitudes necessary to accomplish the challenge(s) provided during the further steps of the procedure. All the skills transferred in this phase will be referred to as *functional blocks*. We assess that students should acquire general capabilities and the ability to abstract problems as early as possible (10–13 years). Then (13–18 years), students can deepen their skills according to personal attitudes. In our experiments, we mainly focus on 13–18 age students, and during the training, we teach them to design an unambiguous high-level solution and technical skills, such as the basic concepts of robotics (e.g., sensors and motors) and coding, and how to assembly robots. While problem-solving is a transversal attitude, the coding and the robotics represent vertical skills (i.e., skills specific to a particular field).
Once all the students learned the useful and necessary functional blocks, we can move to the proper constructionism-based activity. First, a project (also referred to as ***mission***) must be assigned. The mission must be in natural language, at high-level and it should unambiguously detail any desirable sub-goal. The mission should be related to the object-to-think-with, i.e., Roobokart in our case. For instance, a mission can be *implement the line-following feature in Roobokarts.* During the ***resolution process***, each group exploits the problem-solving skills learned during the training. In particular, students have to decompose the mission into sub-problems and design a solution for each of them. In this phase, students experience critical thinking and develop a transversal mindset. The resolution abstraction process can be seen as a tree where the mission represents the root of the tree and the functional blocks are the leaves. The *tree-methapor* is based on the divide-et-impera paradigm: once given a problem, it must be split in immediately solvable or further decomposable sub-problems. If the acquired skills are not enough to solve any sub-problem, it identifies *functional holes*, i.e., skills that must be acquired in the training phase to solve a specific (sub-)problem. By linking the tree root to its leaves, the students empirically experience the CT. During the training phase, students learn to model solutions by automata, since it is unambiguous and be easily experienced by 13–18 age students, but also younger children. Students can model the solution of any (sub)-problem of the mission as an automaton by the following approach:

- detection of automaton states. The convention is to assign to each state the unique name S_x where x is a progressive number, 0-based. It is good practice to take note of the status of the Roobokart in each automaton state, for instance, S_0: Roobokart is turned off; S_1: Roobokart is turned on;
- detection of transition conditions. The convention is to name each transition as $T_{x,y}$ where S_x represents the starting state while S_y the target state. For instance, $T_{0,1}$: Roobokart is switched on and it moves from S_0 to S_1;
- the automaton is graphically represented by modelling each state as a circle attached to its name and each transition as an arrow that connects the starting and the target nodes, and it is labelled by the transition name.

Once designed the solution, it must be ***implemented***. By combing the solution of each state, the entire mission is solved. We propose to implement the automaton by a *switch* on the current state and by modelling each state as a *case*.

The corpus of case corresponding to the state S_x is the implementation of all the transitions that can be performed starting from S_x and the implementation of each transition ends by changing the current state by the target one.

We can resume the proposed approach by the following phases: a training phase, followed by the constructionism-based activities where the project is provided, the solution is designed and, finally, implemented.

4 Experiments and Lessons Learned

In this section, we will present the performed experiments by involving High-school students in using Roobopoli in the formal settings and a preliminary investigation with Middle-school students to test the age-independence of the proposed approach. In conclusion, we will report lessons learned obtained by observing the results of the experiments and the faced challenges.

4.1 Roobopoli at High-School Experiment

Mission. The provided mission is *1) the Roobokart can freely move in the city of Roobopoli, without going out of the lane of the road; 2) it has to read horizontal signs; 3) it has to respect traffic-lights; 4) it can randomly choose the road to take to each road cross; 5) in case of obstacles, the Roobokart has to stop and wait until the obstacle disappears before moving on.*
Sampling. The experiments took place with 300 students with age in the range of 13–18. The evaluation happens in formal education, during the school hours dedicated to *Alternanza scuola-lavoro*, i.e., hours devoted to the learning of skills useful for the job opportunities by a collaboration between companies and the educational system. The whole class is involved, for 120 h in total, 4 consecutive hours a week. We split students into groups of 4–5 components.
Functional Blocks: Provided Skills and Hardware/Software Components. It is assumed the familiarity with a programming language to develop the solution. The provided functional blocks are: basics of coding (e.g., the concept of the cycle); fundamentals in robotics (e.g., sensors and actuators); the design of an automaton as modelling approach. Students learn how to assemble the robot before using it to solve the mission. Therefore, we provide students with a kit consisting of actuators, sensors, motors, and robots' components. The first task is assembling the Roobokart before programming it. The city is provided.
Resolution Design and Implementation. Students analyse the mission, decompose it in sub-problems, and model the solution as an automaton. The states of the automaton represent further decomposable problems or valid status of the robot. It must be refined until all its states correspond to a valid robot status. Transitions between states are guided by the values collected from sensors. While states model robot status, transitions model the evolutionary model of the robot. Once the students designed the automaton, according to the current state, the robot reads the input values by sensors and i) produces output by its actuators, ii) accomplishes a sub-problem modeled by the current stage,

and iii) moves to another state. The automaton can be coded by a switch-case. The code of a possible implementation of the resolution can be inspected at the following link[4], while this video[5] shows the solution performed by a Roobokart.

4.2 Middle-School Experiment

Mission. The mission is *the robot can 1) freely move in the city and 2) randomly choose the road to take to each intersection.* It is enough to formulate less detailed mission to involve also younger participants in the experimentation.
Sampling. The experiments took place with 60 students (25% of whom are female), 10–13 years old. We split the sampling into groups of 4–5 students. The experiments happened after school by students who voluntarily signed up for a course, named *Hello Robot!*. It took place for 16 h, 2 ensuing hours a week.
Functional Blocks. As before, students are provided with the basics of coding and the resolution modelling by an automaton. Because of the participants age, we focus more on transversal attitudes than vertical ones. Students are provided of an already assembled robots. Therefore, students have only to solve the mission. They should be enabled to code the solution by a block-based programming language (e.g., Scratch). In our experiment, we use mBot and its block-based programming language. We consider this experiment as a useful opportunity to collect feedback about the positive and negative aspects of such interface to design an interface for Roobopoli to address also needs of younger students.
Resolution Design and Implementation. Students are spur in designing the solution as an automaton or merely a list of steps. The implementation of the solution will be a program written by the block-based programming language to perform the desired path. The chosen path will be designed according to the starting point, the available roads, and the student's preferences.

4.3 Lessons Learned

Lessons learned are collected by the direct observation of students during the experiments. Here, only qualitative observations are provided, while we plan to collect also quantitative results by questionnaire in the immediate future.

- The *collaboration* implies getting familiar with the confrontation and communication protocols between people. It is beneficial to face challenges by building a network with people interested in meeting the same problems.
- The *mission* should model daily-life scenarios to train students in facing real challenges by experiencing them with incremental steps. For this reason, Roobopoli aims to solve common problems, daily observable in real cities.
- Functional blocks should concern both technical skills (e.g., coding and robotics fundamentals) and transversal soft skills (e.g., unambiguous modeling approach or critical thinking).

[4] https://github.com/Perlatecnica/Roobokart.
[5] https://www.youtube.com/watch?v=g0307OEjOJg&feature=emb_logo.

- The complexity of the mission, the detail level of the functional blocks and the provided equipment should be tuned based on the sampling age.
- Since all the involved students have been successfully engaged in interacting with automated systems and robots, we can assess that the *proposed approach* can be reused in different settings and with a sampling of various ages (by tuning suitable functional blocks).
- The proposed method leads to technologically aware citizenship in automated systems and (by the constructionism) students are actively engaged in experiencing problem-solving and CT by facing real challenges.

5 Conclusions and Future Directions

We have very little time to provide newcomer citizens with all the skills to face the only partially predictable challenges of the future. In particular, they should be provided with a transversal mindset and be technologically aware. We propose to teach technical subjects, in particular robotics, by a constructionism based approach. By involving students 10–18 years old, we observed that they are actively engaged and informed in robotic, while they also develop teamwork competencies and self-direction abilities. As a future direction, we are developing a block-based programming interface for Roobokart, and we will test the effectiveness of the proposed approach on 10–13 age students by using Roobokart rather than third-party robots by questionnaire and quantitative metrics.

References

1. Kurzweil, R.: The Singularity is near: when humans transcend biology. Viking (2006)
2. World Economic Forum: The future of jobs report (2018). https://www.weforum.org/reports/the-future-of-jobs-report-2018. Accessed 20 Jan 2020
3. Wing, J.M.: Computational thinking. Commun. ACM **49**(3), 33–35 (2006)
4. Papert, S.: Mindstorms: Children, Computers, And Powerful Ideas. Basic Books, Inc. (1980)
5. Papavlasopoulou, S., Giannakos, M.N., Jaccheri, L.: Empirical studies on the maker movement, a promising approach to learning: a literature review. Entertainment Comput. **18**, 57–78 (2017)
6. Anwar, S., Bascou, N.A., Menekse, M., Kardgar, A.: A systematic review of studies on educational robotics. J. Pre-Coll. Eng. Educ. Res. **9**, 2 (2019)
7. Bushnell, L., Crick, A.: Control education via autonomous robotics. In: 42nd IEEE International Conference on Decision and Control, pp. 3011 – 3017 (2004)
8. Apiola, M., Lattu, M., Pasanen, T.A.: Creativity and intrinsic motivation in computer science education: Experimenting with robots. In: 15th Annual Conference on Innovation and Technology in Computer Science Education, pp. 199–203 (2010)
9. Burbaite, R., Damasevicius, R., Stuikys, V.: Using robots as learning objects for teaching computer science. In: X World Conference on Computers in Education, pp. 101–110 (2013)

10. Mavridis, N., Al Rashdi, A., Al Ketbi, M., Al Ketbi, S., Marar, A.: Exploring behaviors collaborative mapping through mindstorms robots: a case study in applied social constructionism at senior-project level. In: International Conference on Innovations in Information Technology, pp. 284–288 (2009)
11. Arlegui, J., Pina, A., Moro, M.: A PBL approach using virtual and real robots (with BYOB and LEGO NXT) to teaching learning key competences and standard curricula in primary level. In: Technological Ecosystems for Enhancing Multiculturality Conference, pp. 323–328 (2013)
12. Scaradozzi, D., Sorbi, L., Pedale, A., Valzano, M., Vergine, C.: Teaching robotics at the primary school: an innovative approach. Procedia Soc. Behav. Sci. **174**, 3838–3846 (2015)

Cyber Security Education for Children Through Gamification: Challenges and Research Perspectives

Farzana Quayyum(✉)

Norwegian University of Science and Technology, Trondheim, Norway
farzana.quayyum@ntnu.no

Abstract. With the advancement of technology and the development of new tools, games, applications and social media sites, it is getting difficult every day to keep up with threats and vulnerabilities associated with these tools, apps or websites, especially for children. This paper provides an overview of the challenges in cyber security education for children. The goal of this research is to investigate and further develop new knowledge and tools that will be helpful and effective to teach the children about cyber security by performing various gamified actions in a playful, engaging and motivating manner. The methodology of this research would be both qualitative and quantitative, using interviews, questionnaires, focus group and observations. Initially, this project will define the theoretical and existing practices of cyber security awareness education for children. In the next phase, this research will design and implement interventions based on learning activities like workshops and collaborative tasks; followed by empirical test and evaluation of the proposed interventions.

Keywords: Cyber security awareness · Cyber security education · Internet security · Children · Gamification

1 Introduction

Children nowadays play a lot of online games and browse the internet for several hours every day. On the internet, they meet several opportunities as well as risks; but without relevant knowledge, it is difficult for them to assess the associated risks or threats of using the internet and digital systems. Sometimes they do not even realize the danger of the risks. Children may potentially expose themselves at risks unintentionally in various ways or sometimes may leak personal or confidential information even without knowing. They can also fall victim to cyber security threats like social engineering, cyber bullying, hacking, viruses, and damaging malware, cyber stalking, etc. through search engines, online advertisements and social networking websites such as Facebook, Twitter and lots of other websites [1].

While security practices rely on several factors, one of them is how well people are aware of the threats and how well they can assess the risk and apply their knowledge

Z. Kubincová et al. (Eds.): MIS4TEL 2020, AISC 1236, pp. 258–263, 2021.
https://doi.org/10.1007/978-3-030-52287-2_26

to mitigate threats [2]. Therefore, mitigating the human-related errors or vulnerabilities is a dominant factor for improving security either at a personal or organizational level; and we can do this by raising user awareness on cyber security and privacy issues [3, 4]. Considering the importance of human-related vulnerabilities, we aim to focus on education about cyber security awareness and online etiquette in this research.

Even though many research projects have been carried out so far and multiple applications and platforms have been developed to teach children about cyber security, still the success and impact of these applications on children are not as widespread and dynamic as expected. Thus, building upon previous research studies on motivating children about learning and practicing cyber security knowledge, the purpose of this research is to investigate and further develop knowledge and tools that will be helpful to teach the children about cyber security by performing various gamified actions and experiencing the consequences of those actions in a playful, engaging and motivating manner. Additionally, we also want to investigate and flourish the existing teaching methods of cyber security to become more effective and attractive for the children, enhancing their cyber security knowledge and skills.

2 Background and Motivation

When education starts becoming fun and attractive, children also become more motivated and interested in learning. Using computer games for educational purposes is now very popular. According to [5], there are two categories of games that are used to educate and train users: gamification and serious games. *Gamification's* main goal is to foster more engagement in people by helping to create more robust experiences in everyday life events utilizing game mechanics; while *serious games* are designed to train and are used for stimulation and to educate in virtual environments with previously defined learning objectives [6, 7]. [9] defines gamification as "A way to use game elements to learn but without the entertainment value. It uses game-like features including points and various levels in a way that is not meant to entertain". In recent years, we have seen that gamification of applications is getting a lot of attention from various fields including education. The term was created by Nick Pelling back in 2002, but it was not until 2010 that gamification itself became well known and embraced [8]. Since then the use and popularity of gamification are increasing every year [5]. Considering this popularity and momentum, in this research we will focus on gamification as the technique to educate children about cybersecurity awareness.

Various studies have been previously carried out and many cyber security-based gaming applications have been developed in last few years to educate users, especially for children; for example 'Cybersecurity Lab', which is designed to teach young people basic cyber security skills; 'The Internet Safety', a web-based game on safety in the internet; "Kids game - FBI" is about online safety management and many other games that teach about phishing, password hygiene, malware and other important topics of cyber security. Most of the studies have indicated positive results and impact on the users especially children, in using games and gaming applications as a tool for education on cybersecurity. But that there are still many gaps and issues that need to be addressed by the community. For example, many of the games and applications are often part of research projects and get developed systematically or rapidly, after their evaluation they often disappear and are rarely available to the public [10]. So, in our research, we will focus not only on designing and developing a tool or solution but also on sustainability. We aim to make our proposed solution and its associated contents and functionalities well maintained for a longer period which can be updated and customized for the children time to time along with the evolution of technologies and the changing nature of vulnerabilities.

Alotaibi et al. (2016) conducted a review of various studies focusing on gaming applications and their effectiveness in creating cyber security awareness [11]. From that review, the researcher figured out multiple issues that the future researchers from the industry need to address and consider in their research. For example, some studies including [12] have shown significant improvements in learning using games, but the sample sizes used in the studies were quite small. Other issues that the researchers mentioned include showing positive feedback but not evaluating the impact and effects in terms of learning outcomes or not presenting a conclusive result from the research. So, there is a need for a rigorous study involving large sample populations, well defined and effective methods for evaluation.

From previous research studies like [13–17], we have seen that learning outcomes from games and gaming applications may not be the same for all; the technology and learning preference may vary based on different factors. So, taking these previous research studies into account and be inspired from them, we aim to explore the learning differences from other perspectives also, along with gender; for example, we want to investigate if there is any difference based on culture or social issues in accepting and learning from gamified solutions on cybersecurity awareness education. Thus, building upon previous research studies, the purpose of this research is to address the existing gaps and to investigate and further develop new knowledge and tools that will help to teach a new generation of online users about online etiquette and cyber security.

3 Research Goals and Method

3.1 Research Context

This research targets 8–10th grade students (13–16 years old) of secondary schools. We are targeting this age group because many pieces of research, for example, [18, 19] show that children of this age are more prone to engage in risky internet behavior, as they have easy access to many kinds of digital devices and almost all possibilities that internet provides [18]. We can express the problem foundation of this research by the following research question.

RQ: How can we help children to learn about cyber security, using gamification?

To investigate this overall question and its parameters, we have defined the following sub-questions (SQ) that have been set as a part of this research.

SQ1: What are the trends in gamification of cyber security education?
SQ2: What type of cyber security knowledge and education do children need?

SQ3: How can we make a sustainable solution for cyber security education?
SQ4: How to measure the effectiveness of the proposed solution?
SQ5: How do gender, cultural differences, and social issues influence children's learning on cyber security?

3.2 Research Method, Data Collection and Analysis

Considering the context of this research, focusing on investigating the cyber security related learning experiences of children in informal environments, we have chosen Design-Based research (DBR) as our research methodology. Design-Based research is "A systematic but flexible methodology aimed to improve educational practices through iterative analysis, design, development, and implementation, based on collaboration among researchers and practitioners in real-world settings, and leading to contextually-sensitive design principles and theories" [20].

Following the DBR process, we will perform iterative cycles, starting by surveying existing trends, methods and their limitations for achieving a high impact in cyber security education and awareness. We are currently conducting an in-depth systematic literature review based on the original guidelines as proposed by Kitchenham [21] to answer the first two sub-questions of this research (SQ1 and SQ2). The literature review initially resulted in 64 citations, but after applying some quality criteria, we finally ended up with 26 papers for in-depth review. In the next phase, based on the theoretical understanding, we will design and implement interventions systematically with multiple iterations to refine and improve our initial designs, in collaboration with the participants (SQ3). But before proceeding with the activities, we aim to interview and conduct focus group discussions with experts from the industry and academia to gather necessary information for designing or preparing the contents and activities for the interventions. These interviews and focus group discussions would also help us to signify our answer and finding from SQ2.

We have done a preliminary field study in a workshop called "Kodeløypa" at the Norwegian University of Science and Technology (NTNU) in Trondheim, Norway. Kodeløypa is one of the scientific offers for children at NTNU, which aims to make high school students familiar with programming, especially with robots and games from an early age [22]. The experience of Kodeløypa will help us to design, develop and organize the interventions for our research. Specific learning outcomes will be addressed in each intervention via workshops, collaborative tasks, and other related learning activities. During the interventions, students will receive information about cyber security and will be asked further to participate in various activities with cyber security related learning outcomes.

After each intervention, empirical data will be collected for further analysis and evaluation regarding the students' perception and experience about the activities. Data from this stage will help us to answer the last two questions of this research (SQ4 and SQ5). The researcher will be present in all the interventions to observe the activities and products the participants will be developing during the interventions. At this stage, along with observational data, we also aim to collect the data by conducting pre/post questionnaires and interviews with the participants. The information obtained from the participants will be subjected to numerous statistical analyses and a triangulation among different types of evaluation will be used to ensure the accuracy of the findings and reported results. We believe findings from our studies and measuring learning outcome as well as the engagement of the children will help us to extend and refine our research framework by setting a new starting point.

4 Conclusion

Multiple gaming applications and platforms are available and have been already used to support the teaching of cyber security education for children. Therefore, there are many opportunities for us to improve the existing teaching techniques and methods and also to find out new creative and innovative ways of teaching. We expect these steps towards new techniques and methods along with their evaluation will help future researchers and educators to teach in a more efficient and motivating way. In our research, we will focus not only on designing, developing, and evaluating a new tool or solution but also on how to make the solution sustainable.

Acknowledgements. I would like to thank my supervisors, Letizia Jaccheri (letizia.jaccheri@ntnu.no), Deepti Mishra (deepti.mishra@ntnu.no) from The Department of Computer & Information Science, Norwegian University of Science and Technology (NTNU) and Aida Omerovic (Aida.Omerovic@sintef.no) from SINTEF Digital (Norway) for their support and guidance.

References

1. Hamdan, Z., Obaid, I., Ali, A., Hussain, H., Rajan, A.V., Ahamed, J.: Protecting teenagers from potential internet security threats. In: International Conference on Current Trends in Information Technology (CTIT), pp. 143–152 (2013)
2. Gjertsen, E., Gjære, E., Bartnes, M., Flores, W.: Gamification of information security awareness and training. In: Proceedings of the 3rd International Conference on Information Systems Security and Privacy (ICISSP), pp. 59–70 (2017)
3. Giannakas, F., Kambourakis, G., Papasalouros, A., Gritzalis, S.: Security education and awareness for k-6 going mobile. Int. J. Interact. Mob. Technol. (iJIM) **10**(2), 41–48 (2016)
4. Giannakas, F., Papasalouros, A., Kambourakis, G., Gritzalis, F.: A comprehensive cybersecurity learning platform for elementary education. Inf. Secur. J. Global Perspect. **28**(3), 81–106 (2019)
5. Kim, B.: The popularity of gamification in the mobile and social era. Libr. Technol. Rep. **51**(2), 5–9 (2015)
6. Karagiorgas, D.N., Niemann, S.: Gamification and game-based learning. J. Educ. Technol. Syst. **45**(4), 499–519 (2017)
7. Kim, J.T., Lee, W.H.: Dynamical model for gamification of learning. Multimed. Tools Appl. **74**(19), 8483–8493 (2015)
8. Ypsilanti, A., Vivas, A.B., Raisanen, T., Viitala, M., Ljas, T., Ropes, D.: Are serious video games something more than a game? A review on the effectiveness of serious games to facilitate intergenerational learning. Educ. Inf. Technol. **19**(3), 515–529 (2014)
9. de Byl, P.: Factors at play in tertiary curriculum gamification. Int. J. Game-Based Learn. **3**(2), 1–21 (2013)

10. Roepke, R., Schroeder, U.: The problem with teaching defence against the dark arts: a review of game-based learning applications and serious games for cyber security education. In: Proceedings of the 11th International Conference on Computer Supported Education - Volume 2, pp. 58–66 (2019)
11. Alotaibi, F., Furnell, S., Stengel, I., Papadaki, M.: A review of using gaming technology for cyber-security awareness. Int. J. Inf. Secur. Res. (IJISR) **6**(2), 660–666 (2016)
12. Arachchilage, N.A.G., Love, S.: A game design framework for avoiding phishing attacks. Comput. Hum. Behav. **29**(3), 706–714 (2013)
13. Arnup, J.L., Murrihy, C., Roodenburg, J., McLean, L.A.: Cognitive style and gender differences in children's mathematics achievement. Educ. Stud. **39**(3), 355–368 (2013)
14. Admiraal, W., Huizenga, J., Heemskerk, I., Kuiper, E., Volman, M., Ten Dam, G.: Gender-inclusive game-based learning in secondary education. Int. J. Incl. Educ. **18**(11), 1208–1218 (2014)
15. Lucas, K., Sherry, J.L.: Sex differences in video game play: a communication-based explanation. Commun. Res. **31**, 499–523 (2004)
16. Gorriz, C.M., Medina, M.: Engaging girls with computers through software games. Commun. ACM **43**, 42–49 (2002)
17. Bonanno, P., Kommers, P.A.M.: Gender differences and styles in the use of digital games. Educ. Psychol. **25**, 13–41 (2005)
18. Tsirtsis, A., Tsapatsoulis, N., Stamatelatos, M., Papadamou, K., Sirivianos, M.: Cyber security risks for minors: a taxonomy and a software architecture. In: 11th International Workshop on Semantic and Social Media Adaptation and Personalization (SMAP), pp. 93–99 (2016)
19. Guan, J., Huck, J.: Children in the digital age: exploring issues of cybersecurity. In: Proceedings of the 2012 iConference, Toronto, Canada, pp. 506–507 (2012)
20. Wang, F., Hannafin, M.J.: Design-based research and technology-enhanced learning environments. Education Tech. Research Dev. **53**, 5–23 (2005)
21. Kitchenham, B.A.: Procedures for undertaking systematic reviews. Joint Technical report, Computer Science Department, Keele University (TR/SE-0401) and National ICT Australia Ltd. (0400011T.1) (2004)
22. Papavlasopoulou, S., Giannakos, M.N., Jaccheri, L.: Creative programming experiences for teenagers: attitudes, performance and gender differences. In: Proceedings of the 15th International Conference on Interaction Design and Children. ACM (2016)

Becoming Safe: A Serious Game for Occupational Safety and Health Training in a WBL Italian Experience

Emma Pietrafesa[1](✉), Rosina Bentivenga[1], Pina Lalli[2], Claudia Capelli[2], Gaia Farina[3], and Sara Stabile[1]

[1] Department of Occupational and Environmental Medicine, Epidemiology and Hygiene, INAIL, 00100 Rome, RM, Italy
e.pietrafesa@inail.it
[2] University of Bologna, 40126 Bologna, BO, Italy
[3] University of Padua, 35122 Padua, PD, Italy

Abstract. Through the work-based learning (WBL), students carry out part of their formative curriculum into working contexts, so that the traditional learning, linked to the individual disciplines, is renovating into a new path that includes informal and non-formal competences. Technology is progressively more used to support innovative teaching that allows young people to acquire new skills to meet the needs of the changing labor market. The study describes a participative research and co-design work realized into seven Italian high schools in the agricultural, construction and manufacturing sectors, to co-create an educational tool (videogame) in order to promote occupational safety and health (OSH). The study applied qualitative and quantitative methods to facilitate the participation and the engagement of teachers and students. Learning by playing has always been an activity, today technologies and gaming can offer cognitive and operational elements suitable for recognizing and therefore preventing the occupational risks.

Keywords: Serious game · Occupational safety and health · Work-based learning

1 Introduction

In 2018, in Italy, 16.3 millions of people (aged 6–64 years old) played videogames: the 37% of the Italian population (in particular 54% men and 46% women) [1]. Most players are between 15–34 and 45–64 years old [2]. Videogames brought in revenue of 1.7 billion in 2018, with a growth of 18.6% from the previous year observe, although the vast majority of psychological research on the effects of "gaming" has been focused on its negative impact: the potential harm related to aggression, addiction, and depression [3]. But some studies consider the benefits that videogames could have in several domains: "cognitive (e.g., attention), motivational (e.g., resilience in the face of failure), emotional

Z. Kubincová et al. (Eds.): MIS4TEL 2020, AISC 1236, pp. 264–271, 2021.
https://doi.org/10.1007/978-3-030-52287-2_27

(e.g., mood management), and social (e.g., prosocial behavior)". Games and play have a special quality of social bonding "providing context and motivational aspects that can be used to improve the dynamics and solutions" [4]. Moreover, an increasing number of serious games applied in education are emerging [5]. Easiness of use, elements of surprise in the story-script, open-ended situations have been found as influential factors for the effectiveness in learning outcomes, especially among young players [6, 7].

The OSH Community strategy 2007–2012 [8] considers the prevention culture an important area of action: the purpose is to ensure that students receive risk and OSH education as part of their general education before they start working [9]. A recent study of the French National Research and Safety Institute for the Prevention of Occupational Accidents and Diseases (INRS) found that for young workers who received OSH teaching at school, the occupational accident rate was 50% lower than for young workers who did not [10].

2 Materials and Methods

Starting from international and national literature' review, a participative research and co-design work into 7 Italian high schools – agricultural, construction and manufacturing sectors – was realized to co-create an educational tool (videogame) in order to promote OSH during the WBL programs. This study applied qualitative and quantitative methods:

1. qualitative inquiry through individual interviews and focus group discussions in 4 Italian high schools (12 teachers and 60 students);
2. survey by questionnaire (277 students);
3. four word cafés/focus groups (50 students);
4. ten semi-structured interviews;
5. three WhatsApp groups (12 students) to facilitate the teachers and students participation and engagement.

A beta version of the serious game was tested with 49 students, than a first competition – to encourage the team participation – was organized with 75 students of two Italian high schools. A user' survey, about the satisfaction and commentaries on the videogame, was done by a questionnaire addressed to those 75 first players and by direct observation of their team gameplay. The final version of the videogame, adjusted following the observational and quantitative findings, was realized and tested by a game competition and final award among 39 students of 2 different schools (one of which participated from the beginning of the project).

2.1 The Aim of the Study

In Italy, the paths for transversal skills and orientation (PCTO) are promoted by schools for transversal activities to enhance the formative value of on-going orientation: the focus is on the educational value of the concrete work experience [11]. It is an opportunity to promote WBL paths, which encourage young people to have work activation processes and conscious social identity [12] since the alternation between work and study uses everything' student to acquire skills in the work context which will then be redefined and systematized at school [13].

This study, realized in collaboration with the University of Bologna, was financed by the National Institute for Insurance against Accidents at Work (Inail) through the BRIC 2016–2018 call[1]. In our study' framework the game is considered neither as an idealized way of doing digital education, nor as the stigmatized instrument contributing to the contemporary fashion of gamification and the entertainment market in the society of spectacle that was the object of Débord's critique [14–16]. The challenge more than in the structuring of educational contents is in the amplification of the opportunities to learn, e.g. to facilitate critical awareness about OSH risks, accepting the possibility to agree about taking advantage from prevention. A methodological path was designed, aimed to co-creating a serious game, engaging all the different actors of OSH education: students, teachers, OSH experts. No significant differences in academic achievements were found between digital learning and serious game use, but it is meaningful that "significantly more positive attitudes toward serious game assisted learning were revealed compared with traditional paper-based learning", since they encourage participation [5].

2.2 The Participative Process

The participative process of co-creation [17] follows the following steps:

- a qualitative inquiry in 4 Italian high schools – in 3 different sectors: agriculture, construction and manufacturing – to engage 12 teachers and 60 students in individual interviews and focus group discussions about their representations of risks in general and in the workplace and their learning and teaching experience;
- the creation of a smaller voluntary group of 12 students considered as "peer ambassadors", who had the task to make other interviews to their friends and other acquaintances and to interact with researchers (face-to-face and via instant messaging platforms) about the videogame design;
- a quantitative survey, in collaboration with the health experts, addressed to 277 students (63% boys and 37% girls) of the last three years of the 7 high schools, in order to deepen and verify the knowledge of their representations of risk and learning activities on safety, and evaluate their gained skills in risk prevention;
- a preliminary design of a serious videogame on OSH based on the analysis of the main findings to realize the beta test (a first questionnaire was submitted to 49 students of 2 high schools matching with the teachers and the OSH experts);
- the development of the videogame "Sicuri si diventa"[2] (Becoming safe), in collaboration with the developers of a software house, who were invited to participate in the beta test as well;
- the organization of a first competition encouraging team participation, with 75 students of 2 high schools and a user survey about the satisfaction and commentaries on the

[1] BRIC is a two-years research project in collaboration with University, public research center and other public health institutions financed by the National Institute for Insurance against Accidents at Work.

[2] www.site.unibo.it/sicuri-si-diventa/it/gioca.

videogame, through a questionnaire addressed to those 75 first players and by direct observation of their team gameplay;
- the final version of the videogame, adjusted following the observational and quantitative findings, with a game competition and final award among 39 students of 2 schools, one of which participated from the beginning of the project.

2.3 The Video Game "Sicuri si diventa" – (Becoming Safe)

This study is the result of a participated process that actively involved high-school students in the design of the gameplay mechanics as well as in making the main decisions regarding its educational contents. It was realized in collaboration with a team of developers and designed as a learning tool and a support to mandatory training, to be used in and out the classroom, developing a good practice for OSH culture in accordance with technical and legal issues. The goal is to win a race against time to prevent accidents, protect workers and build a safe workplace. The structure of the game encourages entertainment and interaction to actively engage the players' agentive behaviors. The game is suitable both for individual and team play and therefore can be played cooperatively or competitively. It can also be played on different devices (PCs, tablets, smartphones).

"Becoming safe" is a management simulation game set in a 3D environment with a third-person overhead view, a choice that allows the player to navigate the game easily and move rapidly through the different scenarios. The main scenarios include simplified versions of 3 different work environments – agriculture, construction and manufacturing – graphically translated in the style of some of the most popular videogames among the target users of the project (i.e. cubic and pixelated, similarly to games like Minecraft that are widely played among young people). The overall tone, sound and graphics are left intentionally unobtrusive in order to create an immersive role-playing experience for the player.

Each student can play the role of a junior OSH manager, initially guided by a senior manager who explains the basic rules and regulations that have to be followed to guarantee the workers' health and safety. By enhancing safety and preventing accidents, the safety manager ensures the mental and physical wellbeing of the workers and eventually helps the company become more successful and grow in terms of number of workers and overall production. The focus on the conceptual equation between safety and business growth was one of the key suggestions that emerged from the interviews with students, which often contained references to the commonly held belief that the application of OSH rules inevitably involves a loss of time and money both for entrepreneurs and workers. Therefore, the game includes progressive incentives and prizes, but at the same time, whenever a worker suffers an injury the player loses points and positions in the high-score table. The interaction mode follows the gameplay mechanics typical of a puzzle-platformer game in a role-playing game context – where the player has to drag and drop objects and tap or click to use them.

3 Digital as Opportunity for Education and OSH Training

According to the guidelines – art.1 paragraph 785 of the Law n.145/2018 – despite the specific didactic and training purpose and the limited presence and exposure to risks, the students acquire the status of workers pursuant to article 2, paragraph 1, letter a), of Italian Legislative Decree 81/2008 [18], and therefore they are subject to the obligations required by OSH legislation as training, health surveillance and individual protection devices [19, 20].

3.1 Digital Opportunities and Skills Needs

Today digital skills are a prerequisite to fully participate in the labor market, but benefiting from digital opportunities depends, first, on meeting some skills requirements, and second, on operating in a safe digital environments. These skills include pure digital skills but also the emotional and social skills. Possessing this mix of skills, conveniently labelled as "digital literacy", is a pre-condition for people to harmoniously combine their digital and real lives, and to avoid the mental health problems associated with abuses of digital technologies [21]. Three types of skills needs have emerged in the context of the digital transformation: a) ICT problem-solving skills as well as solid literacy, b) numeracy and problem-solving skills; specialized skills (i.e. cloud computing, big data analysis, block chain and AI are reliant on highly specialized skills); c) additional skills that are complementary to digital technologies, such as creative, social and emotional skills [21]. In the workplace, interpersonal and leadership skills, as well as the ability to navigate and leverage the digital economy, are also becoming more important [22].

3.2 Digital Divide and Education

In this context, besides offering new pathways for learning, schools play an important role in bridging the digital divide and ensure that all children reap the benefits of technological advances. Digital resources in the classroom can serve as an equalizing force between students who do and do not have access to digital technologies at home, allowing the latter to catch up with the digital mastery of the former. The results of digital learning experiences in schools are somewhat mixed, and many studies report limited or no benefits of digital education [23, 24]. This negative effect may be the result of greater distractions in the classroom, when students use Internet connection for chatting or playing rather than learning [25]. Another way digital resources may not necessarily be conducive to improved learning outcomes relates to the lack of digital skills of teachers, in fact, when they are not familiar with digital technologies, digital resources can form a distraction for both teacher and students [21].

4 Results

The main interpretative frames [26] resulting from the focus groups concern: 1) Need to work then to risk; 2) Earning money, which involves accepting risks; 3) The challenge with oneself and with other people; 4) Experience which can but is not always useful to avoid risks; 5) Fate, e.g. "we cannot prevent everything", anyway a minimal distraction or malfunction could make knowledge useless. In fact, interviews suggested that risk at work is often associated with the need to save time, to save money and, consequently, to maximize a company's and workers' profit. In this narrative, the line between acceptable and unacceptable risks becomes progressively more blurred, with some of the students even implying that workers can, after all, accept to sustain minor injuries and that accident, unless they cause permanent physical damage, can be a part of a person's work experience. This perspective also includes an underlying fatalism, which relies on "common sense", rather than on the respect of rules and regulations, to avoid accidents and preserve the workers' health.

The findings of the survey confirm these trends, especially the strong association between risk, getting hurt (30,3%), fate (28,9%) (e.g. the unexpected events) and lack of attention (27,4%). These aspects are underlined also in other control questions linked to the risk situation, where the lack of attention is considered the first most dangerous risk event, followed by rush, convenience and habit. Partially different results arise from qualitative and quantitative data concerning the students' evaluation of the effectiveness of OSH training at school. While a small percentage of the students (21.7%) who attended safety courses generally state that they have been useful, in the focus groups students explicitly point to the excessive abstraction of "boring power point presentations", as opposed to direct experience. In fact, when the questionnaire requires to rank six forms of training (from most to least effective), almost half of the respondents indicate "direct experience" as the most effective form for OSH training, while 39.2% think that traditional lectures - like the ones they are required to attend at school during institutional training - are the least effective.

5 Conclusions

In a rapidly changing world, education and training are called upon to play a key role in the acquisition of skills and competences useful for seizing the opportunities that arise in anticipation of changes in society and in the world of the labor market. In this context, digital technologies can unlock new learning opportunities in the classroom by giving students access to a wider range of resources, by complementing the teacher in learning processes (computer-assisted learning) and by providing other advantages to students, such as access to motivational and informational resources associated with access to tertiary education programs. Learning by playing has always been an activity, today technologies and gaming can offer cognitive and operational elements suitable for recognizing and therefore preventing the occupational risks.

This study confirms that educational elements can be integrated into a gameplay, so that will be subconsciously acquired by the players during the gaming process. The results of the survey, in fact, show a high level of player satisfaction among the students: while in the first group for the beta test only 46% considered the game "adequate" and 31% defined it as "entertaining", the 90% of the second group – who played the final version of the game – love playing with it. According to this second group of students, the game "Becoming safe" is especially useful for learning while having fun (94%) and for acquiring awareness regarding risks in the workplace (86%). It also give the possibility to better remember the OSH rules (83%), in particular those concerning the uses of personal protective equipment (PPE), which play an important role in the game narrative.

This path could be a support for training and usable both in the classroom and in internship, according to the indications for the serious game's design. Evidence suggests that educating individuals to become more flexible, resilient and creative and to pursue personal wellbeing helps them to adapt better to the changing work and life environment [27]. Interactive teaching practices, for example the problem-based learning (PBL), require students to work in groups and use non-cognitive skills in their discussions, to listen to one another when solving problems, or to make choices about their own learning [28]. PBL can contribute to developing a student's cognitive skills [29] and, together with interdisciplinary learning, can facilitate the acquisition of non-cognitive skills by emphasizing the importance of flexibility and innovation for the OSH issues too.

References

1. Aesvi: I videogiochi in Italia nel 2018. Dati sul mercato e sui consumatori, Rapporto Associazione Editori Sviluppatori Videogiochi Italiani (2019). http://www.aesvi.it/cms/download.php?attach_pk=1502&dir_pk=902&cms_pk=3002. Accessed 12 Oct 2019
2. Shudson, M.: How culture works: perspectives from media studies. Theory Soc. **18**(2), 153–180 (1989)
3. Granic, I., Lobel, A., Engels Ruther, C.M.E.: The benefits of playing videogames. Am. Psychol. **69**(1), 66–78 (2013)
4. Schouten, B., Ferri, G., de Lange, M., Millenaar, K.: Games as strong concepts for city-making. In: Nijholt, A. (ed.) Playable Cities, Gaming Media and Social Effects. Springer, Singapore (2017)
5. Zhonggen, Y.: A meta-analysis of use of serious games in education over a decade. Int. J. Comput. Games Technol. **2019**, 8 (2019). https://doi.org/10.1155/2019/4797032. Article ID 4797032
6. Iten, N., Petko, D.: Learning with serious games: is fun playing the game a predictor of learning success? Br. J. Edu. Technol. **47**(1), 151–163 (2016)
7. Wouters, P., van Oostendorp, H., ter Vrugte, J., vanderCruysse, S., de Jong, T., Elen, J.: The effect of surprising events in a serious game on learning mathematics. Br. J. Edu. Technol. **48**(3), 860–877 (2017)
8. EU. Community strategy 2007–2012 on health and safety at work. http://eur-lex.euro-pa.eu/Notice.do?checktexts=checkbox&val=443914%3Acs&pos=1&page=1&lang=en&pgs=10&nbl=1&list=443914%3Acs%2C&hwords=&action=GO&visu=%23texte. Accessed 28 Jan 2020
9. Copsey, S., Debruyne, M., Eeckelaert, L., Malmelin, J., Salminen, S., Buffet, M.A., Bollmann, U.: Mainstreaming occupational safety and health into university education. Publications Office of the European Union (2010)
10. Institut National de Recherche et de Sécurité, INRS: Synthèse étude INRS Accidentologie des jeunes travailleurs. Recevoir un enseignement en santé et sécurité au travail ré-duit le risque d'accidents du travail chez les moins de 25 ans (2018)
11. Gentili, C.: L'alternanza scuola-lavoro: paradigmi pedagogici e modelli didattici. Nuova secondaria **10**, 16–37 (2016)
12. Morselli, D., Costa, M.: Il laboratorio di attraversamento dei confini nell'alternanza scuola-lavoro. Ricercazione. Ricerca educativa, valutativa e studi sociali sulle politiche e il mondo giovanile **6**(2), 193–209 (2014)
13. Marcone, V.M.: La formatività del work-based learning. Tesi di dottorato. Università Ca'Foscari Venezia (2018)
14. Gee, J.P.: Videogames: What Are They Good For? Unpublished ms. Arizona State University (2014). www.jamespaulgee.com/pdfs/What Are Videogames Good For.pdf. Accessed 15 Mar 2019

15. Débord, G.: La société du spectacle, Paris. Editions Champ Libre (1967)
16. Bassetti, C., Teli, M., Murgia, A.: Tin hat games – producing, funding, and consuming an independent role-playing game in the age of the interactive spectacle. In: Briziarelli, M., Armano, E. (eds.) The Spectacle 2.0: Reading Debord in the Context of Digital Capitalism, pp. 167–182. University of Westminster Press, London (2017)
17. Lefebvre, R.C.: Transformative social marketing: co-creating the social marketing discipline and brand. J. Soc. Mark. **2**(2), 118–129 (2012)
18. D.lgs. 9 aprile 2008, n. 81. Attuazione dell'articolo 1 della Legge 3 agosto 2007, n. 123 in materia di tutela della salute e della sicurezza nei luoghi di lavoro. (Gazzetta Ufficiale n. 101 del 30 aprile 2008 - Suppl. Ordinario n. 108) (Decreto integrativo e correttivo: Gazzetta Ufficiale n. 180 del 05 agosto 2009 - Suppl. Ordinario n. 142/L)
19. Accordo tra il Ministro del lavoro e delle politiche sociali, il Ministro della salute, le Regioni e le Province autonome di Trento e Bolzano per la formazione dei lavoratori, ai sensi dell'articolo 37, comma 2, del decreto legislativo 9 aprile 2008, n. 81. (Rep. Atti n. 221/CSR). (GU Serie Generale n.8 del 11-01-2012)
20. Accordo finalizzato alla individuazione della durata e dei contenuti minimi dei percorsi formativi per i responsabili e gli addetti dei servizi di prevenzione e protezione, ai sensi dell'articolo 32 del decreto legislativo 9 aprile 2008, n. 81 e successive modificazioni. (Rep. Atti n. 128/CSR). (GU Serie Generale n.193 del 19-08-2016)
21. OECD: How's Life in the Digital Age? Opportunities and Risks of the Digital Transformation for People's Well-being. OECD Publishing, Paris (2019). https://doi.org/10.1787/9789264311800-en. Accessed 28 Jan 2020
22. Deming, D.J.: The growing importance of social skills in the labor market. Q. J. Econ. **132**(4), 1593–1640 (2017)
23. Bulman, G., Fairlie, R.W.: Technology and education: computers, software, and the internet. In: Handbook of the Economics of Education, vol. 5, pp. 239–280. Elsevier (2016)
24. Escueta, M., Quan, V., Nickow, A.J., Oreopoulos, P.: Education technology: an evidence-based review. National Bureau of Economic Research (2017)
25. Greener, S., Wakefield, C.: Developing confidence in the use of digital tools in teaching. Electron. J. E-Learn. **13**(4), 260–267 (2015)
26. Entman, R.M.: Framing: towards clarification of a fractured paradigm. J. Commun. **43**(4), 51–58 (1993)
27. Cefai, C., Bartolo, P.A., Cavioni, V., Downes, P.: Strengthening social and emotional education as a core curricular area across the EU: a review of the international evidence (NESET II report), Luxembourg (2018). https://www.um.edu.mt/library/oar/handle/123456789/29098
28. WEF: The future of jobs. Employment, skills and workforce strategy for the fourth industrial revolution. In: World Economic Forum (2016)
29. Becker, S.A., Cummins, M., Davis, A., Freeman, A., Hall, C.G., Ananthanarayanan, V.: NMC horizon report: 2017 higher education edition 1–60, The New Media Consortium (2017)

Education Meets Knowledge Graphs for the Knowledge Management

Renato De Donato[1], Martina Garofalo[2], Delfina Malandrino[1], Maria Angela Pellegrino[1(✉)], and Andrea Petta[1]

[1] Dipartimento di Informatica, Università di Salerno, Fisciano, Italy
rended83@gmail.com, {dmalandrino,mapellegrino}@unisa.it, andrpet@gmail.com
[2] ACT OR S.r.l., Rome, Italy
martina.garofalo@act-operationsresearch.com

Abstract. Data are a crucial source for informed decision making. However, their unrestrainable growth requires approaches and tools to learn how to query and identify data of interest. The problem we want to face is how to guide students in going beyond the passive inspection of results returned by a search engine, and in actively searching for the data that best answer their questions. Our final goal is to guide future citizens in actively creating their knowledge supported by data. In particular, we focus on Knowledge Graphs and how they can be used in the knowledge management process in the educational context. We present *ELODIE* and the *datalet* mechanism as a tool to support the knowledge management process. We will describe its features, and we will discuss how teachers and students could exploit it in the educational context.

Keywords: Knowledge management · Data literacy · Information retrieval · Data visualization · Knowledge Graphs · Semantic web · Semantic search

1 Introduction

Data are vertiginously rising [1] and because of their unrestrainable growth, "*data is the new oil, but if unrefined it cannot really be used*" [2]. Therefore, there is the necessity to learn how to read, work with, analyze, and argue with data [3], i.e., we should experience the *data literacy*. It is a crucial skill that future citizens have to acquire by 2030 [4] and it is strongly related to the ability of *searching* and *extracting knowledge* out of raw data. Knowledge is widely studied from several different perspectives, and there is no consensus on its definition. In this article, we consider *knowledge* as *information in action* [5]. It represents one of the phases of the *data* (or knowledge) *pyramid* [6] composed of *data, information* and *knowledge. Data* are observations or facts; the *information* is inferred from data in the process of answering questions such as *who, what, where, how many,*

Z. Kubincová et al. (Eds.): MIS4TEL 2020, AISC 1236, pp. 272–280, 2021.
https://doi.org/10.1007/978-3-030-52287-2_28

when and make data useful for actions or decisions; the *knowledge* is achieved by processing, organizing or structuring information. The process to reach the knowledge starting from data is referred to as *knowledge management* (KM).

The problem we want to face is how to engage students in the KM process actively and to make them leave the position of an indifferent spectator of retrieved results. Usually, users passively accept the list of first results returned by a search engine without further investigating the ones ranked as less important. Ironically, Elon Musk says that the "*safest place to hide a dead body is the second page of Google search results*" since most people stop on its first page. We desire to spur school pupils in assuming the *control* in searching on the Web, in *critically* choosing results of interest, and in *actively* creating their knowledge.

Search engines mainly query documents and look for the occurrence of the searched terms in those files. In this article, we propose a shift from querying documents to semantic searches, i.e., we focus on the exploitation of Knowledge Graph (KG) by the KM process in the educational context. A KG is a knowledge base modeled as a graph, and it combines linked data technologies and ontologies.

We have designed a guided workflow to query KGs and to visualize the retrieved results. While the component to query KG is named *ELODIE*, the results visualization component will be referred to as *datalet mechanism*. In this article, we present the combination of ELODIE and the datalet mechanism as a tool to support the KM process. In particular, it aims to guide students in querying KGs and in replying to questions formulated in natural language to retrieve information. Upon the retrieved results, students can make decisions and acquire knowledge. Our main contributions are the following:

- the proposal of a collection of educational use cases that can benefit from KGs by highlighting potentialities and required skills to query KGs;
- considerations and evidence that the KM process can take advantage of ELODIE and the datalet mechanism [7].

The structure of this article is the following: in Sect. 2, we consider the role of KM and KGs in the education; in Sect. 3, we analyze how the KM process can exploit KGs and in Sect. 4, we point out how ELODIE and the datalet mechanism can be used in the KM process; in Sect. 5, we discuss the potentialities and the required skills to exploit KGs in the educational context; finally, we will conclude with some final considerations and future directions.

2 Related Work

KM is a set of practices that leads to the use of data in decision-making [8]. Rodrigues and Pai [9] defined a list of crucial factors that must be taken into account to develop a suitable KM strategy for schools. Among others, they consider as a critical dimension the *technology and infrastructure*; the *acquisition and learning*, i.e., methods to improve the searching and learning of knowledge; the *dissemination and transfer* of gained knowledge with others. Therefore, students have to learn how to acquire knowledge and mainly how to disseminate it [10], supported by technological solutions. Our proposal, ELODIE and the datalet mechanism[1] [7], could represent a technological solution to support students in the KM process by enabling the knowledge acquisition by querying KGs. About the dissemination, ELODIE and the datalet mechanism are installed in a social platform to share and discuss acquired knowledge.

Our interest in bringing closer KGs and KM is justified by the observation that KGs can facilitate and enhance KM since they explicitly provide a structure to data, and this structure is instrumental in supporting semantic searches and answering more profound and complex questions [11]. KGs and KM are usually combined at the university level [12]. We propose to introduce this approach also in the early stage of the educational plan. This proposal is due to the observation that KGs are gradually applied to teaching and learning in the educational domain to spur students in thinking about entities and their logical relationships [13]. It can promote a more in-depth comprehension of how data are connected among them within a certain domain [13]. In the educational context, they are usually referred to as concept map [14]. It has been proved that concept maps can significantly improve students' learning achievement: it prompts constructive learning, reflective ability, and active interaction [15]. Thus, we assess that KGs can be positively perceived also by young students.

3 Knowledge Management by Knowledge Graphs

Advantages of Knowledge Graphs. An increasing interest is manifested over KG publication: the LOD Cloud [16] (a KG that collects most of the published KGs by academia and industry) counted 12 datasets in 2007 and currently contains 1,239 datasets. Despite the *quantitative* reason to exploit KGs, also the *provenance* is an important aspect: several virtuous institutions invested or are investing in publishing data in the linked format, such as, Europeana[2], Eurostat[3], ISTAT, Beni Culturali[4], the British Museum[5]. Furthermore, it is highly recommended to *interlink* KGs [17] and it implies the possibility to navigate from a KG to another. Because of the extensive range of heterogeneous information stored in KGs, for their easy navigation, thanks to their quantitative and qualitative properties, they could behave as a critical resource for KM.

Challenges Posed by Knowledge Graphs. First of all, it is not always immediately clear how to translate a natural language question in a query over a KG since data modeling can be domain-dependent or domain-agnostic. Therefore, it could be hard to conceptualise data to query and to guess the used terminology.

[1] In the article presented at CSCWD, the same tool is named SPLOD since it was presented as a component of SPOD. Here we consider the tool extrapolated from the surrounding platform, and we refer to it by ELODIE and the datalet mechanism.

[2] https://pro.europeana.eu/page/linked-open-data.

[3] https://ec.europa.eu/eurostat/web/nuts/linked-open-data.

[4] http://dati.culturaitalia.it/.

[5] https://old.datahub.io/dataset/british-museum-collection.

The second aspect is related to the available querying languages. SPARQL is the most common one even if it proves to be too challenging, mainly for lay users [18,19]. It requires technical skills in generic querying languages and in understanding the semantics of the supported operators by SPARQL. Therefore, there is an interest in developing tools able to implicitly compose queries by hiding the underlying complexity to open KGs also to lay users [20].

High-Level Knowledge Management Process. The KM process 1) starts from data that models facts, 2) reaches the information by querying data and collecting results, and 3) realises the knowledge by analysing and structuring achieved information in interpretable and shareable artefacts.

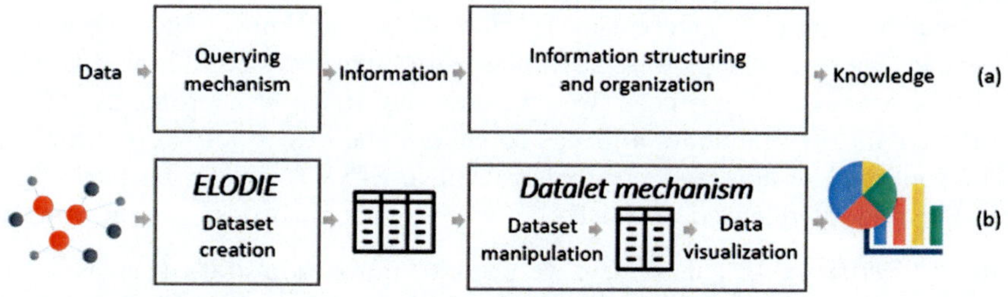

Fig. 1. In (a) we represent the high-level KM approach by KGs, while in (b) we report how we interpreted each phase in our tool. The KM starts from data (KG in our case); data must be queried to retrieve information (represented by a data table); information must be manipulated, and the knowledge can be gained. While the dataset creation phase is realized by ELODIE, the knowledge can be acquired by analyzing and making decisions upon the realized charts.

In Fig. 1a, we design the KM process by KGs, and we report how we interpreted each phase in our tool (Fig. 1b). KGs are the queried data. To move from data to information, we have to define a querying mechanism. As explained in the challenges posed by KGs, we aim to hide the underlying syntactical complexities while raising questions. Therefore, we designed a querying mechanism that leads to 1) the formulation of SELECT queries in natural language and 2) that organises results (i.e., the retrieved information) in data tables. We named this phase as *dataset creation*, and we address it by ELODIE. To move from information to knowledge, users need to process and organise retrieved results. Starting from the first year of their educational plan, children get used to reading tables, creating (simple) charts, performing (basic) manipulation, and interpreting results to reply questions [21]. Thus, we model the gained knowledge as shareable artifacts (e.g., charts) that can justify the decision made. To achieve knowledge, results can be manipulated to be compliant with visualization modes and, then, the desired chart can be realized by the datalet mechanism.

4 ELODIE and the Datalet Mechanism

ELODIE (pronounced elədē) is a guided workflow to extract a tabular representation of data queried by KGs, while the datalet mechanism is a scaffolded approach to visualize the retrieved results and create shareable artefacts. Their combination implements the trialogical learning approach that is articulated into three learning metaphors: knowledge acquisition, participation, and knowledge creation by reusable artifacts [22]. By ELODIE and the datalet mechanism, the user can experience an individual effort and can participate in social discussions. Furthermore, the tool enables the possibility to continue the query of another user by fulfilling the collaborative creation of shared artifacts [7].

Here, we propose ELODIE and the datalet mechanism as a tool for supporting the KM by KGs. While ELODIE realizes the transformation of data in information, the datalet creation leads to acquiring knowledge. We will now analyse each phase of the process as implemented in our tool and how it is related to the KM process. As use case, we consider the desire of a professor to create a library for his/her students and has to choose the best books to collect.

The tool can be accessed via registration in SPOD[6] or can be freely tested here[7]. The code is released on GitHub[8], while quick tutorials[9] are on YouTube.

Dataset Creation. In this phase, we want to move from data (i.e., KGs) to a tabular representation of retrieved results without facing SPARQL challenges. Therefore, we proposed a faceted search interface enhanced by a natural language query. ELODIE organises nodes (referred to as concepts) and links (referred to as predicates) of the KG by facets. Students can click on any provided option to formulate their queries. In Fig. 2, the interface of this phase is visible. Given the need to create a library for children, the professor is interested in retrieving books related to the Children's literature. Furthermore, he/she has to model all the desired parameters he/she wants to take into account in the book selection, for instance, the publication date, the length of the book, books rewarded by critics. By this step, users learn how to actively control retrieved data by choosing step by step, the option compliant with the desired goal. It goads to acquire *independence* and a complete *control* of the search process.

***Dataset Manipulation** and **Data Visualization**.* When the user is satisfied with the retrieved results, we can move on to the knowledge acquisition step, realized by the datalet mechanism. We split this phase in dataset manipulation and data visualization. First, the user has to manipulate the dataset to make it compliant with the desired visualization by aggregating, sorting, and filtering data. Finally, knowledge can be acquired. The creation of a shareable chart realises it. Back to our use case, we can opt for aggregating awards by authors

[6] http://spod.routetopa.eu/.

[7] https://deep.routetopa.eu/deep2/COMPONENTS/controllets/splod-visualization-controllet/demo.html.

[8] https://github.com/routetopa/deep2-components/tree/master/controllets/splod-controllet.

[9] https://youtu.be/e_o32GP-l1c.

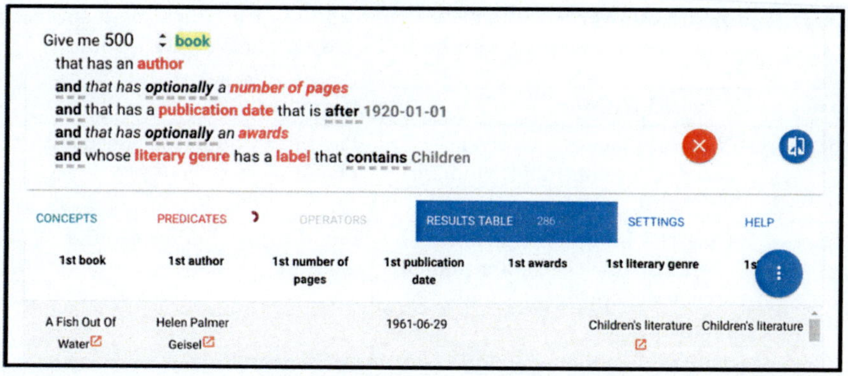

Fig. 2. It represents the dataset creation phase where users can move from data (e.g., books) to information (e.g., books matching the queried filters). By analyzing the retrieved results, the best author/book (i.e., knowledge) can be acquired.

and sort authors by counting the received prizes. We discover that Dr. Seuss (author of *The Cat in the Hat*), Dav Pilkey (cartoonist and author of *Captain Underpants*) and C.S. Lewis (author of the *Chronicles of Narnia*) are the preferred ones by critics. By considering the most recent works, we can cite *River Rose and the Magical Lullaby* written by Kelly Clarkson and *Diary of a Wimpy Kid* written by Jeff Kinney. Then, books of Dr. Seuss and Dav Pilkey are again at the top of the ranking. By considering the book-length, Dr. Seuss is at the top with 32 pages while C.S. Lewis wrote the most verbose stories. By these considerations, we acquire the awareness that Dav Pilkey cannot miss in our library. By this phase, the users learn how to acquire knowledge and how to represent it as a shareable artifact. Each performed analysis so far can be represented as a chart to exhibit evidence and substantiate decisions.

5 Discussion

Potentialities of Knowledge Graphs in Educational Settings. We assess that general-purpose KGs (such as DBpedia) are a useful source in educational contexts because of the heterogeneity of covered topics, from sport to art, from science to geography, and it can be exploited by different educational subjects. In Table 1, we report for each subject a list of useful entities and few examples of queries that can be replied by querying DBpedia (by ELODIE).

Required Skills for Knowledge Management by Knowlegde Graphs. The *KM* requires the ability of *actively* extracting information out of data and representing the acquired knowledge by a shareable artefact. It implies that future citizens must learn how to ***locate and query data*** of interest, *critically* evaluate retrieved information, learn how to ***synthesise*** it to reply the initial query, and to decide how ***represent and share*** the knowledge. To address *data literacy*, students must learn how to interpret artifacts, use them as evidence in

Table 1. It provides an overview of entities covered for each educational subject and few examples of queries whose reply can be retrieved by querying DBpedia.

Subject	Covered entities	Examples
Geography	Natural places (e.g., river), place (e.g., continent, country), mountain, volcano	The most populated continent The longest river
Science	Celestial body, unit of work, chemical substance, anatomical structure, species (e.g., animal, plant)	The orbiting bodies around the Sun
History	Military conflict, royalty, concentration camp, politician	Concentration camps turned into museums
Sport	Sport facility, athlete, activity, coach, league, event, sport season, team	Athlete taller than 2 m
Art	Fictional character, artist, museum, colour	Museums of modern art
Religion	Cleric (e.g., pope, saint)	Saint canonised in XXI century
Technology	Device, software programming language	How programming language influence each other
Music	Musical artist, musical work	Which hard rock artist play piano?
Literature	Writer, written work	Who wrote autobiographies?

discussions while arguing with data. It implies to learn how to ***interpret charts*** and be able to defend an opinion by using it as evidence. In working with KGs, future citizens have to try how data of interest can be abstracted and, then, verify how data are actually modelled. For instance, if I am interested in museums and their paintings, I have to realise that the concepts I am interested in are *museum* and *painting*. Then, I have to think about how this information can be linked: the museum *exposes* a painting or a painting *is located in* a museum. Thus, the ***modelling ability*** plays a crucial role.

6 Conclusions and Future Directions

Data are vertiginously growing and it is crucial for future citizens learning how to exploit them to the best. To actively engage students in the knowledge management, we propose ELODIE and the datalet mechanism that could have the potential to guide users in acquiring knowledge step by step. Our proposal spur students in *locating and querying data* of interest, *critically* evaluating and *synthesising* retrieved information and, finally, making decisions. We have already tested out proposal with high-school students, and we obtained positive results [7]. The next step is to test its effectiveness with younger participants.

References

1. Lynkova, D.: 39+ big data statistics for 2020 (2019). https://leftronic.com/big-data-statistics/. Accessed 14 Feb 2020
2. Humby, C.: Data is the new oil. https://ana.blogs.com/maestros/2006/11/data_is_the_new.html. Accessed 29 Mar 2020
3. Mandinach, E.B., Gummer, E.S.: A systemic view of implementing data literacy in educator preparation. Educ. Res. **42**(1), 30–37 (2013)
4. Organisation for Economic Co-operation and Development (OECD): Core Foundations for 2030 (2019). http://www.oecd.org/education/2030-project/teaching-and-learning/learning/. Accessed 21 Feb 2020
5. O'Dell, C., Grayson, C.J.: If only we knew what we know: identification and transfer of internal best practices. Calif. Manag. Rev. **40**(3), 154–174 (1998)
6. Rowley, J.: The wisdom hierarchy: representations of the DIKW hierarchy. J. Inf. Sci. **33**(2), 163–180 (2007)
7. De Donato, R., Garofalo, M., Malandrino, D., Pellegrino, M.A., Petta, A., Scarano, V.: Linked data queries by a trialogical learning approach. In: 23rd IEEE International Conference on Computer Supported Cooperative Work in Design (2019)
8. Petrides, L.A., Nodine, T.R.: Knowledge management in education: defining the landscape. Institute for the Study of Knowledge Management in Education (2003)
9. Rodrigues, L., Pai, R.: Preparation and validation of km measurement instrument: an empirical study in educational and it sectors. In: International Conference on Knowledge Management, pp. 583–593 (2005)
10. Chu, K., Wang, M., Yuen, A.: Implementing KM in school environment: teachers' perception. Knowl. Manag. E-Learn. **3**, 139–152 (2011)
11. Yan, J., Wang, C., Cheng, W., Gao, M., Zhou, A.: A retrospective of knowledge graphs. Front. Comput. Sci. **12**(1), 55–74 (2018)
12. Mariia, R.: Knowledge graph application in education: a literature review. Acta Universitatis Lodziensis. Folia Oeconomica **3**, 7–19 (2019)
13. Cui, J., Yu, S.: Fostering deeper learning in a flipped classroom: effects of KGs versus concept maps. Br. J. Educ. Technol. **50**(5), 2308–2328 (2019)
14. Chen, P., Lu, Y., Zheng, V.W., Chen, X., Yang, B.: KnowEdu: a system to construct knowledge graph for education. IEEE Access **6**, 31553–31563 (2018)
15. Dias, S.B., Hadjileontiadou, S.J., Diniz, J.A., Hadjileontiadis, L.J.: Computer-based concept mapping combined with learning management system use: an explorative study under the self-and collaborative-mode. Comput. Educ. **107**, 127–146 (2017)
16. McCrae, J.P.: The LOD cloud (2007). http://lod-cloud.net. Accessed 22 Feb 2020
17. Berners-Lee, T.: 5-star OD (2006). http://5stardata.info. Accessed 22 Feb 2020
18. Damljanovic, D., Agatonovic, M., Cunningham, H.: Natural language interfaces to ontologies: combining syntactic analysis and ontology-based lookup through the user interaction. In: Proceedings of the 7th ISWC: Research and Applications (2010)
19. Ferré, S.: SQUALL: the expressiveness of SPARQL 1.1 made available as a controlled natural language. Data Knowl. Eng. **94**, 163–188 (2014)

20. Dadzie, A., Rowe, M.: Approaches to visualising linked data: a survey. Semant. Web **2**(2), 89–124 (2011)
21. Saddiqa, M., Rasmussen, L., Magnussen, R., Larsen, B., Pedersen, J.: Bringing open data into Danish schools and its potential impact on school pupils. In: Proceedings of the 15th International Symposium on Open Collaboration (2019)
22. Paavola, S., Hakkarainen, K.: The knowledge creation metaphor - an emergent epistemological approach to learning. Sci. Educ. **14**(6), 535–557 (2005)

StoryVR: A Virtual Reality App for Enhancing Reading

Federico Pianzola[1,2(✉)] and Luca Deriu[3]

[1] University of Milano-Bicocca, Milan, Italy
federico.pianzola@unimib.it
[2] Sogang University, Seoul, South Korea
[3] PlaySys, Milan, Italy

Abstract. We present a virtual reality app specifically designed for reading/listening to short stories and poems. Empirical studies with VR narrative experiences have proven that the process of embodied simulation enhanced by the VR medium increases users' absorption and engagement. Accordingly, this solution can be effectively used to promote reading and increase motivation for learning. We discuss the design choices adopted to facilitate its widespread adoption and maximise readers' engagement with stories.

Keywords: Virtual reality · Reading · Literature · Audiobook · Narrative absorption

1 Introduction

Digital technologies have been exploited in different ways in relation to literature: by literary artists to create new text forms, by digital humanists for curation and criticism, but no one focused on designing literary texts in virtual reality (VR) to enhance traditional reading experiences. *StoryVR* is the first VR app specifically designed to make literature more appealing by enhancing the level of immersion/absorption experienced while listening to stories. In recent years a few solution for reading in VR have been released [1,3,4], including an official app by one of the leading distributors of VR headsets, the HTC Vivepaper [2], which however is not maintained anymore. All these apps transpose the two-dimensional design of the page into a 3D space replicating the interactive affordance of paper books, that is requiring the reader to turn pages in order to continue the story. With *StoryVR* we decided to take advantage of the increasing popularity of different form of experiencing literature – audiobooks – freeing the user from gestures related to another media and enabling developers to focus on the multimodal integration of audio and visual elements in order to enhance the engagement with the story.

Z. Kubincová et al. (Eds.): MIS4TEL 2020, AISC 1236, pp. 281–288, 2021.
https://doi.org/10.1007/978-3-030-52287-2_29

2 Virtual Reality and Electronic Literature

Virtual reality is a medium used to create experiences that are perceived as highly immersive [6]. VR environments have several positive effects when used as the context in which to perform ordinary activities, e.g. attention and sense of presence are increased [5], and emotional arousal is achieved [10,26]. However, VR has never been exploited to actively promote reading, an activity that could certainly benefit from increased immersion, attention and emotional engagement. Other kinds of digital technology have been used to create new literary formats, which are known as *electronic literature*: hypertexts, multimodal novels, interactive fiction, GPS location-specific narratives, installations in cave automatic virtual environments, etc. [11,14,22]. However, the majority of this kind of works is experimental and cannot be easily accessed by readers. On the contrary, we designed and developed a VR app intended for a widespread standalone VR Head Mounted Display (HMD) – Oculus Go – and with scalable content, since librarians and users are able to upload their own texts and audiobooks.

3 Background Research

The *StoryVR* experience has been designed on the basis of previous theoretical and empirical research on *narrative absorption* [13], *situated reading* [16], and *presence* in VR [17]. The theoretical assumption at the basis of *StoryVR* is that the level of immersion perceived when reading or listening to a story can be enhanced through *environmental propping* [16], that is the process for which features of the reading environment can foster readers' transportation into the story world, and empathy, eliciting a higher aesthetic pleasure. This hypothesis has been experimentally confirmed, showing that the perceived shift of embodiment from the actual world to the VR space facilitates a further shift of perceived embodiment into the story world [24]. The same study also found that VR can effectively be used to promote reading, since people who read in VR were more willing to continue reading. A similar research showed that when the VR experience is designed to let users embody the audience of a historical event, they feel a stronger sense of presence and are more willing to later seek more information about the event to which they participated [28]. However, although there is a widespread enthusiasm about the use of VR in education [29],VR does not seem to autonomously encourage learning in all fields: application to mathematics and science learning proved to be controversial [18,19] if not supported by complementary learning strategies [23]. This process is in line with what predicted by cognitive theory of multimedia learning [8]. Overall, VR works well as a tool to enhance motivation and enthusiasm towards a topic, especially with narrative experiences that enable embodied simulation.

4 Design and Implementation

The app is specifically designed for reading/listening to stories, creating distraction-free moments dedicated to this activity. For this reason, the user experience and interface have been designed with purpose-driven limitations. It is a seated experience with movements limited to 3-degrees-of-freedom, no interaction with the physical environment is possible, and the user interface (UI) is intuitive and requires only a few clicks to start reading/listening to the story.

4.1 User Interface

The UI is a non-diegetic panel in which users can choose an environment and a text/audiobook from two menus (Fig. 1). The panel with the two menus can be accessed by clicking on a 3D object floating in the scene (spatial interface), a closed old book that opens and reveals the panel. Users can pause the audiobook, change background scene and activate/deactivate the audio reverb (see below). By clicking again on the 3D book the panel disappears.

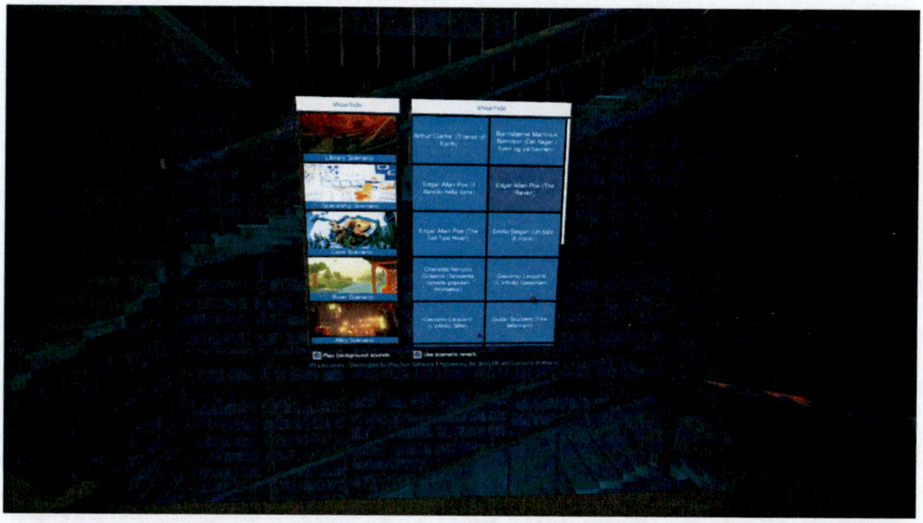

Fig. 1. Home scene with user interface for the selection of story and environment.

4.2 Graphics

A series of environments have been designed and manually modelled in 3D with the aim of creating an atmosphere that can help to focus on the reading experience and increase the absorption in the story by matching the environment with the story's theme or atmosphere. The home environment is a mysterious wooden library, in which users are surrounded by books. Here they can select a story or poem and one of four different environments (Fig. 2), designed to offer visual and audio stimuli that will be "peripherally perceived" during reading/listening, thus supporting imagery and transportation into the story world

[16]. The scenes do not represent the story content, therefore we believe they will not be an impediment for imagination, since users will still be able to picture as they want the situation evoked by the story. On the contrary, having a visual prompt can help people who do not have a high mental imagery ability [12,20]. A 19th century foggy alley at night is suitable for horror and mystery stories; a spacecraft, for science fiction; an underwater cave, for adventure and pirates stories. The fourth environment is a hut by the river, which is more neutral and has been conceived – based on users' feedback – as a relaxing place suited for a variety of stories (Fig. 3). The environments have been created with a low-poly graphic style, which has been chosen to make the experience pleasant for both children and adults, while keeping the graphic computing power required to a minimum, so that the app works smoothly even in low-specifications HMDs like the Oculus Go, which are more economically affordable for schools.

Fig. 2. Details of the four animated scenes that users can choose as environment while listening to audiobooks.

4.3 Sounds

Ambient sounds have been carefully selected and mixed to increase immersion. Every environment has its own specific background sounds and audio reverb

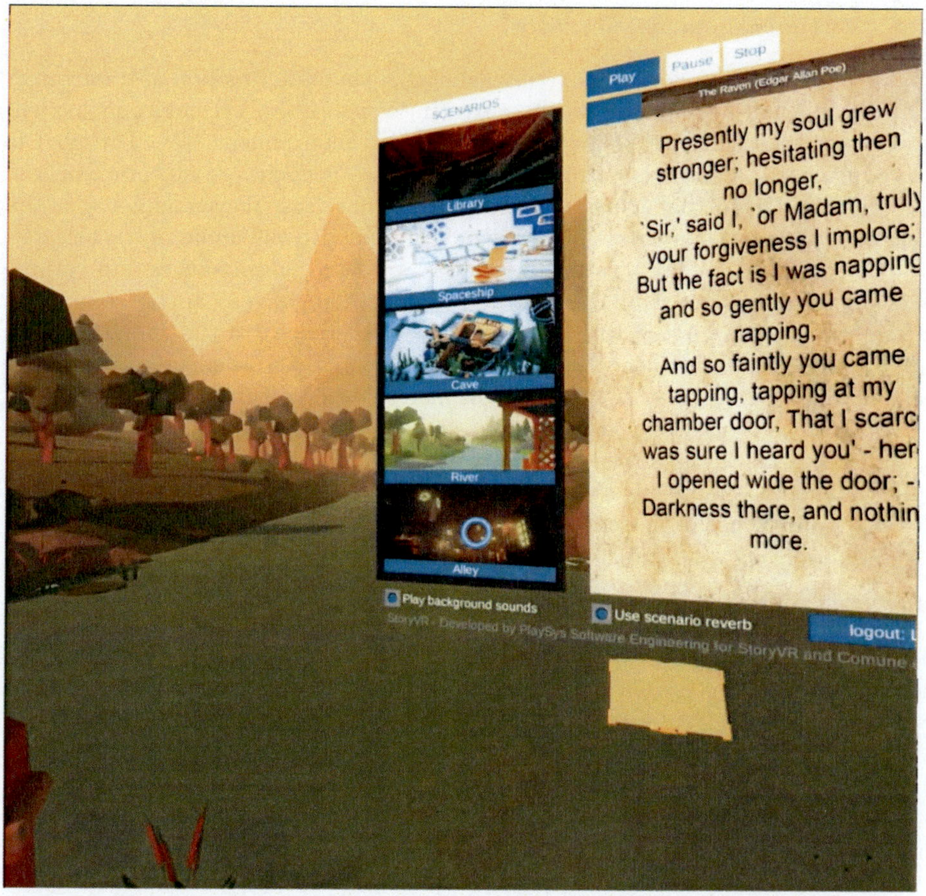

Fig. 3. Relaxing environment with the menu panel and text visible.

zone that simulates the different spatial reflection of sound in each environment. Moreover, we considered that when we read engaging narratives there is an unconscious narrowing of the attentional focus to the story, which consequently limits our *epistemic awareness* [27], that is the reduction of sensitivity to external stimuli. However, in VR the ambient sounds are perceived as part of the mediated experience and it is difficult for the user to "filter off" the background noise. Therefore, to simulate the perceptive attentional focus we used a script that progressively reduces the volume of the ambient sound when the audio of the story start or the text is displayed. This feature facilitates immersion into the story world.

4.4 Virtual Agent as Narrator

We are currently developing a virtual agent [25] that can function as a narrator. Having an agent that embodies the narrating voice in the VR environment will likely increase the sense of presence in the VR environment [15]. The goal is to reduce users' mind wandering by helping them keeping the attention on the story [9,30]. The virtual agent is programmed to have casual eye-contact with the user and the movement of the lips is synchronised with the audio of the selected story. Mouth shapes that create lip-syncing are based on phonemes and various sets of phonemes are available so that different languages can be simulated. Facial expressions and gestures are automatically synced to the speech to give appropriate emphasis to the story and create a more positive effect on users [21]. The implementation of this solution will increase the verisimilitude of the sense of co-presence of user and virtual agent and the feeling of a positive experience [7]. Previous research showed that, in a narrative context, embodied simulation of the audience is more effective than first-person embodiment of the protagonist for increasing sense of presence, engagement, and motivation to learn [28].

4.5 User Tracking

For research purposes, we included in the app analytics tools to track the users' choices of scenes and stories. Data collected regard the amount of stories read/listened, the rate of abandonment, and the session time. This information can be used to understand reading preferences but also to investigate more in depth the relationship between reading and environmental stimuli. Used in combination with more sophisticated HMDs – e.g. with eye-tracking or EEG sensors – *StoryVR* can be used for neuropsychological research on reading. The advantage will be a greater control on environmental variables during experiments compared to laboratory settings, which do not usually offer a relaxed environment encouraging reading for enjoyment.

5 Conclusion

In the context of a no-profit project promoting reading, the *StoryVR* app is first being distributed in public libraries together with HMDs, where specific spaces ("VR corners") are designed to host this new service. *StoryVR* is being used to attract young people to the library and to collect data about library users' preferences in matching environments and stories. In order to encourage the lifelong improvement of reading skills, using tasks that focus only on the development of reading comprehension is a solution with major shortcomings. In this respect, helping people to become passionate and motivated readers is of tantamount importance to sustain their future autonomous reading activities. By leveraging an attractive technology like virtual reality, whose learning and entertainment potential based on embodied simulation has been widely proved, the *StoryVR* app can be used in educational context to engage students, offering a new point of access to literary cultural heritage. We are currently doing experiments with quantitative and qualitative methods in Europe and South Korea, investigating how the environment influences the reading experience and whether cultural differences influence narrative absorption and the acceptance of the hybridisation of literature with VR technology. Users who tried the *StoryVR* app so far reported it to create both an "intimate" and "cultural" experience. Overall, the app proved to have a great potential for the promotion of reading and could be used for educational purposes as well as for leisure. Future improvements will be focused on adding sound effects and animations based on the progression of the story, in order to increase narrative absorption even further.

References

1. Chimera reader (2016). https://www.oculus.com/experiences/gear-vr/901261583298725/?locale=en_GB
2. Vivepaper (2016). http://www.vivepaper.com/en.html
3. immersionVR reader (2017). https://www.oculus.com/experiences/go/1397503436990351/
4. VirtualBook Viewer (2018). https://www.oculus.com/experiences/go/2341468065880359/
5. Baños, R., Botella, C., Alcañiz, M., Liaño, V., Guerrero, B., Rey, B.: Immersion and emotion: their impact on the sense of presence. Cyberpsychol. Behav. **7**(6), 734–741 (2004). https://doi.org/10.1089/cpb.2004.7.734
6. Bailenson, J.N.: Experience on Demand. What Virtual Reality is, How it works, and What it can do. Norton & Company, New York (2018)
7. Bailenson, J.N., Beall, A.C., Loomis, J., Blascovich, J., Turk, M.: Transformed social interaction, augmented gaze, and social influence in immersive virtual environments. Hum. Commun. Res. **31**(4), 511–537 (2005). https://doi.org/10.1111/j.1468-2958.2005.tb00881.x
8. Colvin Clark, R., Mayer, R.E.: e-Learning and the Science of Instruction: Proven Guidelines for Consumers and Designers of Multimedia Learning, 3rd edn. Pfeiffer, San Francisco (2011)
9. Corballis, M.C.: The Wandering Mind: What the Brain Does When You're Not Looking. University of Chicago Press, Chicago (2015)
10. Diemer, J., Alpers, G.W., Peperkorn, H.M., Shiban, Y., Mühlberger, A.: The impact of perception and presence on emotional reactions: a review of research in virtual reality. Frontiers in Psychology **6**(26) (2015). https://doi.org/10.3389/fpsyg.2015.00026
11. ELMCIP: Electronic Literature Knowledge Base. https://elmcip.net/knowledgebase
12. Green, M.C., Kass, S., Carrey, J., Herzig, B., Feeney, R., Sabini, J.: Transportation across media: repeated exposure to print and film. Media Psychol. **11**(4), 512–539 (2008). https://doi.org/10.1080/15213260802492000
13. Hakemulder, F., Kuijpers, M.M., Tan, E.S., Bálint, K., Doicaru, M.M. (eds.): Narrative Absorption. John Benjamins, Amsterdam (2017)
14. Hayles, N.K.: Electronic Literature : New Horizons for the Literary. University of Notre Dame Press, Notre Dame (2008)
15. Kuzmičová, A.: Literary narrative and mental imagery: a view from embodied cognition. Style **48**(3), 275–293 (2014)

16. Kuzmičová, A.: Does it matter where you read? situating narrative in physical environment. Commun. Theory **26**(3), 290–308 (2016). https://doi.org/10.1111/comt.12084
17. Lombard, M., Biocca, F., Freeman, J., Ijsselsteijn, W., Schaevitz, R.J. (eds.) Immersed in Media: Telepresence Theory, Measurement & Technology. Springer, Cham (2015). OCLC: 907095808
18. Madden, J., Pandita, S., Schuldt, J.P., Kim, B., Won, A.S., Holmes, N.G.: Ready student one: exploring the predictors of student learning in virtual reality. PLOS ONE **15**(3), e0229788 (2020). https://doi.org/10.1371/journal.pone.0229788
19. Moreno, R., Mayer, R.E.: Learning science in virtual reality multimedia environments: role of methods and media. J. Educ. Psychol. **94**(3), 598–610 (2002). https://doi.org/10.1037/0022-0663.94.3.598
20. Neumann, D., Moffitt, R.: Affective and attentional states when running in a virtual reality environment. Sports **6**(3), 71 (2018). https://doi.org/10.3390/sports6030071
21. Oh, S.Y., Bailenson, J., Krämer, N., Li, B.: Let the avatar brighten your smile: effects of enhancing facial expressions in virtual environments. PLOS ONE **11**(9), e0161794 (2016). https://doi.org/10.1371/journal.pone.0161794
22. Organization, E.L.: Electronic Literature Directory. http://directory.eliterature.org/
23. Parong, J., Mayer, R.E.: Learning science in immersive virtual reality. J. Educ. Psychol. **110**(6), 785–797 (2018). https://doi.org/10.1037/edu0000241
24. Pianzola, F., Bálint, K., Weller, J.: Virtual reality as a tool for promoting reading via enhanced narrative absorption and empathy. Sci. Study Lit. **9**(2) (2020)
25. Qu, C.: Talking with a Virtual Human: Controlling the Human Experience and Behavior in a Virtual Conversation. Ph.D. thesis, TU Delft, Delft (2014). http://resolver.tudelft.nl/uuid:b4f77e2f-b6b8-493f-bc6e-e536dada300e
26. Riva, G., Mantovani, F., Capideville, C.S., Preziosa, A., Morganti, F., Villani, D., Gaggioli, A., Botella, C., Alcañiz, M.: Affective interactions using virtual reality: the link between presence and emotions. Cyberpsychol. Behav. **10**(1), 45–56 (2007). https://doi.org/10.1089/cpb.2006.9993
27. Schwitzgebel, E.: Do you have constant tactile experience of your feet in your shoes? or is experience limited to what's in attention? J. Conscious. Stud. **14**(3), 5–35 (2007)
28. Slater, M., Navarro, X., Valenzuela, J., Oliva, R., Beacco, A., Thorn, J., Watson, Z.: Virtually being lenin enhances presence and engagement in a scene from the Russian revolution. Front. Rob. AI **5**, 91 (2018). https://doi.org/10.3389/frobt.2018.00091
29. Slater, M., Sanchez-Vives, M.V.: Enhancing our lives with immersive virtual reality. Front. Rob. AI **3**, 74 (2016). https://doi.org/10.3389/frobt.2016.00074
30. Varao Sousa, T.L., Carriere, J.S.A., Smilek, D.: The way we encounter reading material influences how frequently we mind wander. Front. Psychol. **4**, 892 (2013). https://doi.org/10.3389/fpsyg.2013.00892

Ph.D. and Master's Student Competition

Ph.D. and Master's Student Competition

The goal of the competition is to create a setting in which students can present their work and receive valuable feedback from both established researchers and attending students. Overall, the competition is an outstanding opportunity to showcase the student's creativity to leaders in the field, turn their ideas into reality, and win fabulous prizes that will foster the development of their scientific interests and will help them with networking in the research community.

To encourage peer learning, participants are invited to interact with their colleagues, sharing the most significant and innovative aspects of their research, in terms of both content and methodology. Thus, each participant will be assigned a mentor, expert researcher in the field to help them interact with the research community, exchanging ideas and comments before the conference/workshop. The mentor will introduce the student and will provide guidance throughout the event.

Organization

Organizing Committee

Giovanni De Gasperis	University of L'Aquila
Fernando De la Prieta Pintado	University of Salamanca
Alessandro De Leonardis	Armundia Group
Tania Di Mascio	University of L'Aquila
Rosella Gennari	Free University of Bozen - Bolzano
Beatrice Ligorio	Università di Bari
Vittorio Scarano	University of Salerno
Pierpaolo Vittorini	University of L'Aquila

Designing IVR Serious Games for People with ASD

An Innovative Approach

Federica Caruso(✉) and Tania Di Mascio

University of L'Aquila, 67100 L'Aquila, Italy
federica.caruso1@graduate.univaq.it, tania.dimascio@univaq.it

Abstract. This paper aims to articulate and discuss the problem statement, motivations, goals, challenges, and preliminary results of an ongoing doctoral research project. We would argue the serious games potentiality as TEL-solutions for autistic people, especially if these adopt immersive virtual reality technology; on the other hand, we would like to point out critical issues hidden in existing design approaches found in the related literature. We present an overview of the adoption of serious games and immersive virtual reality technology in the autism spectrum disorder domain; the goal of the research is then framed starting from the literature evidence. Moreover, the current status of the project is presented and finally, we will describe the future direction of the PhD project.

Keywords: Serious game · Immersive virtual reality · Design methodology · Autism spectrum disorders

1 Introduction

Continuing advances in the field of Information and Communication Technologies (ICTs) and the concomitant system-cost reductions have supported the development of more usable, useful, accessible and promising tools, especially in the domains of education, therapy, and rehabilitation (e.g., Autism Spectrum Disorders - ASD) [1]. The use of ICT-tools for ASD treatment is encouraged by the literature since the behaviour of autistic children concerning technological devices is the same as their typical development peers [23]. Moreover, since the 1970s, several papers have shown that the engagement and motivation of autistic people who used ICT-based teaching tools are greater than no-ICT-based [21]. This is since ICT-tools could provide a digital environment in which autistic people could practice and acquire new and difficult skills (e.g. reading, emotion recognition, and social skills) in a safe and non-threatening manner [3, 4, 23].

From the intervention viewpoint, according to [19], Serious Games appear the most promising tools. When well-designed [23], serious games balance the pedagogical and the entertainment aspects. They are able: (1) to motivate and promote the learning object acquisition, (2) to promote its generalisation, defined as far-transfer to real-world social interaction [9]. In summary, designing serious games means merging learning theory

Z. Kubincová et al. (Eds.): MIS4TEL 2020, AISC 1236, pp. 291–295, 2021.
https://doi.org/10.1007/978-3-030-52287-2_30

and empirical findings with principles of game design to create a unique intervention tool to improve outcomes beyond the context of the game. In order to create a right intervention tool, it is mandatory taking into account the right design pattern; it is then clear that designing a serious game for ASD people could be a challenge. In fact, taking existing design patterns for game design and forcing their adaptation in the ASD context did not even lead to relevant results in the general case [23]. The introduction of pedagogical elements has led to solutions which appear in some cases more oriented to the entertainment than to the learning, in other cases the solutions appear too oriented to the learning without considering entertainment aspects [15].

From the technological viewpoint, ICT-tools for the ASD people have been developed with different solutions. There are tools realised as simple apps [6] or tools realised as complex Virtual Reality (VR) environment [3]. A particular type of VR is the Immersive VR (IVR), where the level of immersion provided by the virtual environment is moderate-high; in this regard, as asserted by [17], this property could influence the learning of people with ASD and their engagement. Nowadays, the most used devices for IVR are Head-Mounted-Displays (HMDs) or Cave Automatic Virtual Environment (CAVE) systems. When social skills are the focus of learning outcome, the most appropriate ICT-tools for the ASD treatments appear to be those realised through IVR [13, 16]. The immersion can increase the effectiveness of learning by eliminating environmental distractions and supporting autistic people in better maintaining the focus on the main task to perform [5]. Besides, these technologies can improve the capacity of autistic people to deal with anxiety and phobias in social situations. It is worth noting that, in general, ICT-tools based on IVR technology have been evaluated with small groups of participants and often these experimental groups did not include a control group with typical developing people [13]. Nevertheless, if the studies are considered as aggregate [16], they show a promising nature of IVR as a training technology for autistic people, in particular for social skills learning and training [5, 7, 13].

Starting from this literature analysis, it is then clear that there could be a new generation of ICT-tools for ASD people: *IVR-serious games*. To properly develop a system, we should start with its right design; at this point, the key question is *What is the most appropriate approach to design an effective serious game for autistic people using immersive virtual reality technology?*

2 Related Literature

To try to find an answer, a review of available literature on approaches to designing SGs for autistic people was carried out. Unfortunately, the literature has revealed only general guidelines about game elements that could have a positive impact on the learning and the treatment of ASD [23] or frameworks which provide a list of game-elements which could enhance the learning of some specified skills by children with ASD [12, 20]. As a result, this has allowed making some preliminary considerations; despite our best efforts, our analysis of the literature regarding the design of serious games for autistic people seems not reveal any methodology that could be adopted *as-is*. Moreover, as confirmed by [8], the available literature shows some limitations, such as the lack of description of the approach adopted during the game design.

In this vein, we enlarge our analysis by investigating literature about serious games design approaches in a broader sense. It emerges that design approaches that we found are generally sequential and not iterative [11, 14, 15, 25]. More specifically, few works make some considerations about the role which experts should play in each step of the design of the game elements [14, 15]. Moreover, it emerges that few frameworks explicitly take into consideration which final users will be [12, 20] and/or the purpose/s of the serious games [2, 13]. Last but not least, we observed that few frameworks take into account the specificity of the technological device will be adopted in the implementation [6, 14] or the level of immersion to involve users as much as possible [24].

It is clear that the reviewed literature lacks in some important aspects which could directly influence the learning outcomes: (1) The serious game design process specification, (2) The role that experts plays in the design process, (3) Explicit characterization of final users, (4) The explicit skills that define the purposes of a serious game to implement, (5) The technological aspects of the serious game to implement, (6) The produced level of immersion of the serious game to implement.

3 Preliminary Considerations

The aforementioned aspects, rather than others in the literature, are more important since they could directly influence the learning outcomes provided by playing with a serious game. In fact: point (1) is strictly related to the quality of the realised serious game as well highlighted in [15]; point (2) is related to the in-depth nature of serious game which should be well balanced between learning and entertainment aspects and only a multidisciplinary team composed of different expertise is able to realise such games [15]; point (3) is related to the nature of ASD people who see and feel the world around them in a particular way, do not consider their characteristics means do not use their potentialities [22]; point (4) is strictly related to the efficacy of the serious game, i.e. if the skills to support are not explicitly defined the purposes of the serious game could be confused and then the serious game could result not effective [18]; point (5) is related to a positive or negative impact on the learning outcome of autistic people [22]; point (6) is related to the learning of autistic people and their engagement [17], i.e. the immersion can increase the effectiveness of learning by eliminating environmental distractions and supporting autistic people better maintaining the focus on the main task to perform [10].

It is evident that an innovative design methodology would be useful: (1) it should be able to support the whole design process of IVR-serious games for ASD people, (2) it should take into account the needs and the attitudes of final users, (3) it should bear in mind the technology adopted and the related level of immersion, (4) it should define the people that could be involved in the whole design process, from the idea to the final product.

The PhD research is going in this direction; it is being conducted according to an Action-Research methodological approach called *deejay* [6] to ensure equal attention to research and real-world problem objectives. The aim is dual: (1) define a sustainable, ecological, iterative, adaptive, and participative methodology for designing serious games for ASD people able to meet the above-mentioned considerations, and (2) implement an IVR serious game for the treatment of ASD people.

In this vein, some research activities have already been conducted: (1) wide study of literature about ICT tools dedicated to autistic people, and (2) deep investigation of existing design approaches for serious games. The Introduction and Related Literature sections of this article briefly report the main outcomes of these activities.

4 Future Works

Currently, we are conducting a systematic review including both clinical and technical databases with the following keywords: "Serious Game", "Immersive Virtual Reality", "Autism Spectrum Disorders". The aim is to investigate the design approach adopted by each research group and to deduct lessons from their hands-on experiences. Once this activity is over, we have planned multiple interviews sessions with people from different research domains (i.e., doctors, psychologists, therapists, educationalists, computer scientists, game designer, etc.) who have experience in the design of systems for and with autistic people, to appreciate this special context from the respondents' perspective and to explore the significance of their experience.

The activities that will follow will be planned according to the outcomes of each *deejay* iteration cycle until the research and real-world problem objectives are achieved.

References

1. American Psychiatric Association: Diagnostic and statistical manual of mental disorders (DSM-V®). American Psychiatric Pub (2013)
2. Andreoli, R., Corolla, A., Faggiano, A., Malandrino, D., Pirozzi, D., Ranaldi, M., Santangelo, G., Scarano, V.: A framework to design, develop, and evaluate immersive and collaborative serious games in cultural heritage. JOCCH **11**(1), 1–22 (2018)
3. Aresti-Bartolome, N., Garcia-Zapirain, B.: Technologies as support tools for persons with autistic spectrum disorder: a systematic review. Int. J. Environ. Res. Public Health **11**(8), 7767–7802 (2014)
4. Boucenna, S., Narzisi, A., Tilmont, E., Muratori, F., Pioggia, G., Cohen, D., Chetouani, M.: Interactive technologies for autistic children: A review. Cogn. Comput. **6**(4), 722–740 (2014)
5. Bozgeyikli, L., Raij, A., Katkoori, S., Alqasemi, R.: A survey on virtual reality for individuals with autism spectrum disorder: Design considerations. IEEE Trans. Learn. Technol. **11**(2), 133–151 (2017)
6. Di Mascio, T., Tarantino, L., Cirelli, L., Peretti, S., Mazza, M.: Designing a personalizable ASD-oriented AAC tool: an action research experience. Adv. Intell. Syst. Comput. **804**, 200–209 (2019)
7. Di Mascio, T., Tarantino, L., De Gasperis, G., Pino, C.: immersive virtual environments: a comparison of mixed reality and virtual reality headsets for ASD treatment. Adv. Intell. Syst. Comput. **1007**, 153–163 (2020)
8. Grossard, C., Grynspan, O., Serret, S., Jouen, A.L., Bailly, K., Cohen, D.: Serious games to teach social interactions and emotions to individuals with autism spectrum disorders (ASD). Comput. Educ. **113**, 195–211 (2017)
9. Grynszpan, O., Weiss, P.L., Perez-Diaz, F., Gal, E.: Innovative technology-based interventions for autism spectrum disorders: a meta-analysis. Autism **18**(4), 346–361 (2014)
10. Halabi, O., El-Seoud, S.A., Alja'am, J.M., Alpona, H., Al-Hassan, D.: Immersive virtual reality in improving communication skills in children with autism. Int. J. Interact. Mob. Technol. **11**(2), 146–158 (2017)

11. Ibanez, B.C., Boudier, V., Labat, J.M.: Knowledge management approach to support a serious game development. In: 9th IEEE International Conference on Advanced Learning Technologies, pp. 420–422 (2009)
12. Khowaja, K. Salim, S.S., Al-Thani, D.: Components to design serious games for children with autism spectrum disorder (ASD) to learn vocabulary. In: 5th International Conference on Engineering Technologies and Applied Sciences, pp. 1–7 (2018)
13. Lorenzo, G., Lledò, A., Arràez-Vera, G., Lorenzo-Lledò, A.: The application of immersive virtual reality for students with ASD: a review between 1990-2017. Educ Inform Tech **24**(1), 127–151 (2019)
14. Marfisi-Schottman, I., George, S., Tarpin-Bernard, F.: Tools and methods for efficiently designing serious games. In: 4th European Conference on Games Based Learning, pp. 226–234 (2010)
15. Marne, B., Wisdom, J., Huynh-Kim-Bang, B., Labat, J.M.: The six facets of serious game design: a methodology enhanced by our design pattern library. In: European Conference on Technology Enhanced Learning, pp. 208–221 (2012)
16. Mesa-Gresa, P., Gil-Gòmez, H., Lozano-Quilis, J.A., Gil-Gòmez-J, A.: Effectiveness of virtual reality for children and adolescents with autism spectrum disorder: an evidence-based systematic review. Sensors **18**(8), 2486 (2018)
17. Miller, H.L., Bugnariu, N.L.: Level of immersion in virtual environments impacts the ability to assess and teach social skills in autism spectrum disorder. Cyberpsychol. Behav. Soc. Netw. **19**(4), 246–256 (2016)
18. Mitgutsch, K., Alvarado, N.: Purposeful by design? a serious game design assessment framework. In: International Conference on the foundations of digital games, pp. 121–128 (2012)
19. Noor, H.A.M., Shahbodin, F., Pee, N.C.: Serious game for autism children: review of literature. WASET **64**(124), 647–652 (2012)
20. Park, J.H., Abirached, B., Zhang, Y.: A framework for designing assistive technologies for teaching children with ASDs emotions. In: CHI 2012 Extended Abstracts on Human Factors in Computing Systems, pp. 2423–2428 (2012)
21. Pennington, R.C.: Computer-assisted instruction for teaching academic skills to students with autism spectrum disorders: A review of literature. Focus Autism Other Dev. Disabil. **25**(4), 239–248 (2010)
22. Valencia, K., Rusu, C., Quinones, D., Jamet, E.: The impact of technology on people with autism spectrum disorder: a systematic literature review. Sensors **19**(20), 4485 (2019)
23. Whyte, E.M., Smyth, J.M., Scherf, K.S.: Designing serious game interventions for individuals with autism. J. Autism Dev. Disord. **45**(12), 3820–3831 (2015)
24. Winn, B.M.: The design, play, and experience framework. Handbook of Research on Effective Electronic Gaming in Education, pp. 1010–1024 (2009)
25. Yusoff, A., Crowder, R., Gilbert, L., Wills, G.: A conceptual framework for serious games. In: 9th IEEE International Conference on Advanced Learning Technologies, pp. 21–23 (2009)

Improved Feedback in Automated Grading of Data Science Assignments

Alessandra Galassi(✉) and Pierpaolo Vittorini

Department of Life, Health and Environmental Sciences, University of L'Aquila, P.le S. Tommasi, 1, 67100 Coppito L'Aquila, Italy
alexandra2g2016@gmail.com, pierpaolo.vittorini@univaq.it

Abstract. The paper deals with the automated grading of assignments made up of R commands, their output and comments written in natural language. Compared to other tools presented in the literature, our tool supports both students and teachers, uses static source code analysis for the code-snippets and a supervised classifier based on sentence embeddings for the open-ended answers, and provides a feedback to students and includes the instructor review. After more than one year of use, improvements in terms of the feedback provided to the students are discussed in the paper, that in turn should offer manifold benefits to both students and teachers. Finally, the paper proposes a study finalised to refine and test the effectiveness of the proposal.

Keywords: Automated grading · Formative assessment · Structured feedback · Data science assignments

1 Introduction

The paper deals with the problem of the automated grading of assignments made up of R [9] commands, their output and comments written in natural language, which are common in the data science domain. Several solutions have been proposed in the past to perform automated grading of open-ended and code-snippets answers [4,10].

Regarding the open-ended answers, the common task is to assign either a real-valued score (e.g., from 0 to 1) or to assign a label (e.g., correct or irrelevant) to a student response [4]. These attempts have mainly focused on English, while our courses are in Italian language. In connection with code-snippets questions, the first attempt towards automated assessment of student programming assignments can be dated back to the sixties [7]. Nowadays, the available tools use different approaches, either centred on the teacher, or on the student, or on both [10]. In terms of the programming languages, most of them focuses on Java, C, and C++ [10]. Unfortunately, no system still focuses on R, which is the programming language we are interested in. Therefore, compared to other

Z. Kubincová et al. (Eds.): MIS4TEL 2020, AISC 1236, pp. 296–300, 2021.
https://doi.org/10.1007/978-3-030-52287-2_31

tools presented in the literature, our tool [2,3,5] has the following characteristics: it provides an automated grading of the assignments, supports both students and teachers, uses static source code analysis for the code-snippets and a supervised classifier based on sentence embeddings for the open-ended answers, provides a feedback to the students and includes the instructor review [2,5].

In detail, the assignments we take into account are similar to that depicted in Fig. 1. A solution can be considered as a list of triples containing the command, its output and the possible comment. In our approach [2], such a solution is compared with the correct solution given by the professor. So, a student may: (i) give a correct command that returns the correct output; (ii) give a command with a wrong output, because the command has a mistake either in the call (e.g., a `t.test` command without the required `paired=TRUE` option) or in the passed data; (iii) miss the command; (iv) interpret the result in the right/wrong way. The higher the number of missing commands, commands with different output and wrong comments, the larger the distance between the two solutions. For a formal definition of such a distance, the reader may refer to [2]. The next section describes the research proposal in terms of the current limitations and the possible improvements, whereas the concluding section discusses the expected benefits.

Let us consider the following dataset:

Subject	Surgery	Visibility	Days
1	A	7	7
2	A	5	7
	...		
10	B	16	12
	...		
20	C	19	4

The data regards 20 subjects (variable Subject) that underwent three different surgical operations (variable Surgery). We observe the scar visibility (variable Visibility) in terms of ranks ranging from 1 (the best) to 20 (the worst). We also measure the hospital stay (variable Days). You are required to:

1. calculate the central tendency (with confidence intervals) and dispersion of the hospital stay;
2. calculate the distribution (with confidence intervals) of the surgical operations;
3. verify if the hospital stay can be considered as extracted from a normal distribution;
4. comment on the result;
5. calculate the central tendency and dispersion of the hospital stay for the different surgical operations;
6. verify if the aspect of the scar is different among the different surgical operations;
7. comment on the result.

Submit as solution a text containing the list of R commands with the respective output, as well as your interpretation of the analyses 3 and 6.

Fig. 1. Sample assignment

2 Research Proposal

2.1 Current Limitations

So far, the system has been used with more than one hundred students in the Health Informatics course of the Medicine and surgery degree. During the use of the system as a formative assessment tool (i.e., the students used the system solve exercises and receive immediate feedback), we noticed some issues that can be classified into two categories: (i) technical improvements; (ii) feedback structure.

About the technical improvements, the issues follows. Let us take into account points 1 and 5 of Fig. 1. For both points, means are required. If the student solution contains the means in the reverse order (i.e., first the mean for point 5, then for point 1), the commands are seen as having an incorrect output, because the system expects the answers in the order in which they are asked in the assignment (let us call this issue T_{I1}). Furthermore, students sometimes remove the R prompt ">", that is used by the parser to recognise the commands (T_{I2}).

About the second class of issues, the most important one is that the feedback is more technical than didactic, i.e., it reports that the command is wrong, or has a wrong output, or that the command is not correct, without stating why (F_{I1}). Furthermore, the automated grade returned by the system – even if close to the manual one (an $R^2 = 0.740$ was measured in [2]) – may lead to false pass/fail outcomes (a Cohen's $K = 0.52$ was measured in [2]), especially when the automated grade is close to the sufficiency (F_{I2}). Finally, both in the calculation of the distance and in the feedback, an output wrong either because of the syntax or because of the passed data, is reported in the same way (F_{I3}).

2.2 Proposed Improvements

The improvements we propose are the following:

- T_{I1} The tool will have to scan all commands in the teacher's solution, then – for each command in the student's solution – firstly search if a command with the correct output exists, even if at the end of the solution, then search for a command with a wrong solution, before marking the command as missing;
- T_{I2} An heuristic approach can be followed: the system could check all words contained in the student solution. If the first word on a new line is in the list of the possible commands and is not preceded by a ">" symbol, the system may add it. After having checked all words, the tool can proceed normally;
- F_{I1} Referring to descriptive statistics, errors could be an improper choice of the command for central tendency, dispersion or distribution (e.g., considering Fig. 1, point 1, a `median` instead of `mean`). So far, the system simply reports a "Wrong command", whereas a more detailed feedback like "You used the median to calculate the central tendency. However, since the variable is numeric, the proper central tendency indicator is the mean" could be more helpful for the student. Referring to the hypothesis testing, a mistake could come from a misunderstanding about the variable type (quantitative rather than qualitative) or about the type of study (e.g., paired rather than independent samples) or the non-execution of a normality test (which may lead to use a parametric instead of a non-parametric test). All these cases are – again – all labelled as "Wrong command" instead of producing more wordy (and potentially more useful) sentences like "You use the analysis of variance. Since the outcome variable is qualitative and not quantitative,

you should use a Kruskal-Wallis test" (which can be a possible mistake in solving point 6 of Fig. 1);

F_{I2} The quantitative grade (i.e., the distance) can be substituted by a qualitative evaluation. To this aim, the proper thresholds must be defined so to reduce the number of false fail/pass cases. Accordingly, given that in [2] a perfect agreement between the manual and the automated grader was obtained by removing the cases with grades in the range [16, 20][1], we could experiment the following qualitative evaluation: "Very poor" = [0–8), "Poor" = [8–15), "Average" = [15–21), "Good" = [21–27), "Very good" = [27–31].

F_{I3} A long term improvement could be to revise the current definition of the distance, by separating the two cases of a wrong output (i.e., if the mistake is caused by a wrong syntax or by wrong data), which in turn may lead to a system able to provide a differentiated feedback to the students.

3 Expected Benefits

As known, a formative assessment tool should help the students to identify their strengths and weaknesses, as well as to target areas that need further work, encouraging their self-evaluation. As a results, the system should improve the students' understanding of the subject and allow them to pass the exam in a more profitable way. It is commonly used for keeping student engaged and to help the teacher to have a full understanding of how students are learning as he/she teaches a subject [6]. In our proposal, the feedback plays a fundamental role. In such a context, we will organise and perform a usability test as follows [11]:

Purpose The purpose of the test is twofold. First, to verify if the feedback is considered useful by students and to improve it accordingly. Second, to evaluate if the use of the system as a formative assessment tool helps students to achieve better learning outcomes;

Sessions To achieve the goals specified above, we will organise:
- Before implementing the novel feedback, focus groups with students who already used the current system, so to discuss with them the aforementioned improvements for F_{I1}, F_{I2} and F_{I3};
- After the implementation of the novel feedback, remote or in-person testing with students during their homework.

Metrics Both qualitative and quantitative measures will be collected:
- During the focus groups, structured notes will be taken, which will guide us to define an improved and refined feedback structure;
- For the remote or in-person testing, at the end of each homework we will administer (i) the User Assessment Engagement Scale (UAES) [1] to measure the learners' engagement, (ii) the After-Scenario Questionnaire (ASQ) [8] to measure the students' user experience, plus (iii) questions regarding the students' previous skills in computer science, statistics and software for data analysis;

[1] Please note that a grade in the Italian system ranges from 0 to 30 "cum laude", customarily considered as 31.

– At the end of the course, we will collect the final grades achieved by all students, so to compare the average grades of the students that used the system with the others, by taking into account also the possible confounding factors previously listed.

References

1. Angelone, A.M., Vittorini, P.: A report on the application of adaptive testing in a first year university course. In: Communications in Computer and Information Science, vol. 1011, pp. 439–449. Springer, Heidelberg (2019)
2. Angelone, A.M., Vittorini, P.: The automated grading of R code snippets: preliminary results in a course of health informatics. In: Proceedings of the 9th International Conference in Methodologies and Intelligent Systems for Technology Enhanced Learning. Springer, Heidelberg (2019)
3. Bernardi, A., Innamorati, C., Padovani, C., Romanelli, R., Saggino, A., Tommasi, M., Vittorini, P.: On the design and development of an assessment system with adaptive capabilities. In: Methodologies and Intelligent Systems for Technology Enhanced Learning, pp. 190–199. Springer, Cham (2019). http://link.springer.com/10.1007/978-3-319-98872-6_23
4. Burrows, S., Gurevych, I., Stein, B.: The eras and trends of automatic short answer grading. Int. J. Artif. Intell. Educ. **25**(1), 60–117 (2015)
5. De Gasperis, G., Menini, S., Tonelli, S., Vittorini, P.: Automated grading of short text answers: preliminary results in a course of health informatics. In: ICWL 2019: 18th International Conference on Web-Based Learning, LNCS. Springer, Magdeburg (2019)
6. Harlen, W., James, M.: Assessment and learning: differences and relationships between formative and summative assessment. Assess. Educ. Principles Policy Pract. **4**(3), 365–379 (1997)
7. Hollingsworth, J.: Automatic graders for programming classes. Commun. ACM **3**(10), 528–529 (1960). http://portal.acm.org/citation.cfm?doid=367415.367422
8. Lewis, J.R.: Psychometric evaluation of an after-scenario questionnaire for computer usability studies. ACM SIGCHI Bull. **23**(1), 78–81 (1990). http://portal.acm.org/citation.cfm?doid=122672.122692
9. R Core Team: R: A Language and Environment for Statistical Computing (2018). https://www.R-project.org/
10. Souza, D.M., Felizardo, K.R., Barbosa, E.F.: A systematic literature review of assessment tools for programming assignments. In: 2016 IEEE 29th International Conference on Software Engineering Education and Training (CSEET), pp. 147–156. IEEE (2016)
11. Tullis, T.T., Albert, B.W.: Measuring the User Experience: Collecting, Analyzing, and Presenting Usability Metrics. Elsevier, Amsterdam (2013)

Smart Object Design by Children as Protagonists

Eftychia Roumelioti(✉)

Faculty of Computer Science, Free University of Bozen-Bolzano,
Piazza Domenicani 3, Bolzano, Italy
eftychia.roumelioti@stud-inf.unibz.it

Abstract. Making children the *protagonists* of a design process by allowing them to express themselves through technology and mixing this knowledge with design competence and reflection opportunities has been seen in the CCI community as one way to empower them. This paper presents a card-based board-game and the way it was used in design workshops with children in order to involve them as *protagonists*.

Keywords: Children · Protagonist · Empowerment · Design · Making · Smart objects · Cards · Game

1 Introduction

The maker movement and the arrival of programming languages and prototyping toolkits for children opened new horizons in children's empowerment through making and learning. Lately some attempts have been made to include a "designerly" approach to digital fabrication and empower children through design and reflecting thinking as well (e.g. [2,3,9]). Considering this way of empowerment, a novel perspective on children's roles in design emerged, the role of *protagonist* [7,10]. Design toolkits, besides prototyping, can support this direction. However, there is still a lack of research dedicated to toolkits specially for children that can guide them across design parts and stir reflections.

This paper reports on a card-based game, *SNaP*, created to offer an engaging and inspirational toolkit that provides to children a structured way to design technology, smart objects in particular, and opportunities for convergent, divergent and reflective thinking. The paper also presents the design workshops organised using *SNaP* in which children were involved in the design of smart objects, with the aim to empower them through this process.

2 The SNaP Game

SNaP is a collaborative card-based board-game, for designing smart objects for specific environments. It is played by 2 to 4 players, from 11 years old onward,

Under the supervision of Dr. Rosella Gennari and Dr. Alessandra Melonio.

Z. Kubincová et al. (Eds.): MIS4TEL 2020, AISC 1236, pp. 301–304, 2021.
https://doi.org/10.1007/978-3-030-52287-2_32

who have the role of Junior Designers. The game is facilitated by a moderator, with the role of Senior Designer. During the game-play, each player has a mission to accomplish and all the players "win" the game collaboratively; the game ends when each of them has designed at least a smart object fulfilling his/her mission.

SNaP is divided into 3 levels, each mirroring steps of the ideation stage:

1. Divergent ideation: players, by dice rolling and card drafting on a game board, collect coins and "conquer" cards of 3 different kinds, corresponding to the parts necessary for ideating a smart object, i.e. object, input and output cards (see examples in Fig. 1).
2. Conceptualisation of ideas and self reflection: each player receives a conceptualisation idea sheet and reflects on ideas starting with the collected cards. Each sheet consists of colorful spots to organise the cards that players want to use from the collected ones, mark down extra cards they might need and write down a description of their smart objects.
3. Convergence into a single idea and group reflection: each player, in turns, presents the idea of his/her choice to the other players. The rest of the players play the roles of expert reviewers. By reading cards, they pose questions and give feedback on the presented idea concerning design heuristics, in order to further reflect on it collectively. Subsequently, the Senior Designer gives the final approval and the player, if needed, can buy extra cards with the coins (see Fig. 2).

The reflection lenses of the expert reviewers concern: (i) the feasibility of the idea, (ii) the consistency of the idea with the mission and (iii) the elaboration of the idea regarding clarity and complexity. Two versions of *SNaP* have been designed so far for two different, familiar to children, contexts, i.e., a park and a school classroom.

Fig. 1. Cards (input, object, output) of *SNaP* for classrooms

Fig. 2. Third level board of *SNaP* for classrooms

3 Workshops with Children

Six design workshops have been organised so far with both versions of *SNaP* involving overall 39 children, from 11 to 14 years old. In the first four workshops, children played with previous versions of *SNaP* and ideated smart objects [4–6]. In the last two workshops, 27 children in total experienced a full design circle, from ideating to prototyping. Children first explored and tinkered with Micro:bit micro-controllers [1] and the programming environment of MakeCode [8]. After playing with *SNaP*, they were given the input and output devices that corresponded to the chosen input and output cards used in their SNaP idea conceptualisation sheets. Subsequently, they moved on assembling the components to the Micro:bit micro-controllers and programming their ideas, according also to the description on their sheets. Finally, they made a physical prototype of their smart objects (see Fig. 3).

Data related to children's design competence, reflections, learning and engagement have been collected during these workshops. A brief description of the results is following. So far, all the children involved in workshops, guided by *SNaP*, managed to design at least one smart object. Recurrent practise of similar design activities, in particular, showed that they can also become independent in doing so, without adult help. Moreover, the majority of children reflected on theirs and others' ideas, collaboratively, making them evolve. However, reflections were mostly observed during the ideation stage, supported by the "tangible" reflections lenses of *SNaP*, and fewer in the programming and prototyping stage. Finally, the majority of children seemed to have learned the main programming concepts explained and experienced during the workshop and showed rather high engagement in playing with *SNaP* and prototyping.

Fig. 3. Children ideating (left), programming (middle) and presenting (right) a *SNaP* prototype.

4 Conclusions and Future Work

The paper reports on the *SNaP* game and its usage in design workshops with children. Results regarding children's design competence, reflections, learning and engagement are overall positive. They indicate that the game-based structure of the workshops can help empowering children in developing the skills and reflexivity for critically shaping digital technology. Future work will include more tangible reflection stimuli, through *SNaP*, also in the programming stages, as well as stimuli that will foster also the creativity of children's ideas.

Acknowledgements. Sincere thanks go to my supervisors, Dr. Rosella Gennari and Dr. Alessandra Melonio, for their continuing guidance and support. I also thank Prof. Maristella Matera who has closely collaborated on this research, Dr. Mehdi Rizvi for his technical assistance, as well as all the participating children.

References

1. Micro:bit educational foundation—micro:bit (2019). https://microbit.org. Accessed 06 Sept 2019
2. Bekker, T., Bakker, S., Douma, I., van der Poel, J., Scheltenaar, K.: Teaching children digital literacy through design-based learning with digital toolkits in schools. Int. J. Child-Comput. Interact. **5**, 29–38 (2015). Digital Fabrication in Education
3. Eriksson, E., Iversen, O.S., Baykal, G.E., Van Mechelen, M., Smith, R., Wagner, M.L., Fog, B.V., Klokmose, C., Cumbo, B., Hjorth, A., et al.: Widening the scope of fablearn research: integrating computational thinking, design and making. In: Proceedings of the FabLearn Europe 2019 Conference, FabLearn Europe 2019. Association for Computing Machinery, New York, (2019). https://doi.org/10.1145/3335055.3335070
4. Gennari, R., Matera, M., Melonio, A., Roumelioti, E.: A board game and a workshop for co-creating smart nature ecosystems. In: Proceedings of the 9th International Conference in Methodologies and Intelligent Systems for Technology Enhanced Learning (mis4TEL 2019). Springer, Heidelberg (2019)
5. Gennari, R., Matera, M., Melonio, A., Roumelioti, E.: A board-game for co-designing smart nature environments in workshops with children. In: Malizia, A., Valtolina, S., Morch, A., Serrano, A., Stratton, A. (eds.) End-User Development, pp. 132–148. Springer, Cham (2019)
6. Gennari, R., Matera, M., Melonio, A., Roumelioti, E.: Snap 2: the evolution of a board game for smart nature environments. In: Extended Abstracts of the Annual Symposium on Computer-Human Interaction in Play Companion Extended Abstracts, CHI PLAY 2019 Extended Abstracts, pp. 405–411. Association for Computing Machinery, New York (2019). https://doi.org/10.1145/3341215.3356281
7. Iversen, O.S., Smith, R.C., Dindler, C.: Child as protagonist: expanding the role of children in participatory design. In: Proceedings of the 2017 Conference on Interaction Design and Children, IDC 2017, pp. 27–37. ACM, New York (2017). https://doi.org/10.1145/3078072.3079725
8. Microsoft: Microsoft makecode (2019). https://makecode.microbit.org. Accessed 06 Sept 2019
9. Smith, R.C., Iversen, O.S., Hjorth, M.: Design thinking for digital fabrication in education. Int. J. Child-Comput. Interact. **5**(C), 20–28 (2015). https://doi.org/10.1016/j.ijcci.2015.10.002
10. Södergren, A.C., van Mechelen, M.: Towards a child-led design process a pilot study: when pre-schoolers' play becomes designing. In: Proceedings of the 18th ACM International Conference on Interaction Design and Children, IDC 2019, pp. 629–634. ACM, New York (2019). https://doi.org/10.1145/3311927.3325330

Author Index

Z. Kubincová et al. (Eds.): MIS4TEL 2020, AISC 1236, pp. 305–306, 2021.
https://doi.org/10.1007/978-3-030-52287-2